Classical Recording

Classical Recording: A Practical Guide in the Decca Tradition is the authoritative guide to all aspects of recording acoustic classical music. Offering detailed descriptions, diagrams, and photographs of fundamental recording techniques such as the Decca Tree, this book offers a comprehensive overview of the essential skills involved in successfully producing a classical recording. Written by engineers with years of experience working for Decca and Abbey Road Studios and as freelancers, *Classical Recording* equips the student, the interested amateur, and the practising professional with the required knowledge and confidence to tackle everything from solo piano to opera.

Caroline Haigh grew up with equal love for music, maths, and physics and combined study of all of them on the Tonmeister course at the University of Surrey. She began her career in classical post-production working for Decca during her final year at university, and she stayed there for several years before moving around the corner to Abbey Road Studios (EMI) in 1996. At both Decca and EMI, she gained experience with countless major classical artists and became a sought-after and skilful editor, working on several Grammy award–winning albums, including *Les Troyens* (Decca – OSM/Dutoit), Best Opera 1995. Having enjoyed giving guest seminars at the University of Surrey during her time at Abbey Road, Caroline was recruited to teach on the Tonmeister course on a permanent basis from 2009. She currently teaches recording techniques, production/post-production skills, and electro-acoustics (microphones), and she continues to work as a freelance classical editor.

John Dunkerley is one of the world's most highly respected and emulated classical recording engineers. Throughout a long career at Decca and then as a freelancer, his recordings have been renowned for their ravishingly beautiful sound and attention to detail. He has worked with almost all the major artists of the last 40 years and has made over a thousand CDs, and his recordings have earned over 15 Grammy awards. John is one of the last engineers alive to have learnt his craft from the great Kenneth Wilkinson, the inventor of many of the techniques that underpin the classical recording art. John teaches workshops at the University of Surrey, at the Banff Centre, and at the Abbey Road Institute.

Mark Rogers studied on the Tonmeister course at the University of Surrey. He began his career working with John Dunkerley at Decca, and then spent nine years around the corner at EMI's Abbey Road Studios, where he was the chief technical engineer for Studio One, famous for its orchestral and film score recordings. Here he worked with hundreds of different producers and engineers and gained unique insight into the huge variety of techniques used in classical recording. In 2000 he moved to a management role at Warner Music, and after four years left to become a freelance recording producer, engineer, and musician. Since then he has worked for a wide variety of clients, including the Royal Opera House, Covent Garden, and back where he started at Decca. His recordings have won many accolades, including a Grammy award in 2009. Mark is a visiting lecturer at the University of Surrey.

AUDIO ENGINEERING SOCIETY PRESENTS...

www.aes.org

Hack Audio: An Introduction to Computer Programming and Digital Signal Processing in MATLAB
Eric Tarr

Loudspeakers: For Music Recording and Reproduction
Philip Newell and Keith Holland

Beep to Boom: The Development of Advanced Runtime Sound Systems for Games and Extended Reality
Simon N. Goodwin

Intelligent Music Production
Brecht De Man, Ryan Stables, and Joshua D. Reiss

Women in Audio
Leslie Gaston-Bird

Audio Metering: Measurements, Standards and Practice
Eddy B. Brixen

Classical Recording: A Practical Guide in the Decca Tradition
Caroline Haigh, John Dunkerley, and Mark Rogers

The MIDI Manual 4e: A Practical Guide to MIDI within Modern Music Production
David Miles Huber

For more information about this series, please visit: www.routledge.com/Audio-Engineering-Society-Presents/book-series/AES

Classical Recording

A Practical Guide in the Decca Tradition

Caroline Haigh, John Dunkerley,
and Mark Rogers

LONDON AND NEW YORK

First published 2021
by Routledge
2 Park Square, Milton Park, Abingdon, Oxon OX14 4RN

and by Routledge
52 Vanderbilt Avenue, New York, NY 10017

Routledge is an imprint of the Taylor & Francis Group, an informa business

British Library Cataloguing-in-Publication Data
A catalogue record for this book is available from the British Library

Library of Congress Cataloging-in-Publication Data
Names: Haigh, Caroline, author. | Dunkerley, John, author. | Rogers, Mark (Sound engineer), author. | Audio Engineering Society, editor.
Title: Classical recording : a practical guide in the Decca tradition / Caroline Haigh, John Dunkerley, and Mark Rogers.
Description: Abingdon, Oxon ; New York, NY : Routledge, 2021. | Series: Audio Engineering Society presents... | Includes bibliographical references and index.
Identifiers: LCCN 2020018477 | ISBN 9780367321338 (hbk) | ISBN 9780367312800 (pbk) | ISBN 9780429316852 (ebk)
Subjects: LCSH: Sound—Recording and reproducing. | Decca Music Group. | Music.
Classification: LCC TK7881.4 .H35 2021 | DDC 621.389/32—dc23
LC record available at https://lccn.loc.gov/2020018477

ISBN: 978-0-367-32133-8 (hbk)
ISBN: 978-0-367-31280-0 (pbk)
ISBN: 978-0-429-31685-2 (ebk)

Typeset in Sabon
by Apex CoVantage, LLC

Visit the eResources: www.routledge.com/9780367312800

To the memories of Kenneth Wilkinson and Jimmy Lock for their wisdom, patience, and understanding, and to all the Decca family who went before to show us the way.

Also to our families: Alan, Jonathan, Anna, Claire, Isaac, and Milly with thanks for their support.

Contents

Acknowledgements

We would like to offer our grateful thanks to the many friends and colleagues who have helped us out with fact-checking, opinions, conversations, photographs, drawings, and cups of tea.

Particular thanks go to Jonathan Allen; Peter Cobbin and Kirsty Whalley; Richard Hale; James Shannon; Simon Eadon; Jonathan Willcocks; Edward Weston (Decca); Elena Turrin (Fazioli); Dr Russell Mason, Alan Haigh, and Prof. Dave Fisher (University of Surrey); Hannah Fitzgerald and Carlos Lellis (Abbey Road Institute); Anja Zoll-Khan, Steven Zissler, Giacomo Carabellese, Ross Hendrie, and Mark Thackeray (Royal Opera House, Covent Garden); Martin Schneider (Neumann GmbH); Richard Evans (Trinity School, Croydon); Emma Button, Crispin Ward, and Simon Growcott (University of Chichester); Paul Mortimer and Dominique Brulhart (Merging Technologies); Jeremy Powell (Vectorworks); Shannon Neil and Hannah Rowe (Focal Press); our copy editor, Jennifer Fester; and production editor, Abigail Stanley.

Some of the drawings in this book include images that are representations created in Vectorworks software by Vectorworks, Inc., and some include images used and adapted with kind permission of the Royal Opera House, Covent Garden.

Introduction

Our philosophy of recording

The authors of this book have known each other for over 30 years, and having all spent years at both Decca and EMI at different times, it is perhaps unsurprising that our philosophy of recording classical music has become something that we have in common.

At the core of this is the belief that the recording itself should serve and enhance the music at all times. A good recording should work with the performers and composer to excite and engage the listener and enable the sort of emotional involvement that is experienced in a live performance. The whole enterprise should have musical communication at its heart: we all love music; it is why we do what we do.

To pursue only the accurate spatial reproduction of a concert as experienced from the 'best seat in the house' is to miss opportunities to make the listener sit up and listen. A recording is best regarded as a different medium to a live performance, and why should the recorded experience not aim to be better than sitting in the concert hall with our eyes closed? At a concert, we have many visual cues which affect our perception of the music; we can see who is playing, where they are sitting, we can see the soloists, how big the hall is – all these things serve to engage us in the experience. When we shut our eyes, these visual cues are no longer available and our perception of the sound is altered. If we want to make a recording that draws the listener in, we have to find ways to recreate these cues in audio form and reactivate the energising sense of engagement that we experience in a live performance – the upper strings should soar, the basses should be warm with rosin flying off their bows, the brass should sparkle and thrill, and the tone colours of the woodwinds should come through as beautiful, individual highlights.

There are plenty of books dealing with recording one instrument at a time or with recording instruments that are well isolated from one another, as in a pop studio. In these situations, avoidance of spill of one source onto another's microphone is a central concern so that instruments can be processed separately. But when recording classical music, with all the players together in a live space, the key to success is in embracing spill, accepting its inevitability, and learning to blend and balance the different microphones to create the sound that you want. In this way, it is more akin to watercolour painting than colouring in a line drawing.

At the core of this recording approach is the creation of a sense of a real space in which the performers are situated. This does not mean it has to be simply a reproduction of a real space exactly as we find it, but it can be enhanced or created by artificial means as long as it has the characteristics of a real space. This includes achieving a sense of depth with larger sources such as orchestra or choir, where some performers feel closer to the listener than others. We are aiming for an impression of space, depth, and clarity so that the music can speak as the composer and performers intended. The result must be believable even if it contains a great deal of artifice.

In a good classical recording, we never want to become aware of individual microphones at work, dragging a single instrument too far into the foreground and reducing it to a point source, or of artificial reverb added like a layer of thick varnish over a painting, obscuring detail and tone colours. Blending several microphone sources and reverb is a key part of the craft, and no amount of expensive gear will make up for a lack of skill in balancing. In an interview given to Andrew Achenbach for *International Classical Record Collector* (ICRC), Richard Itter (Lyrita Records) said:

> We tend to forget today that the great art of the engineer is in the balancing . . . the equipment – be it analogue, digital or come what may – is secondary; it is the skill in the placing of those microphones and balancing on the desks that is the true art, and it was an art that the folk at Decca had down to a tee.[1]

The aims of this book

In this book we will introduce a number of guiding principles for the recording of classical music, and we will also reassure the reader that excellent results can be obtained using good technique with relatively inexpensive equipment. Our aim is to cover some good starting points for the recording of all the common classical ensembles; we have chosen not to cover every single possible technique but to present a selection that is inevitably subjective. It is not our intention to be prescriptive but rather to help the engineer understand what works and why, and how to avoid the most common pitfalls. We will suggest core techniques for many scenarios, with the proviso that this might have to be adapted in the hall you find yourself in or with the players that you have. At this point, we will then suggest what you should be listening for and what steps might be taken to correct things. The overall aim is to give practical, achievable advice for both live and studio scenarios, with consideration given to being visually discreet when a concert is being filmed. At all times the emphasis will be on principles and technique and not on specific pieces of expensive equipment. Following on from recording, the chapters on post-production work will guide you through the pitfalls of getting your recording from the studio takes to a final polished product, including editing, noise removal, and mastering.

Our target readers

The book is intended for readers who already have some recording knowledge, although not necessarily in the classical field. These would include students at the undergraduate and post-graduate levels, home recording enthusiasts who might want to also get out and record larger-scale acoustic music such as choral society events and school orchestral concerts, and professionals who want to explore other approaches to recording.

What is not covered

It is assumed that the reader is familiar with microphone types (condenser, dynamic, and ribbon) and directivity patterns (cardioid, omnidirectional, figure of eight, etc.) and is able to plug in microphones, activate phantom power when needed, and get a signal into and through a mixing desk or recording DAW.

eResources

Spotify playlists for each chapter containing both referenced recordings and illustrative listening can be found at www.routledge.com/9780367312800

Note

1 *International Classical Record Collector (ICRC) No 18, Autumn 1999, pp34–40*
Article: Lyrita, 40 Years On
Author: Andrew Achenbach interviewing Richard Itter

Glossary of terms, acronyms, and abbreviations

0 dBFS 0 dB full scale – the highest level on a digital meter, representing full digital modulation.

AB pair A pair of microphones for recording stereo that are spaced apart.

AFL After fade listen – a means of listening to an individual audio signal at the level it is being used in the mix without disturbing the mix bus output. Useful for fault finding while in record.

Cardioid A microphone that picks up sound primarily from its front axis and discriminates against sounds from the rear with reduced output at the sides. The name is from its 'heart-shaped' polar response pattern.

Condenser A microphone that uses a variable capacitor as its transducer and requires phantom power or another power supply. They are the most common microphone type used for classical recording (ribbon microphones are the second) because of their smoother and more extended frequency responses and better transient responses when compared with dynamic microphones (which are not normally used for classical recording).

DAW Digital audio workstation – a computer, audio interface, and software programme for recording and editing.

EQ Abbreviation of 'equalisation'

Fig of 8 Microphone with a figure-of-eight polar response: it picks up sound from sources in front and behind in equal amounts whilst discriminating against sounds arriving from the sides. The electrical output produced by a sound source located in front of the microphone will be of the opposite polarity to that produced by the same source located behind the microphone. This is because the diaphragm will move in the opposite direction, causing the transducer to produce a positive voltage in one case and a negative one in the other.

HPF High-pass filter – a filter that removes low frequencies. They are frequently used in classical recording to remove room rumble, to clean up low-frequency spill from a higher-pitched soloist's microphone(s), and to avoid the muddying build-up of layers of lower frequencies when using a lot of microphones.

LUFS Loudness units relative to full scale – loudness units are a measure of the perceived 'loudness' of a whole programme. See section 19.4.3.

NOS Nederlandsche Omroep Stichting – a stereo recording technique using spaced and angled cardioids; see section 3.3.

Omni Omnidirectional microphone – one that picks up sound equally from all around. It will show gradually reduced output at the rear and sides at higher frequencies due to the microphone itself forming an obstacle to shorter wavelengths.

ORTF Office de Radiodiffusion Télévision Française – a stereo recording technique using spaced and angled cardioids; see section 3.3.

PFL Pre fade listen – a means of listening to an individual audio signal when it is not currently faded up but without disturbing the mix bus output. Useful in live work for checking that a signal is the correct one before fading it up.

Proximity effect The increase in low-frequency output from a microphone as it is moved closer to a source. It is not exhibited by omnis but only by directional microphones, such as cardioids and fig of 8s.

Ribbon A microphone that uses a ribbon of aluminium suspended in a magnetic field as its transducer. They have a smooth frequency response with a gentle, early high-frequency roll-off and a figure-of-eight polar response. They are easily damaged by air movement and wind.

Solo-in-place A means of listening to an individual audio signal in a mix that works by cutting all the other channels. It will affect the mix bus output, and should not be used while in record or when broadcasting.

SPL Sound pressure level relative to a specified reference level. Measured in decibels (dB), SPL is the objective level of a sound source. The subjective perception of 'loudness' is related not only to SPL but also to frequency content and duration.

XLR connector A universal professional connector for microphones and other audio equipment. Most commonly found in a 3-pin form for balanced lines.

XY pair A pair of microphones for recording stereo that are arranged to have their capsules as close together as possible (co-incident pair).

Glossary of recording attributes

Describing aspects of audio with words can be quite tricky, but here goes:

Focus Image stability and concentration in one place. If you shut your eyes, can you locate the instrument very precisely, more generally but in a particular direction, or not at all?

Bloom The important sense of the depth and size of the real space the player is in. The early reflections and the reverb time will convey information about the nature of the room or hall and the space around the player. This translates into a sense of the instrument not being a pinpoint location in the recorded image but having some size and feeling of 'air' around it.

Depth This is the perception of distance from the front to the back of the ensemble. Performers at the rear will tend to sound more reverberant, with some loss of high frequency, and produce less 'close' instrumental noise such as key noise, bow noise, and breathing. They will also have a narrower image width if they are an instrument with any sort of size or a section taken as a whole.

Detail Details of the instrument's sound/noises, such as you get from placing a microphone closer. There needs to be an *appropriate amount of detail* for the distance from the listener: the instrument should not be pulled out of its context in an ensemble by picking up too much in the way of close noises.

Width The amount of space between the loudspeakers that is filled by the recorded image. This works very differently on headphones, and loudspeaker monitoring is needed to judge this well.

Part I

Before recording

Chapter 1

Acoustics and the recording venue

Capturing the musicians' performance in a suitable acoustic space is central to the art of recording classical music. A good sounding space will enhance both the recording and the performance itself; playing music in a rewarding acoustic is a much more enjoyable experience than playing in a deadened room or a space with distinctive echoes or other problems. Reverberation from the room enables the players to hear their own playing better and thus helps with tuning and other performance attributes. Commercial record companies will spend a lot of time searching for halls and churches with a suitable acoustic characteristic, and as a result, many venues are used by multiple companies and freelancers for making recordings. However, it is not always possible to pick and choose the venue in a non-commercial situation, and so we need to consider what can be done in a less than ideal environment. The use of additional artificial reverb is an essential skill in classical recording, and this will be considered in Chapter 17.

1.1 Brief introduction to room acoustics

This book is intended to have a practical approach, and so for detailed discussion of theoretical acoustics and mathematical modelling of spaces, you should seek out an acoustics textbook. However, there are a few basic concepts that are worth including here.

Reverberation is built up from the repeated reflections of sound waves within a building, and after the very early reflections, it quickly becomes a blend of thousands of reflections and interactions that are no longer individually discernible. At every reflection of a soundwave, some energy will be absorbed and some will be reflected. This behaviour varies with the surface involved, and most importantly, it is frequency dependent. In a room where most of the sound energy is reflected at each interaction with a boundary, the sound will continue bouncing around the room for a longer time and produce a longer reverberation time. In a room where most of the sound energy is absorbed at each boundary interaction, the reverb will die down very quickly. After the first early reflections in the room, the reflections quickly multiply to form a characteristic reverb signature which is determined by a number of factors (see sections 1.1.1–1.1.4).

1.1.1 The materials making up the surfaces in the space

Any very smooth, hard surface will reflect more high frequencies (HF) than a rougher surface, which tends to scatter reflections and dissipate energy. Stone and glass will generate the most HF reflections, followed by painted, modern plastered walls and wood with a hard modern varnish. The more HF reflections there are, the more HF content will be in the reverb, and the space will have a bright reverb characteristic. Rougher surfaces such as oiled and waxed wood finishes will reflect less HF energy at each reflection, and carpets, curtains and upholstered seating will absorb higher frequencies very quickly.

Low frequency (LF) energy is absorbed very little by soft furnishings, but is taken up by structures with low resonances such as sprung floors and partition walls that are not solid. A lot of the energy from the LF waves will be taken up by these structures, making them move slightly, and will then not be reflected to form part of the reverb. A room with a sprung floor or partitioned walls can tend to sound bass-light when compared with one that has a solid floor and walls. An interesting example of a sprung floor was Brangwyn Hall in Swansea that had a floor with adjustable springiness so that it could be used as a dance hall. When Decca recorded there, the floor was adjusted to its firmest point to minimise LF loss.

Kingsway Hall (which was situated at 75 Kingsway, London WC2) was another very well-used recording venue. It had a sprung floor with a large void underneath that affected the sound in a different way. The void was used to store large rolls of theatrical fabric which were very absorbent, and they reduced the sound of the underground trains that ran beneath. This became apparent when the void was cleared out in the 1980s and the train sounds became more intrusive and amplified by the resonance of the now empty space. The floor and void in combination had a resonance in the 600 Hz region that actually enhanced the viola section.

Figure 1.1a and 1.1b show images from Kingsway Hall.

1.1.2 The size of the space

The larger the distance from the performers to the first boundary (a wall, ceiling, or floor), the longer the time taken for the first early reflections to arrive back to the players and any microphone placed near them. This timing gives a characteristic signature to the sound that enables us to judge the size of the space. A large space will also tend to have a longer reverb time because of the distances involved between reflections. A bathroom or a stairwell is likely to have a longer reverb time than a large softly furnished sitting room because of all the hard surfaces, but it will not sound like a large concert hall because the first reflections happen after a very short space of time. A useful rough guide is that sound takes about 3 milliseconds (ms) to travel 1 m (3′4″), or roughly 1 ms per foot. As noted in Chapter 17, artificial reverb is generally better at modelling the reverberation tail than the early reflections.

1.1.3 The geometry of the space

The geometry of the space includes both the shape and size of the room. Surfaces might be flat or curved, parallel or divergent, and these affect how sound waves behave in the room. Simple-shaped, small rooms with dimensions of around 3–10 m with parallel walls are susceptible to LF standing waves because the dimensions involved tie in with the wavelengths of frequencies in the 30–100 Hz regions. Parallel walls that are very reflective can produce a distinct series of echoes as the sound is bounced back and forth between them. These 'flutter echoes' are easy to perceive using a transient sound such as a hand clap to set them off, but they will be an unpleasant

Figure 1.1a A view of Kingsway Hall taken from the stage looking towards the balcony, which could be used for placing a chorus of up to 300 for recording. This image shows Anatole Fistoulari conducting the LPO with Mado Robin.

Photo: Courtesy Decca Music Group Ltd.

colouration on most sound sources placed between the walls. Rounded wall sections, as are sometimes found behind a church altar, can focus sound or can reflect it around the edges as in the whispering gallery at St Paul's Cathedral in London. Temple Church in London has a rounded west end which produces some interesting focussing effects on the organ pedal notes, which are particularly noticeable just before the point where the rounded walls start. (See Chapter 14.)

Domed ceiling areas can cause greater focussing of sound as they are curved in all directions, and it is usually a good idea to avoid setting the players up under a clearly defined domed area. Low ceilings will send very early reflections back to the players and will impart a rather boxed-in feeling to the recording, so any high ceiling would be preferred.

The effect of arched ceilings that are often found in older concert halls and churches is dependent on their materials; those that are made of a combination of ornate wood and plaster can work

Figure 1.1b A view of Kingsway Hall towards the stage with the balcony behind the camera. This image shows Sir Georg Solti conducting the LPO and Vladimir Ashkenazy playing the piano for Bartók sessions.[1]

Photo: Courtesy Decca Music Group Ltd.

really well as they disperse the sound energy in a complex way that enhances the reverb time. Both Kingsway Hall (now demolished) and St Eustache Church, Montreal (used for many Decca Orchestre Symphonique de Montréal (OSM) recordings) have ornate arched ceilings. Where an arched ceiling is undecorated and made of hard stone, the effect is not nearly as pleasant.

Figure 1.2 shows an image from St Eustache Church, with the seating removed and flooring in place for recording.

Figure 1.2 St Eustache Church and the Orchestre Symphonique de Montréal

Photo: Courtesy Decca Music Group Ltd.

Another interesting feature of churches are side chapels; if large enough, these areas can have their own distinctive reverb which only becomes audible when the players are playing loudly, and so their effect is unpredictable and intermittent. To locate the source of the problem you will have to go and walk around the church while the performers are playing.

1.1.4 The contents of the space

Soft furnishings, curtains, carpets, and padded audience chairs will absorb HF and mid frequency (MF) reflections and shorten the reverb time. The presence of an audience seated on hard chairs will shorten the reverb time in comparison with the unoccupied hard chairs (which is something to remember when moving from rehearsal to concert). Likewise, areas of wall surface that are uneven, such as areas of ornate wooden decoration or cases full of books, will disperse and absorb reflections and are very good for taming echoes. When running a recording session, two of the simplest things you can do to change the acoustics is to add or remove any upholstered audience chairs and open or close any large curtains.

1.2 What to be aware of when looking at a venue

When you first visit a venue that you are considering using (or having to use) for a recording, it would be useful to be able to get a feel for whether the space is going to produce difficulties or whether it is going to give you a usable reverb signature. Background noise levels must also be taken into account, and Chapter 2 contains further advice on visiting a potential venue.

Things to pay attention to immediately are:

- Traffic noise and aircraft noise. (Is there a quieter time of day? Can you record at night?)
- Buzzing lights. (Can you get them fixed? Work without them?)
- Air conditioning noise.
- Noises and cracking sounds from the heating system pipes (many venues have old heating systems that have to be run for quite a while before the temperature settles down and the noises stop).
- Broken panes of glass that might let birds in. Noisy birds within a church roof space are a common recording problem.
- Identify anywhere that that has some degree of acoustic isolation from the main space that you might be able to use as a control room.

In terms of assessing the acoustic suitability of the space, there are a few things you can do.

Firstly, visual clues:

- What's on the floor? The surface most likely to produce good results for classical recording is a wooden floor over a solid foundation, with a natural finish such as spirit varnish, wax or oil. This will give some nice reflections and slightly more warmth as some of the HF is absorbed. A shiny polyurethane varnish or a stone floor will give a more brittle sound as it reflects more HF. Carpeted floor will absorb a lot of HF and MF and will be less enjoyable for the players if it covers the whole floor, although a small area of carpet around the performer can be a help if the floor reflections are very bright.

- Is it a sprung floor? If so, it could act as a bass trap and remove some of the LF reverb spectrum, depending on its exact properties.
- What's on the walls and other surfaces? Stone will give brighter reflections, and too much HF can make the strings in particular rather tizzy and over-bright. Wooden areas are good, especially if they are rough or have a lot of architectural detail. This will disperse sound and mitigate against strong, distinctive single reflections. A minimalist glass and steel building with little in the way of architectural decoration inside will have a lot of distinctive HF reflections and very little to disperse sound energy, and as a result will probably sound coloured and horrible.
- Is there an arched ceiling? Again, if this contains a lot of wooden beaming, or panelling, it is likely to be beneficial, but if it is very unadorned stone, it will create a lot of HF reverb.
- Is there a dome? Are the players planning on sitting right underneath it? See if you can talk them out of this.
- Is the ceiling low? Be aware that a space that is too small in volume can become acoustically saturated quickly if the source is large and/or loud such as a brass band or orchestra. Playing something loud in too small a room is likely to be unpleasant for the players, and the recording will also suffer as the sound will not feel open with its full dynamic range but compressed and closed in.
- Are there any side chapels the size of small rooms that are enclosed on three sides? These will have their own reverb characteristics that might be triggered by loud sounds in the main space.

Then, some aural clues:

- Clapping at various locations within the building will give you an idea about the timing and quality of early reflections. Are there any distinctive echoes where the sound seems to slap straight back at you? Are there any areas of flutter echoes? You will also get a sense of what the HF decay sounds like, and whether the reverb is bright.
- Stand next to a wall and talk or sing at it – this will give you some useful information about what sort of frequencies the walls are reflecting. Hard plaster will reflect differently to older horsehair-type plaster.
- Making a variety of different frequency sounds around the building – hooting, squawking, and so forth; use your chest voice and head voice and see how the building responds to a variety of stimuli.
- To check for LF standing waves, you will need to be producing a substantial volume of lower-pitched sounds. See if you can set any off within the building. If you are recording something with organ, see if you can ask the organist to play some lower notes to check for any unusual acoustic behaviour.
- Stand at the position within the building where the performers are likely to be situated. Do sounds that are emitted here set off any unpleasant echoes, flutters, or standing waves? Do they activate any side chapel reverb if loud enough?
- Ask someone to walk down into the building and talk back to you from various places. They should do this without shouting or raising their voice so that the room is not excited to any great extent. The intelligibility of the speech will give you a helpful indication of the nature of the early reflections, and how much clarity you are going to get from the musicians at

various distances. If you find that speech intelligibility is compromised at 3–4 m away (9′10″ to 13′), you should not consider using the venue at all. If it is becoming difficult to understand at 9 m (30′) away, the recording might be more challenging but achievable; 9 m is roughly the depth of an orchestra, and the back players should at least be able to understand the conductor.

1.3 What can you do to help with poor acoustics?

Without the ability to completely remodel the space you have to work in, there are still some things that you can do to help where the acoustics are less than ideal. Below are some of the most common problems, and some suggestions for helping with them.

1.3.1 The space is boxy sounding with a short reverberation time

Short reverb time is one thing; this can be helped by the addition of artificial reverb (see section 17.5). The boxy quality is down to the room being a little small, possibly with a low ceiling, and simple in shape, and the early reflections are what gives this quality to the recorded sound. The amount of room colouration that you are picking up needs to be reduced, and this is best done by using curtains everywhere possible to reduce reflections as much as possible, and then adding in artificial reverb later on. Adding reverb to a very dry sound is discussed in section 17.5.7; it can be difficult to do realistically and might involve layering more than one reverb. If adding curtains to the space is not possible, moving microphones a little closer to the musicians and using more directional microphones will reduce the proportion of reverb to direct sound. The danger of this approach is obtaining too close and localised sounds from the players, so should be used with care. Again, the resulting recording will need artificial reverb.

1.3.2 The space has an extremely long reverb time

This would be typical of a large cathedral, especially if the surfaces are stone and marble. Reverb times in the range of 6–12 seconds are not unusual, but make the recording of non-liturgical music quite difficult. Given the size of the space, there is likely to be little you can do to significantly reduce the reverb time, although keeping as many padded chairs out as possible is sensible. Wooden pews and chairs will not have the same beneficial effect. As in section 1.3.1, getting closer with microphones and using directional microphones so that the direct sound dominates the signal will be the best way forwards. The reverb tail will still be there, but at a lower level, and careful addition of some artificial reverb of a shorter length to fill in will help improve the reverb tail to make it feel more natural. Figure 1.3 is an illustration of what you might try to achieve. (See also Chapter 17.)

1.3.3 The space has prominent flutter echoes

There are two approaches. One would be to avoid placing the players between the walls that are causing the problem; this might of course be impossible. The second approach will be to treat the walls by hanging something absorbent, such as curtains, over at least one of them to prevent the repeated bouncing back and forth of the sound.

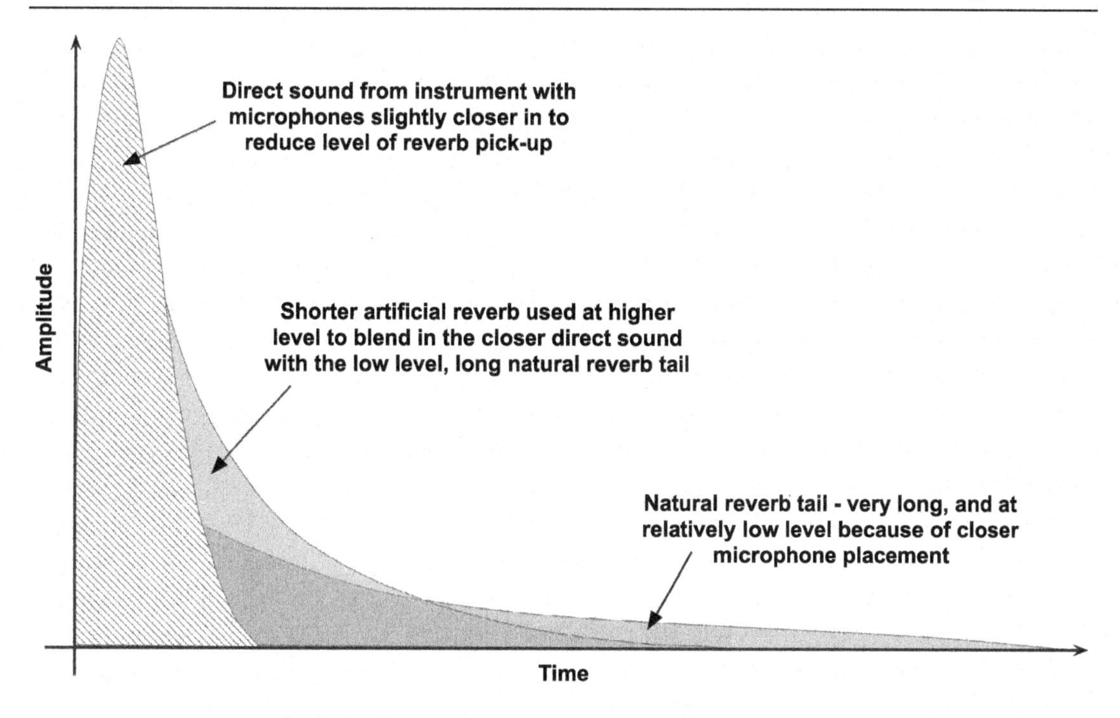

Figure 1.3 Shows a loud sound with a long, fairly low level reverb tail, and how shorter artificial reverb can be used to fill in the gap

1.3.4 The floor is too reflective

If the immediate floor reflections are very bright, then placing the players on a small piece of carpet (about 1.5 m (5′) square for an individual player) will help this. Annoying floor reflections are much more of a problem with chamber music where there are very few players and a large expanse of shiny floor. Where there is an orchestra, there are more bodies in the room that will help to absorb some of the immediate sound and break up floor reflections. This effect can be partially replicated when performing chamber music by adding randomly distributed upholstered chairs around the players to reduce and disrupt floor reflections. They can also be used to good effect in the corners of the room to break up general early reflections; the addition of a lot of upholstered chairs will reduce the overall reverb time.

1.3.5 There are distinctive low to mid standing waves

LF standing waves are not something that can easily be prevented without big architectural alterations as LF energy is not absorbed by materials such as curtains and carpets. If it is possible to avoid placing your microphones where the particular frequencies are exaggerated or diminished, you should do so. Sometimes, moving the room set up around by 90° can help with standing waves.

Any decisions about changing aspects of the performance space should be made with the involvement of the musicians who will need to know what you are doing and why and be given a chance to try out the results. If they strongly dislike the effect on their playing of you having curtained off the entire room, you will have to accept a compromise position. Acoustics affects both the performance and the recording, and we need both for a successful outcome.

Note

1 BARTÓK/Piano Concertos 1 & 2/LPO/Solti/Ashkenazy/DECCA (1980) LP SXL 6937.

Chapter 2

Studio techniques and working on location

This chapter covers the essential practicalities involved in running a classical recording session on location and coming away from it all with plenty of usable recorded material, venue staff who would be happy to see you return, and no third-party insurance claims. Much of this chapter is about preparation, getting to know your equipment well, and devising an efficient workflow in advance. Making recordings is essentially about communication and working with the fragility of musical performance, and the more time and attention you are able to give to the musicians, the happier and more productive the session will be. This means making your technology as quick and simple to use as you can so that it does not get in the way of the flow of the session.

It is strongly advised that you make a detailed visit to a venue that you have never used before. By introducing yourself to the venue staff (and making yourself look organised and professional in their eyes) you can work out all the practical details in advance that make the difference between a successful or stressful experience. A reconnoitre is particularly valuable prior to a church or cathedral organ recording; see sections 14.2 and 14.9 for discussion of problems particular to churches.

2.1 Equipment

Preparation of a standard checklist that you take out every time you go is essential. Make a signal flow diagram of the whole process from microphones back to control room, including monitoring, talkback, and other communications, and then go through it methodically to make your list. Once you have done this, try to think what could go wrong at every stage of setting up and decide what you could do, or what you might need to ameliorate this. Take spares of everything; if you plan on using 10 microphones, take 14 with you.

For example, the list of things you might need to be sure of being able to rig a microphone could be:

- Microphone.
- Clip – make sure clip has a thread adapter (3/8″ to 5/8″).
- Shock mount.
- Microphone stand.
- Boom (and counterweights).
- Stand weights or sandbags.

A cable run planning process might look like this:

- How long are the cable runs? Measure them, don't guess.
- How many cables?
- Multi-core or single cables?
- Will the cables need to go through a door at any point? Find out.
- Will cables need to go over a door (fire exit)? Find out.
- What sort of protection will you use for cables placed in vulnerable positions?

Cables are something that need looking after, and they will last a very long time if you do. They will need protection anywhere that they might be trodden on, have vehicles drive over them, pianos wheeled over them, or even get chewed on by rodents. (Rodent-proof fibre cable is available; it is about 10 mm thick, but is worth having if you have to leave a long cable run through the outdoors for any length of time.) Cable covers are used both to protect cables and prevent a trip hazard. In reverse order of effectiveness, the options are:

- Gaffer tape – can be used to stick cables down and prevent tripping. Cheaper makes in particular can leave sticky floor and sticky cables – and the venue might not allow you to use it.
- Carpet – will prevent tripping, but will not protect the cables if a piano goes over them.
- Rubber cable covers with a channel on the underside – will prevent damage to cables unless a vehicle drives over them.
- Heavy duty cable channel covers (such as Adam Hall defenders) will protect cables from damage by vehicles, so they are for use outdoors.

Fibre cable can be protected from damage in door hinges and so forth by using some cheap, extruded foam pipe insulation that comes with a long tube slit down the side and is easily cut to length.

Apart from the obvious things like microphones, stands, cables, mixing desk, and DAW, there are several additional essentials:

- Tape measure or a cheap laser measurer – for measuring cable runs, room dimensions and microphone locations. Logging your microphone positions is a good discipline; it allows you to accurately add delays in mixing, or reproduce the session again another time, or to learn from your experience.
- White insulation tape and a permanent waterproof marker – these can be used for labelling cables and mixing desk channels.
- Black insulation tape – for discreet marking of microphone stand positions.
- Torch – always useful for when you have to plug something up, find the essential widget you dropped in the dark, or read something in a dark corner of an old, badly lit building.
- A heater for the control room (see section 2.2.4 for heating and lighting the musicians' areas.)
- Sandbags to use as stand weights – especially for lightweight stands such as Manfrottos.
- Plug-in circuit tester and Residual Current Devices (RCDs) – you will find out if there is mains-borne hum and if the sockets are safe to use.

2.2 Practicalities at the recording venue

As noted in the introduction, doing a reconnoitre of the venue beforehand with a member of staff is immensely valuable, if not essential. You will need to plan cable routes, find power points, locate any potential control room, and assess parking and building access. This will tell you what, where, and who your problems are going to be. Background noise levels are often a particular problem in churches and cathedrals; see section 14.9 for further discussion.

2.2.1 *Venue staff*

Your first job on arriving at the venue for a reconnoitre is to introduce yourself by name to all the officials, to give them your phone number, and to get their names and phone numbers if you can. Be nice to them, and be open about your plans; you want to avoid surprising them with anything that you want to do, as many people's reaction to surprise is to say, 'No'. Make them a cup of tea/coffee when you are having a break, bring them mid-morning pastries if you are having them, and never underestimate the power of goodwill to smooth the path if you have to negotiate something tricky.

Someone will be in charge of keys and letting you in and out of the building. If you are going to be there for a few days, and have made a good impression by being organised, safety-conscious, courteous, and all-round good guests, they might even be persuaded to let you have a set of keys for the duration. If you are leaving equipment in the building overnight, locking up securely is very much in your own interests. There will also be someone in charge of health and safety (H&S) and fire regulations, and it is extremely important that you do as they ask with good grace. They have the power to inspect your equipment and to put a stop to your session at any time if they are so minded, so try not to get involved in any altercations. There will also be rules about whether or not you can use gaffer tape, especially if it is a historic building. Many venues do not allow it at all or have a specific brand that you must use, and it is up to you to ask about this beforehand and adhere to the rules.

2.2.2 *Health and safety: fire exits, cable runs, and electricity supply*

At all times, remember that you are a guest in someone else's premises, and what you consider to be safe might not be the same as that which the H&S officer considers as safe. The fire exits are sacred; they should be the first thing you look at when you go into a location in order to work out how you are going to manage your cable runs around them. You will not get away with running cables across the floor in front of a fire exit – so do not even try – and if the fire officer tells you that a piece of equipment can't go somewhere, then it can't go there. Cables that have to cross a doorway have to pass above the doorway on brackets fitted for that purpose; they cannot be taped around the door frame as this would fail in the event of a fire.

All cable runs should be as neat as possible; it always looks better to run cables together as far as possible and branch out with individual cables as required rather than having them splaying out at random. Tidy cable runs make it safer for the players in the stage or orchestra pit area and also make it easier to troubleshoot and replace a faulty cable where required. Leave the coil of cable next to the stand in case you need to move the microphone; if the session is being filmed or it is a live concert, the coil should be located by the stage box once microphone stand positions are confirmed.

Colour coded cables can work on a small session, but labelling both ends of a cable (with insulation tape and a marker, or masking tape and pencil/ballpoint) is the most comprehensive way of dealing with a large number of cables, and all cables will have to be black if it is being filmed or is a live event. Insulation tape is also good for marking the positions of microphone stands that need to be moved and replaced in the same position. Place a marker under the centre pole of the microphone, and include an arrow for the direction in which the microphone should be facing. If you are coming back in a few days to continue with a recording, it is worthwhile making sure that the cleaners know not to remove your microphone stand markers.

The electricity supply should be tested using a socket tester, and then you should use an RCD circuit breaker on each of a number of distribution boards, and feed everything from these. As modern recording equipment does not consume a great deal of current, the only things that should have an individual supply would be large, class A power amplifiers.

2.2.3 Security and insurance

You should of course have adequate insurance for your equipment and professional indemnity insurance to cover accidental damage to buildings, musical instruments, and people. Using a specialist insurance broker will usually be more cost-effective than going directly to a mainstream musical instrument and equipment insurer that is primarily aimed at gigging musicians and their individual instruments. It is easy to lose track of the value of what you have, especially if you have built it up over a number of years, so make sure your insurance is adequate for your needs, in terms of value and locations in which you might work.

Once you are working in a venue, be aware that there are a lot of expensive instruments and equipment lying around, and there should always be someone in the hall if any doors to the building are unlocked. Security in older churches can be particularly difficult, as they are sometimes harder to secure completely and can be located in large urban areas with a higher risk of opportunistic crime. You need to find out if you can leave your set-up overnight if sessions are to run into another day; ideally, you would get a set of keys and make sure that the keyholders treat locking up extremely seriously. Avoid advertising your presence in the building by keeping any control room blinds down, even during the times that you are working. It is easy for someone to spot that you are working there and decide to come back later when all is quiet. Make your vehicle as anonymous as possible for when it is parked up outside the venue; choose a van without windows, and if you have to leave equipment in it, see if you can arrange for others to block the van in with other closely parked vehicles to make opening doors impossible.

2.2.4 Heating, lighting, air conditioning, and ventilation

All these are aspects of making the location a comfortable space for the musicians to give of their best. It can be easy for an artist who is feeling a little stressed to channel that feeling into complaints about draughts, temperature, and lighting. You should make every effort to pre-empt these and address any complaints that arise as best you can. Do not be drawn into an argument that tells the artist that they are wrong to be feeling too hot/cold/unable to see the score. They have to perform, and your job is to make that easier for them. To get a good idea about additional heating or air-conditioning needs, do your recce visit at an appropriate time of day; if you are planning an overnight session, bear this in mind if you visit the church on a warm day.

The temperature in an old church is almost inevitably too hot or too cold; the building's heating system (often hot air based) is likely to emit a lot of noise while it is getting up to temperature or cooling down. There might also be localised hot spots near to scalding heating pipes and hot air vents, and extremes of temperature are bad for both performers and acoustic instruments. The best plan is to get the heating switched on early, and let the temperature settle to avoid 'cracking' sounds. If you are working in a really large space, you might need additional heating to create an area that is warm. Gas heaters are a common solution, but if they are a bit hissy, you will have to move them away from the microphones, and then prevent the musicians from repositioning them. You could switch them off for when actually recording as they will still retain a lot of latent heat and should radiate this efficiently. Calor gas heaters will produce a lot of moisture and some fumes, so ventilation will also be important. Electric heaters will involve a lot of current, so you need to be sure that the electrical system in the venue is able to cope with this. If they are fitted with a thermostat, they will switch on and off at intervals with an acoustic click, and very possibly an electrical click that can turn up on your recording. Making sure that the recording gear is on a different circuit is one way of dealing with this, but it will not solve the acoustic noise.

Few recording venues other than studios have air conditioning built in (in the UK), but if they do, you can at least hope that it is both quiet and effective. The stand-alone portable air conditioning units that can be brought in tend to make a lot of noise, and they generate heat and water that have to be channelled into a different room or to the outside. The only way to use these is to turn them on in the breaks only. A room full of musicians, such as choir and orchestra will generate a lot of heat and moisture, and will lead to air quality loss over a long session. This increased concentration of carbon dioxide can lead to creeping tiredness amongst the players and loss of performance. This can happen even in a large space, such as St Eustache, Montreal, if there are enough people, so opening any possible ventilation in the breaks is a good idea. The Royal Albert Hall has air vents in the top of the roof for this purpose. The same problem might manifest itself with a smaller ensemble in a small space and is also likely to affect your control room, so bear it in mind.

Lighting is very important for classical players when they need to be able to read their score or see the conductor clearly, and it can also be used to create a comfortable session ambience, such as by the use of table lamps in a chamber music session. It needs to illuminate the right things and not dazzle anyone, so diffuse lighting is best. Barn doors fitted to lamps can be very useful in directing light, and solving problems of lights shining into people's eyes. Different bulb types have advantages and disadvantages, but LED bulbs are the best overall solution. Halogen lighting can be very easily damaged, especially if the musicians move them around to suit themselves. Old-fashioned bulbs will give out heat as well as light, which is potentially useful in a cold church, but not good in a room that has struggling air conditioning. Being able to set the colour temperature of the room is psychologically advantageous, and if you can find a set of LED lights that will crossfade between a set of warm and cold white colours, this would be a very flexible solution. Lights can be mounted on most microphone stands, and the tall aluminium stands, such as Manfrottos, were originally designed for lighting but have been adopted for microphone use as they have fittings for both. If you have a heavy light mounted on a tall stand, make sure it is absolutely safe by weighting the bottom of the stand.

2.3 Rigging microphones and running cables

2.3.1 Planning signal routing

Unless it is a very small session, draw a set-up diagram and write a plug-in list (which microphone, what it is pointing at, which channel of multi-core it is going down, any intermediate routing, and track number on the DAW). Prepare this in advance and make multiple copies for anyone who is helping. This helps very clear communication and avoids time-wasting with assistants. Label microphone cables at both ends, and for a complicated set-up or live show, label the stands with microphone type, what it's pointing at, and which line it is going down. This enables assistants to go through the whole orchestra or pit band and set up efficiently by following instructions, without needing to have a full overview.

2.3.2 Stands

Making stands safe is essential. Lightweight stands are strong, but need weighting at the bottom as they can more easily tip over. Where you are using a boom arm, it should be properly counterbalanced with weights, and a long, extended boom arm should stay where it is even if you undo the clutch. A friction clutch is not sufficient to hold a full Decca Tree rig in place, and it is dangerous practice to rely on this to prevent it from swinging down onto the orchestra or conductor. Stand legs can be a trip hazard in a tight space, and places like the orchestra pit at the Royal Opera House will usually use heavy, round-based stands so that they can be fitted in more neatly.

2.3.3 Stereo bars

It is worth saying a few words about stereo bars here. Many of the spacings discussed in the book require wider stereo bars than the standard narrow ones which are about 15 cm (6″) wide. It is possible to buy wider stereo bars at around 25 cm (10″) wide, and even wider still at around 70–80 cm (27″ to 30″). Some of the widest stereo bars can be expensive, but look for one that will allow for mounting several microphones on the single bar. Cheaper versions of the wide stereo bars made by DPA and Manfrotto are coming onto the market (e.g. the Superlux MA90). If you find yourself without a wide enough bar, two or three K&M narrow stereo bars can be screwed together using the microphone clip 3/8″ screws that are attached, but they can be completely unscrewed if necessary.

2.3.4 Slinging microphones

The first thing to say about slinging microphones is that it is a serious business, and you need to be certain of what you are doing. Do not do anything without a detailed discussion with the technical manager of the hall; the decision to go ahead will be a joint one, and you cannot expect to turn up on the day and start suspending microphones from parts of the building without permission. Any large venue is likely to specify what you can use to sling microphones, and all rules must be followed; for example, the Royal Albert Hall has very detailed technical specifications for what can be used, and who can access any gantries in the high roof spaces. When someone is working at height, there should be a system for clearing the floor below. Even if it is a small venue that is more relaxed, you should be vigilant and safety conscious. Do not take risks – you will be liable for damages if anything happens as a result of your microphone rig descending from the ceiling by accident. If you can avoid slinging, you should; it will take a lot longer than you think, and once microphones are slung, it is time-consuming to move them if you need to do so.

Never try to sling microphones alone as it will take hours of adjustment and will probably not be safe while you do so. You will realistically need three people to make it more efficient. Two will deal with the mounting positions on the building structure (attachment to either side of a balcony, for example) and the third needs to be in the middle to watch over the microphone rig and to take control of the positioning instructions. To give an idea of the time-consuming nature of safe slinging of multiple microphones in a large space, it takes the BBC about a week to rig the Albert Hall for radio and TV broadcasts from the Proms season each year.

One of the key principles of slinging microphones is that there should be redundancy in your rig; that is if something breaks, something else should still be holding the microphones in place. Microphone cables are not to be used for suspending microphones, but additional steel wire (preferred) or ropes need to be securely attached to both the microphone stereo bar and the building support. Microphones need to be firmly attached; the clip must be screwed properly onto the stereo bar, and the microphone tightly held by the clip. Spring loaded clips are not suitable for slinging, so friction clips that can be tightened around the microphone should be used. This is to protect both your microphones and anyone below the rig. It should also be noted that when you suspend microphones on a stereo bar, they will be mounted hanging down below the bar rather than sitting above it, as would be the case if it were on a stand. This keeps the centre of gravity of the assembly as low as possible.

Manufacture of materials that are suitable for slinging is a specialist undertaking, and you should look to a theatrical chandlers (such as Flints in the UK) for advice and stocks of black steel wire and ropes that will not melt under hot lights. These companies will also finish off steel wires to your own requirements, including properly made loops on the ends that will hook into a shackle or other clip. For tying off steel wires to an appropriate length, there is an excellent range of cable grippers made by Reutlinger. These come in various sizes for different thicknesses of wire and weight limits and are spring-loaded, enabling quick adjustment of wire length by quickly pulling it through. The spring-loaded grip ensures that the cable won't slip back while it is being adjusted, and once the microphones are in the right position, the mechanism can be screwed up securely. Give some forethought to the equipment that you might need, remembering that if you look professional, you will inspire more confidence in your clients.

When suspending microphones on a stereo bar or a lightweight Decca Tree frame, you will have to deal with running microphone cables out along the supporting wires. The cable can be attached to the wires at intervals with cable ties, and this needs to be done before you hoist the whole rig into the air. The weight of the cables needs to be evenly distributed to avoid overbalancing the rig, and making it tilt to one side. If you rig a stereo bar with a steel wire on each side, the cables will have to be run out from the bar one to each side. If you rig a wide pair of omnis plus a central ORTF pair (see Chapter 3) onto a single wide stereo bar, the cables from both left microphones will need to be tied along the left wire and the cables from the right microphones along the right wire. This is less convenient for keeping your cable runs together, but will make it possible to hoist the rig without it tilting.

2.4 The control room

When you commandeer a room to use as a control room on location, it will almost certainly be unsuitable in some way, but you will have to make the best of it, and keep it as tidy and organised as possible. If you are in a theatre or hall, you might end up with a dressing room that is covered in mirrors and is an awkward shape and too small. It is a good idea to come prepared with fabric that you can use to drape on mirrors to dampen off any HF reflections. You won't be able to affect much else about the acoustics, but this is worth doing. On the practical front, you need somewhere for microphone cables to come into the room, and if you can avoid putting them through the same door that the artists will use to come in for playbacks, you should do so. Trip hazards are something you should be aware of in general, but particularly in the areas of the control room that will be used by you and the musicians, who will be fresh from performing and not looking out for cables.

The control room has three zones:

1 A desk surface for computers/fader controllers/mixing desks/producer scores.
2 Area in front of the desk where engineer/producer and musicians are moving around.
3 Behind the desk where the loudspeakers are (if you have them).

Any mess belongs behind the desk and loudspeakers as it is assumed that no one will be walking in that area unless they are fixing something.

2.4.1 Monitoring arrangements: loudspeakers and headphones

If you can monitor on loudspeakers, you should do so, as there are many aspects of classical recording that can't be well judged on headphones, such as lateral panning and reverb choices. (See section 2.6.6 for discussion of monitoring on headphones versus loudspeakers as it affects workflow.) Most control rooms will not have enough space to set up full-size studio monitors in a large equilateral triangle, and the acoustics will probably not be suitable for this. Smaller active monitors that can be used in the near to mid field will be the most practical solution, and will give you a better idea of the overall sound than using headphones alone. They also enable better communication between artists and producer during playbacks because closed-back headphones in particular make it difficult for people to listen and exchange comments at the same time. If you do make allowance for several pairs of headphones in the control room, make sure that each has its own volume control that is isolated from any main loudspeaker level control. If you have some stands for the loudspeakers, then bring them; if they are not very substantial or weighty, see if you can use sandbags both to damp them and to prevent them falling over.

2.4.2 Aids to monitoring: phase scopes

A very useful tool for indicating stereo image width, particularly when monitoring on headphones, or on loudspeakers in a poor quality control room (such as a church vestry), is a phase scope – also known as a vector scope or Lissajous display. Originally these were built around

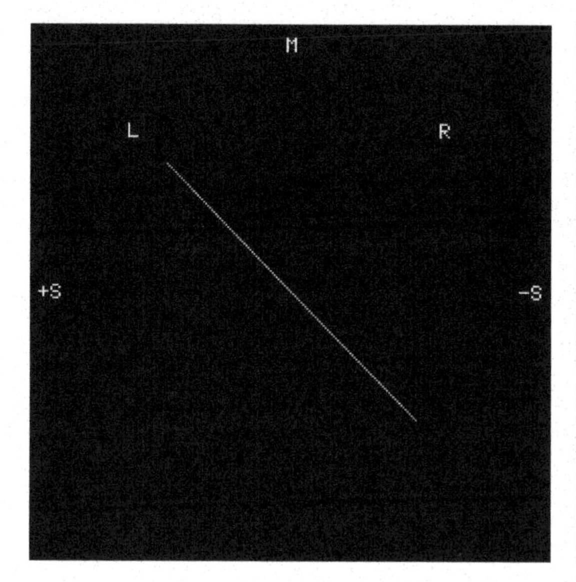

Figure 2.1 Signal on left channel only

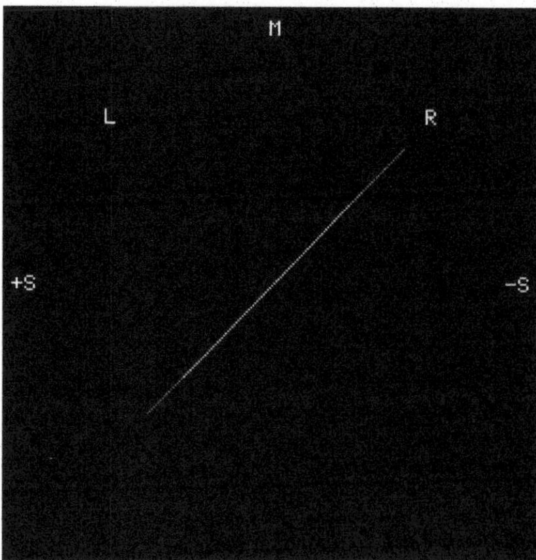

Figure 2.2 Signal on right channel only

modified cathode ray tube oscilloscopes, but these days it is much easier (and cheaper) to use a software plug-in or application. There are numerous freeware and shareware phase scopes available; the images in Figures 2.1 through 2.4 are taken from the Vector Audio Scope included in RME's free Digicheck software, which will work with any RME audio interface.

The phase scope takes an incoming stereo signal and uses the two channels to generate a display using a moving point that leaves a trace on the screen. If only a left signal is present, then the result is a line (formed by the dot moving back and forth at speed) that is tilted over to the left by 45°, as shown in Figure 2.1. The length of the line corresponds to the amplitude of the signal. Likewise, if there is only a right signal, the display shows a line tilted over to the right by 45°, as shown in Figure 2.2.

When there are signals on both channels, the resulting display depends on the difference between the two signals. If they are identical (i.e. mono) the result is a vector sum of identical left and right signals (i.e. a vertical line), as shown in Figure 2.3. If they are identical, but one is phase inverted, the result is a horizontal line, as shown in Figure 2.4.

A display taken from actual music will look like those shown in Figure 2.5. The shape of the 'ball of scribble' gives information about the stereo image width, L-R balance and degree of phase coherence between the left and right channels. The more vertical the display, the more components of the sound are in phase; the more horizontal the display, the more they are out of phase. The 'fatness' of the display gives a very good indication of stereo width, and when its width exceeds its height, the amount of out of phase components will make a stereo image that is over-wide, spacey, and uncomfortable to listen to. See Chapter 5 for more practical examples.

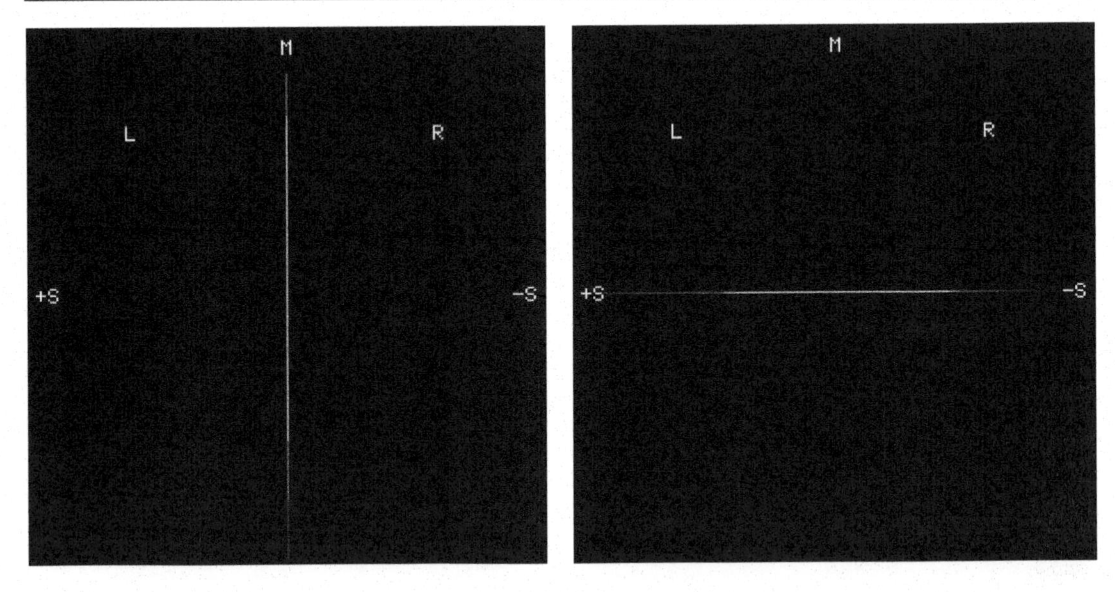

Figure 2.3 Identical signal on both left and right channels (mono)

Figure 2.4 Identical signal on both left and right channels, one with inverted phase (reversed polarity)

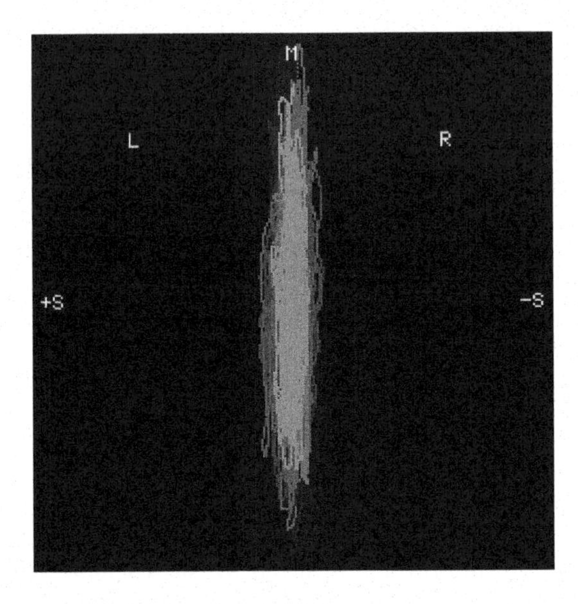

Figure 2.5a–c Stereo images of increasing width

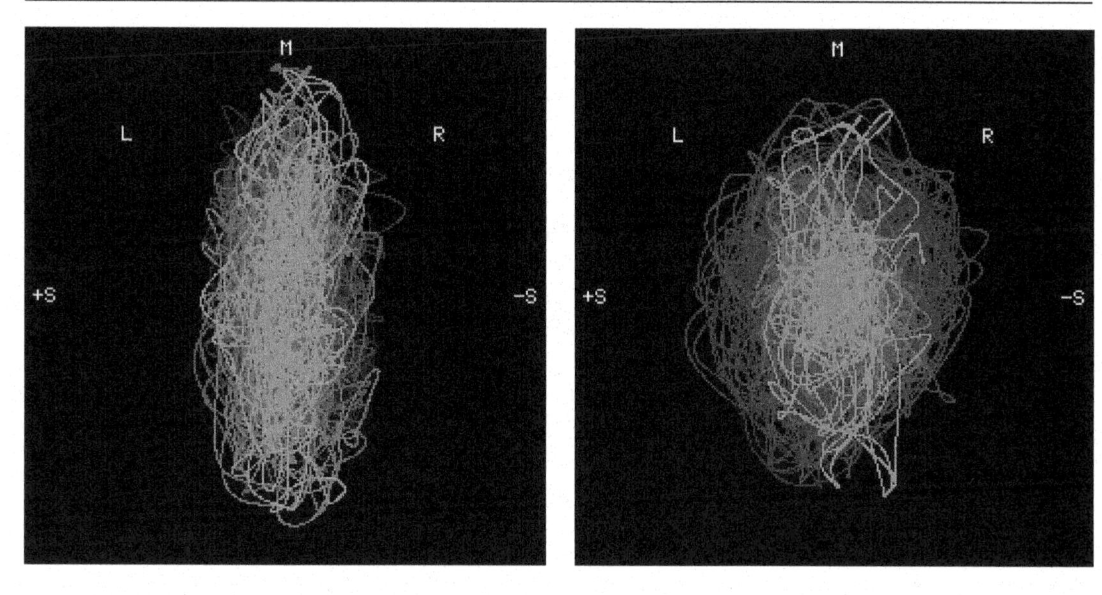

Figure 2.5a–c (Continued)

2.4.3 Control room seating layout

Different layouts of the control room encourage different styles of communication between the engineer and the producer (unless one person is performing both roles). Emerging from the 1950s and 1960s, Decca and EMI developed different traditions in this respect; at Decca, the engineer sat centrally with the producer alongside, which reflected a collaborative relationship; at EMI, the producer sat behind the engineer in a central listening position at a separate desk. Both approaches have different merits; side-by-side seating aids communication but places the producer in a less good stereo listening position. A desk area is always needed for the producer to spread out scores and notes; placing this behind the engineer gives the producer a good workspace, but it also requires a deep control room. Your control room positioning might be a matter of compromise as well as personal preference.

Finally, control room heating and lighting needs to be considered, as both producer and engineer will be sitting down for a long time, and a cold church in the winter will be quite miserable in that regard. You will need an electric heater, without a fan, to keep the background noise level down. Run it from a separate circuit to your equipment if you can, or make sure that the thermostat is not set so that it keeps clicking in and out.

2.5 Studio communications: talkback, telephones, and cue lights

During a classical recording session, none of the performers is wearing headphones as a rule, (there is usually no click-track or backing track), and a small loudspeaker is used to relay the producer's comments to the studio floor. Where there is a larger ensemble and a conductor, a talkback telephone is also used. This enables a more private discussion to take place if necessary, although the producer should avoid this going on for too long as it can cause stress to any soloists who naturally assume that the conversation is about their performance, and also that it is not complimentary. If the telephone is used all of the time, the performers can feel out of the communication loop, but if it is used too rarely on a session, their immediate assumption is that it is bad news, so a balance needs to be struck.

A basic talkback (TB) system consists of a microphone in the control room and an active loudspeaker in the studio area (requiring a power supply) connected by a microphone cable, or a built in microphone tie line if there are any that you can use. Ideally, the loudspeaker should be on the same electricity supply as the control room to avoid potential for hum on the talkback loudspeaker, but using an earth lift XLR barrel will solve this. The TB microphone needs to be switchable, preferably dimming the control room monitoring when it is engaged. This function will be included in any mixing desk with a dedicated talkback. A more complicated TB system will include a selection of destinations for the producer to talk to, so they do need to be certain of which buttons to press to talk to whom. This is only likely to come into play on a large session with chorus, orchestra, soloists, and offstage performers requiring additional conductors.

A bell (rather like a small fire alarm bell) is sometimes included in talk-back systems and should be operable from the control room. It has two functions; in a busy session, it can be used to gain everyone's attention if they are chattering and can't hear the talkback, and it is also a useful source of sound on the studio floor that can be triggered from the control room for testing microphones.

In order that the artist or conductor can talk back to the producer, they will also need a microphone. There are, of course, many microphones set up on the studio floor, but they are not all necessarily near enough to the performers themselves, and a dedicated talk microphone might be needed. This will always be the case for pianists and organists who are seated some distance from the microphones on their instruments. Make sure that any talk microphones are cut during recording, or are not sent to the main mix bus if you are able to arrange separate monitoring for them.

Some sort of visual communications such as a video link is a very good idea, and some would say essential. Being without a window into the studio area means that you are left guessing what is happening if something isn't going to plan, and means that you cannot see if someone is not standing where they should be in relation to their microphone. Something to beware of is the latency of modern digital video monitors, and this can only be avoided with an analogue camera and CRT monitor. Latency is not a problem if the images are just for general use, but if you are using remote cameras for communication with a conductor who needs to cue in some offstage performers, this can be a problem. Line whistle from a camera in the studio is another thing to look out for; even if you are old enough not to be able to hear it (15.625 kHz for PAL video), you should be able to check for it using a spectrum analyser on your DAW.

Cue lights are traditionally red, but there is an interesting school of thought that suggests using a green light might be less psychologically stressful for the musicians. Whatever the colour, everybody in the room needs to be able to see the light to know when you are in record, and you might need to place more than one light at the back of the orchestra. Mounting a bulb on a tall stand (with a diffuser if it is very bright and directional like an LED) will mean it is generally visible. It is also useful to have a separate light outside the studio door that can either be tied in to switch with the studio floor lights or treated separately so that it is on all the time during a session to discourage interruptions. Switching of lights (which is under the control of the engineer or producer) is best implemented by use of low voltage DC control lines that can be used either to drive an LED or to operate a mains relay near to where the mains powered lights are located. This avoids long runs of mains cable that can induce hum onto microphone cables and is safer than switching mains voltage directly. There are several companies that supply control panels for this purpose.

2.6 Optimising recording workflow

This section deals with engineering work practices aimed at achieving a smooth and quick operation that minimises flow disruption in a session; section 2.7 deals with the production aspects of running a session.

2.6.1 *Headroom and microphone gain*

Working at 24-bit resolution is standard in the classical world, and this does allow you to leave enough headroom when setting microphone levels so there is no risk of accidental clipping. A degree of under-modulation of between about –5 dB and –10 dB will not produce any really significant noise problems; the ambient noise of a typical, practical room will be down at around –60 dBFS.

Keep notes of your microphone gains on each session and you will get to know your microphones and your microphone amps. Typical gains will be around 30 dB for condenser microphones and 50–60 dB for ribbon microphones. (Dynamic microphones are not usually used in classical recording as their off-axis colouration in particular will detrimentally affect the quality of spill.)

2.6.2 *Capturing clean mic feeds plus a stereo mix*

There are a several methods of achieving the desired result: clean microphones feeds plus a stereo mix. One is to use an in-line console arrangement, and make the stereo mix (along with any EQ) from the monitor paths. The stereo mix is then sent to the DAW to be recorded alongside the clean microphone feeds from the channel paths. Another is to use a straight to stereo mixer to provide the stereo mix, but record the channel direct outputs after the microphone amps as additional tracks. Finally, if you are working completely inside the DAW without a mixing desk (but perhaps with a hardware fader controller), the mix can be created within the DAW and then be sent via an internal bus to another pair of tracks to be recorded alongside the microphone feeds. This has the advantage that all the fader moves can be retained as automation within a single project, and small amounts of remixing can be just a matter of trimming the fader moves from the original mix. Where you are working with an external mixing desk, microphone gains, fader levels, panning and EQ should be logged to enable you to address any remixing as efficiently as possible.

2.6.3 *Using faders versus mixing with a mouse*

Using real faders, whether on a mixing desk or on a control surface for the DAW, enables the engineer to react immediately to the need for a live, real-time level change, and control more than one level at once. The experience of balancing a few sources against one another simultaneously is very important when learning to mix and, although it is not impossible to mix with a mouse alone, it is harder and it takes longer. Using a mouse inevitably changes the workflow from making balance decisions at the time of recording to deferring them until there is more time to manipulate channels singly. In a commercial situation, using less time is very important, and although it feels more pressurised to make decisions during the session, greater experience will make this easier, and ultimately workflow will be considerably faster.

2.6.4 *Mixing to stereo on session: why use this workflow?*

The traditional classical workflow was to mix everything live to stereo on the session and produce a series of two-track master takes that were then edited together. However, to allow for the possibility of remixing the edited material at a later date, it is more flexible to record the stereo mix alongside the clean microphone feeds as part of the same multi-track file. As noted in 2.6.2, this is easy to achieve, and the multitrack files can be edited across all the tracks whilst monitoring the stereo mix. If the session mix is good enough, it can be used for the production master and the edited multi-track files ignored. Budget constraints do not always allow for remixing later on, but this method is less time-consuming than the old-fashioned method of remixing all the individual takes and editing the whole project again if a remix was required.

Apart from time or budget constraints, other factors point to the need to spend time on obtaining a good, working balance at the start of a session:

1 Microphone placement needs to be assessed as part of the mix; if the microphones won't blend together well, the answer will be to move some of them. If you don't take the mix seriously on session in the belief that anything can be 'fixed in the mix' later, you might find the microphone placement is not good enough to make a really good sound.
2 Doing the best mix you can on session (while there are conversations and input from the producer and musicians) means that you will not be starting from scratch at a later date without this memory and experience.
3 You need to have something good to play to the musicians during session playbacks, and to send to them as part of the post-production schedule. It makes the post-production workflow quicker if the edited master can be sent directly to the musicians at any stage without the need to spend time on creating a mix.
4 Editing with a monitor mix that is going to be significantly changed is likely to cause audible edits once remixed.

For further discussion of managing post-production workflow, please see section 18.2.1.

2.6.5 *Obtaining a good monitor mix quickly*

Achieving a good balance quickly is increasingly important in a commercial context. This is due mainly to increasing financial pressures but also partly to some engineers' egos and the somewhat erroneous belief that technological advances mean that all problems can be sorted out after the session.

The UK and US systems of running a session gave the engineer different amounts of time to obtain a balance. In the UK, the 'rehearse then record' system gave the engineer about 20–40 minutes of session time to get a good balance, so this gave time to move microphones if necessary. In the United States, the engineer had about 10 minutes at the start of the session, and the beginner should be reassured that short a time is only possible if the engineer is experienced and knows the hall and the orchestra. Additionally, as noted in 2.6.3, using faders rather than a mouse will make for much faster balancing.

Eighty to ninety percent of the overall balance comes from the main microphones, with the rest being finishing touches, so it is worth prioritising these and taking time to get them in the right place. Speed of balancing will come with experience, but a good balance can only be achieved on the back of good microphone placement. If you have the opportunity to set up and try out your balance in a rehearsal environment, you should take it. This gives you the chance to take your time, to listen to what the band actually sounds like in the hall, to listen to how the room behaves, and to observe any strange squeaks and noises that might be coming from piano stools, pedals, or other instrumental sources that need to be addressed. Spending time in the hall listening also tells the musicians that you care about what they are doing and makes for a greater degree of trust in your work.

You should *avoid changing balance and microphone or player positions part way through a session*. Changing any of these things makes editing very, very difficult, and you can end up with two sets of mutually incompatible takes. This does make the responsibility to get it right at the start quite terrifying, but it is ultimately much better to live with your initial decisions. The only exception to this should be if the request for a change during the session comes from the artist; this is unusual, and it has to be for a good reason. The producer should be aware that any takes recorded before the change may need to be discounted.

2.6.6 *Monitoring: loudspeakers and/or headphones*

The most flexible arrangement on location is to take both loudspeakers and headphones, but you should avoid spending too much time on headphones if you can, as reverb and lateral panning judgements made on headphones do not translate well onto loudspeakers. It does, however, work perfectly well the other way around, so a mix made on loudspeakers will translate onto headphones. Headphones are useful for listening to small details in a compromised listening environment such as on location, and most producers like to wear them during takes. Semi-open back designs will often give a slightly better spatial sense that is closer to the loudspeaker sound, but you will have to dim or cut the loudspeaker monitoring when you use them. Closed back headphones will shut out the sound from the loudspeakers and any other sources, including spill from the studio floor.

In practice of course, there will be times when you have to monitor on headphones because there is no isolation between your control room and the main room, or you are actually in the room with the artist. If this is the case, then you will not really know how the recording is sounding at the time; the best thing to try is to record part of the rehearsal and play it back during a break so you can make some judgements while you have the chance. If you are working on headphones alone, panning the headphone feed inwards to about 75% can help a little with compatibility with the expected sound on loudspeakers. Under these circumstances, you should multi-track the session if at all possible and avoid adding any high-pass filters to microphone sources. Despite the monitoring limitations, you should create and record a monitor mix which will give you some idea of whether your microphone placement is working and give you something to use in playbacks (see section 2.6.4). If you are not able to multi-track but can only capture a mix, keep any additional reverb low in level as it is easy to add too much – you can always add a little more later if you need to. See also Chapter 18.

2.7 Running the session

There are many organisational and logistical aspects to running a session quickly and smoothly, all of which will need some advance planning and preparation. The time when the musicians are with you is precious and should not be wasted. From the players' point of view, the session needs to flow without pauses for technical hitches or administrative tasks such as logging takes.

2.7.1 Keeping musicians happy: home comforts

People involved in the session need to be looked after, which means that someone needs to have sorted out refreshments and toilets. If this is an orchestral session, the orchestra should have the means of supplying hot drinks, although you might have to be involved with agreeing where to set up the water urn, kettles, waste bins, and so forth, and locating the water supply if the venue does not have a kitchen area. If your session is a smaller group, you should provide the refreshments, including snacks as well as drinks. Be generous in your provision and it will not go unnoticed; the recording session is a special event for the artist, and treat food in the form of nice biscuits or similar will usually be welcomed. Tea break frequency and timings will be decided according to local musicians' union rules for a large ensemble and in accordance with the session flow for a smaller group. Comfortable levels of lighting and heating were discussed in section 2.2.4.

2.7.2 Session durations and session breaks

Most UK/US recording sessions are three hours long and run for three sessions a day, usually 10.00–13.00, 14.00–17.00, and 19.00–22.00. If you are working with a soloist or small group, the time can be split up in whatever way works, but for a larger group, a system is needed. The current arrangements might seem rather strict, especially for those who imagine that musicians are simply doing something they love, and so will be happy to play for additional minutes without payment. However, it must be remembered that they are highly trained professionals, and the system evolved to put a stop to record company abuse in the early days. Unions have worked hard to maintain co-operative session discipline without which recording an orchestra would be very difficult, given than any one player can halt the session. The strict discipline on timings is there because it is a team operation that requires every member of the team to be present at the right time.

A 10.00 start time means that the first downbeat can be at 10.00, so everyone has to be set up, tuned, and ready to play at that time. This also means that the engineering team has to be ready to begin, and being late is considered extremely discourteous to all the other people involved. If you are late, it suggests that you consider yourself important enough to waste everybody else's time.

In the UK, the standard recording session length as defined by the Musicians' Union is three hours, with five minutes' break per hour (or part hour) to be taken at once (15 minutes) approximately halfway through the three-hour session. During this time, the maximum duration of repertoire to be recorded (once it has been edited together) is 20 minutes. This excludes breaks taken for playbacks, and the players are free to use the break time to do anything they like. Overtime has to be agreed by all the players in advance of the end of the session (some players might have a booking elsewhere and be unable to stay on), and so all are entitled to leave at the end of the allocated time, and they will do so. There is an agreed minimum unit of overtime (15 minutes) that will become payable if you overrun the session by a second. For up-to-date rules and rates in the UK, contact the Musicians' Union, 60–62 Clapham Road, London SW9 0JJ (www.musiciansunion.org.uk).

In the United States, the basic regular session as set out by the American Federation of Musicians is of the same length, but has to contain two 10-minute breaks, and a maximum of 15 minutes of repertoire can be recorded. The basic unit of overtime is 30 minutes, but this can be 15 minutes if the purpose of the overtime is to complete a piece of repertoire. However, different rules apply where the recording is by a named 'symphonic orchestra'. For symphonic orchestra recordings, the basic session is three hours but can be extended to four by agreement. Break time is 20 minutes in each hour, but this can be grouped together, with players playing for a maximum of 60 minutes without a break, so a session can be run as 'one hour on, one hour off, one hour on', and the break used for playbacks. Seven and a half minutes of music can be recorded in 30 minutes of session time, which means that 45 minutes of repertoire can be recorded in a three-hour session. For up-to-date rules and rates in the United States, contact the American Federation of Musicians, 1501 Broadway, Suite 600, New York, NY 10036 (www.afm.org).

Because the rules are strictly applied, the producer really needs to be on top of the session planning and timing, and there are many occasions where an orchestral recording has had to be halted with a few bars to go and the whole piece tackled again during the next session. It is not unknown for players, particularly in the rear string desks, to set wristwatch or phone alarms to go off a few seconds before the end of the session.

2.7.3 Over-exertion of artists, especially singers

It is very important that you remain aware of rising tiredness levels, especially where you are running a session that does not have set break times. All players, but soloists in particular, are giving very deeply of themselves, and if they are new to recording, they are likely to find this particularly draining. For the artist, making a recording is a really important and special experience, even if it is a routine experience for the engineer. Singers are particularly prone to becoming tired, and you should also be aware that solo singers will usually give of their best in afternoons and evenings, so morning sessions are best avoided.

2.7.4 *Page turns and scores*

Page turns are noisy, and you should plan in advance how to deal with them. The most prominent will come from a choir who will all be turning the page at the same time, whereas orchestral parts all have their page turns at different times. Most orchestral players will be sharing a stand so that one player can turn while the other plays. For large choral page turns, you might need to record small sections from memory to cover the page turn in editing. Ring-bound scores should be avoided as they cannot be turned quietly.

Soloists will need to prepare scores before the session with the aim of minimising page turns. Pianists will often ask for a page turner, but if the turner is inexperienced, they might not be completely silent. Both the actual page turns and the movements of the page turner as they stand up and sit down have to be inaudible. Someone with experience should do an excellent job, but if no one experienced is available, it is often preferable for the piano score to be photocopied and spread across the music stand, which will take five sheets on a grand piano.

The producer also needs to prepare scores, and this will sometimes mean photocopying to make quick repeats easier to deal with. Many producers prefer to work with a pile of single-sided photocopied score sheets than a bound score; in this case, make sure that page numbers are included. It is important to make sure that everyone is working from the same edition, and it is advisable to put in page and bar numbers before the session, as these might be the only common reference point between the artist and control room. If the artist has been asked to provide a score for the producer, it should be made clear to them that this score will need to be kept after the session for post-production, and that it will be written on in pencil and ink as part of the production and editing process.

2.7.5 *Approaches to recording takes*

In order to allow the musicians to settle into the room and get into their playing zone, it is usual to do about three complete takes at the start of the session. Record all of these; the first take is rarely the best, but it might contain useful short segments of material that can be used to patch another take. It is usually on the third time through that the performance will start to gel properly. After this, the producer can start on patches, sometimes working through the piece from the start, some preferring to work backwards through the piece, making sure that any flaws that exist in the complete takes are covered. Where you need a short patch, take a longer one as it will work better for editing. Having told the musicians where the problem lies, ask them where they would like to pick up from so they can pick a comfortable starting point. Once all mistakes have been covered (and covered more than once for preference, and more editing choice), it is often a good idea to run another complete take. The aim should be to get relaxed musicians for this take, so tell them that you have everything that you need already, and that they should just enjoy the performance without worrying about mistakes. With any luck, you will get another good, complete take.

Flaws that you are listening for include intonation, wrong notes, missing notes, wrong rhythm, poor ensemble (not playing together), poor dynamics, noises in the room such as page turn noises and bow hits, and so forth.

2.7.6 *Pitch and tempo consistency*

Maintaining tempo consistency is a factor in all recording sessions, and maintaining pitch is a problem when recording unaccompanied choir. Most experienced musicians playing repertoire that they know well tend to be very consistent in tempo between takes; the performance tends to settle into its own pace and feel rather like breathing. The times that tempo can become unstuck are when recording shorter sections, when the overall feel can get lost. In this situation, you should have a metronome to hand which you can use over the talkback if needed. Most metronomes of phone apps now have the means to measure a tempo if you tap along, and this can be a useful double check between takes.

When it comes to choir pitch, it is useful to have a realistic assessment of the choir's consistency; the choirmaster or mistress should have a good idea from rehearsals. A pitch reference needs to be given before each take, either from a piano or keyboard (which can be played over the talkback if needed.) Anything you can do to avoid having to correct pitch later on in post-production will save time and money. (See section 18.5.4 for discussion of managing pitch correction after the session.)

2.7.7 *Logging takes*

Keeping a record of the takes that have been recorded and which bars of the piece they cover is an essential part of the session. The session will move quickly, and your primary task is to concentrate on the artists and the music. Any take logging system needs to be very simple and should not involve you having to make complicated decisions about naming takes. If the session record-keeping is not good enough, expensive post-production time will be wasted. Most typically, this will mean trying to reconcile the producer's notes with an incompatible list of timings produced by the engineer or assistant. Inconsistencies make it difficult to work out which takes are actually meant to be used in the edit.

The requirements of any system are:

1 Everyone must agree on which take number is currently in use (this avoids the producer's take numbers not matching up with engineer's take numbers in post-production).
2 The take number should appear in the audio file name, as a marker on the project timeline, or both.
3 The take durations should be readily available during the session (useful for a rough guide to the speed of performance).
4 The numbers of the bars covered by each take should be recorded.

If the recording session is a simple one, with a single producer-engineer, it is relatively straightforward to log takes and avoid confusion about which take number is current. Where you have both a producer and an engineer (or assistant) operating the DAW, it is usual for the DAW operator to log take timings and name audio files correctly. They will also take charge of deciding on the take numbering because they are starting and stopping the recording process; many workstations will display the take number on a large display if required. The producer will note musical flaws in the score and keep notes of which bars of the piece are covered. The coverage of each take might also be noted by the engineer or assistant as a backup, especially if the producer is known for keeping sketchy notes.

For take logging to work smoothly, you need to get to know the marker system and audio file naming system on your DAW inside and out. Evolve a plan for working with the markers so that you can use it at speed on session, and know what will happen to the markers under every circumstance:

- Do they increment each time you add one?
- Will it add one every time you drop into record?
- Will they all get renumbered if you insert a marker?
- Are the markers attached to the timeline or the recorded media?
- Do you have to drop out of record at the end of every take?
- Are the audio files named incrementally?
- If you have to stop and start unexpectedly, how will you deal with this?
- Can you name a file at the end of recording to enable you to drop-in quickly?

It can be very useful to be able to leave the DAW in record through a section of short retakes that happen one after another – can you split these up and rename them easily afterwards?

Most DAWs do not really adequately manage the nuanced business of logging takes on a classical session, and as a result, many professional engineers use additional pieces of software of their own devising. One important facet of these is their ability to differentiate between the act of dropping into record and marking the start of a take. It is often useful to drop into record some time ahead of the take start, and so any bespoke system will allow the user to use keystrokes to log take starts, ends, false starts, breakdowns and restarts, all independently of dropping in and out of record. Accurate start and stop times also allow for the automatic calculation of take durations, which is a good guide to tempo consistency. Some systems will allow the user to leave the DAW in record throughout and will automatically split up the long audio file into shorter clips and rename each according to the take logging marker information. Systems of this kind superseded the manual jotting down of time codes at Abbey Road during the 1990s.[1]

Where an artist makes a start on a take, but stops again well short of the planned end point, it becomes very important to have a single person in charge of take naming, and to decide if the next attempt is a new take number, or whether the aborted first attempt (which could be several bars long) is just a 'false start' to the current take number. A take might have several false starts, and the producer will log them all in this way: F/S, F/S, F/S. This material might still be used as part of an edit sequence, and would be indicated as 'take 56 F/S 2' for the second false start to take 56.

Another frequent occurrence is where the players stop a take, and then quickly pick up again a few bars earlier. They will do this of their own accord after a very short discussion so that the producer and engineer have no time to intervene. In this case, the take will have 'BDRS' noted, along with a bar number and/or time code to show that there was a 'break-down and restart' in the middle of the take. Soloists can manage a BDRS almost imperceptibly as they need no discussion and can stop and start a few bars earlier with minimal pause in the music. The producer needs to keep an ear open for this happening and make a note of it.

A final note on the subject of takes is whether to announce the take number to the studio floor or not. Announcing it to the DAW is an excellent way of adding a means of unambiguous take identification to an audio file should the file naming go astray, and will be valued in post-production. However, announcing the take number to the studio is not universally practised because it can be stressful for artists to hear that they are on a very high take number. This requires a talkback system that can send to the DAW independently of the studio floor.

2.7.8 Playbacks

Artist playbacks are an essential part of the recording session, and they are an opportunity for them to comment on the sound as well as the performance. It is usual to invite them into the control room after the first few complete takes to assess what has been recorded so far, and what still needs to be done. This also gives the producer an opportunity to review the recorded material and plan the next moves in the session.

The producer will be in charge of what is to be reviewed, and the engineer must locate the material quickly and without hesitation. The process needs to be seamless, fast, and efficient. You should get to know the fastest way to locate to a particular marker, and how to mute the playback until the right section of audio comes along. Artists really do not like to hear out-takes, false starts, and chatter between takes. Listening to their own mistakes makes them more stressed, which is the opposite of what is required for good session performance. As noted earlier, headphones during playbacks will reduce interaction between performers and the producer, or they will all start shouting at one another over their closed back headphones.

2.7.9 Recording room tone

The final word in this chapter belongs to room tone, or 'ambience' as it is sometimes known. This is a recording of the sound of the room with nothing happening in it, and needs to be done with the microphone set-up as you have it for each session that you do. Room tone is essential in classical recording, as it is used for the gaps between pieces on the final master and is useful for editing repairs and extending pauses. It will make editing and post-production much easier if you capture a few minutes of quiet room tone on the session; if you forget to do this, you will have to spend a lot of time scraping together about 20 seconds of room tone constructed from half-second snippets taken from between takes (if the artists stop talking, making noises or tuning for long enough). Record several minutes of the artists sitting silently if they can manage it or leave the DAW in record when you go for lunch. If you do this on every session, you will eventually build up a collection of room tone; this sounds very unexciting, but it is a tremendously valuable resource that is hard to fake.

Note

1 The system introduced at Abbey Road in the 1990s was 'Takelog', which was based on an earlier Decca system.

Part II

Recording

Chapter 3

Basic two-microphone stereo techniques

The focus in this chapter will be on the two-microphone stereo techniques that are frequently referred to throughout the book for non-orchestral recordings such as piano, chamber music, choirs, and organs. These fundamentally useful techniques are co-incident directional microphones, spaced omnis, and spaced and angled cardioids (of which the ORTF pair is the most well-known example.)

3.1 Co-incident microphone techniques

When two microphones are mounted as close together as possible (or when a dedicated stereo microphone with two capsules is used), the microphones are said to be co-incident, meaning that they are located at the same point in space. In theory, there are no timing differences between the two signals generated by such a pair, as sound waves from any source will arrive at both microphones at the same time. The resulting stereo image is produced by level differences between the two signals; these level differences are created naturally by the use of directional microphones pointing in different directions. Given that it is not really possible to mount two microphones in exactly the same place, there will be some very small timing (and therefore phase) differences, but these can be disregarded in practice.

To make the terminology clear, Figure 3.1 shows a general co-incident pair of unspecified directivity pattern (the patterns most used will be fig of 8, cardioid, and hyper-cardioid). The microphones are mounted with an angle between their front axes, and depending on this mounting angle and the directivity pattern, the pair will have a characteristic L-R 'stereophonic recording angle'. This is the angle between the positions of sources that will appear fully left and right in the stereo image produced by the pair, and it is not the same as the mounting angle. (It is assumed that the pair of microphones are fully panned.)

The L-R stereo recording angle is dictated by the level differences between the microphone signals that arise due to the directivity of the pair, and an individual instrument will only be perceived to be fully left or right when the level difference between its signal on each microphone reaches about 18 dB.[1] The width of the final image of the source will depend on how much of the pair's stereo recording angle it occupies. If the pair being used has a very wide stereo recording angle, and the source is a piano that only occupies a small segment of it, the recorded piano image will be narrow even with the pair fully panned. To make the piano wider without moving closer, the L-R stereo recording angle needs to be reduced so that the piano occupies more of it.

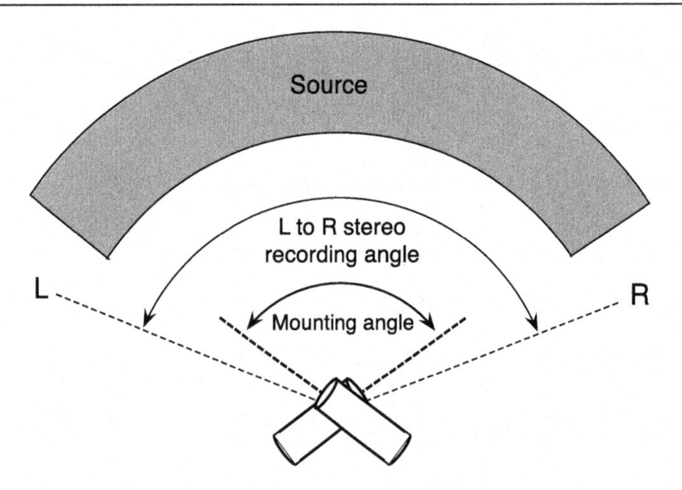

Figure 3.1 Generic co-incident pair

In general, an increase in the mounting angle results in a decrease of the stereo recording angle of the pair. This means that increasing the mounting angle will also increase the image width of the source.

The most well-documented co-incident microphone technique is the Blumlein pair, which consists of a pair of fig of 8 microphones mounted with their front axes 90° apart. This pair has a stereo recording angle that is also 90°, but care must be taken not to place any sources outside this angle to the sides of the pair, as the source will be picked up on the front lobe of one microphone and the rear lobe of the other, meaning that one signal will be phase inverted. When reproduced over left and right speakers or headphones, this phase inversion is a very unpleasant listening experience. Co-incident fig of 8 microphones are almost invariably used mounted at 90°.

When cardioids are used as a co-incident pair, the problem of a phase-inverted region does not arise, but some mounting angles will produce significant angular and level distortion in the stereo image. Angular distortion means that some parts of the image are compressed into a narrow arc, and other parts are stretched out to occupy a wide section of the image. This occurs because of the way that the level difference between the microphones changes with the angle of incidence. Level distortion means that some parts of the stereo image are louder than others (the areas of the image that are compressed into a narrow range are also louder.) The practical upshot is that cardioids mounted at 90° will produce an image which is centre heavy; most of the image is squashed into a narrow central range, with the rest stretched out at the sides. Conversely, cardioids mounted at 180° will produce an image that is stretched out and quieter across the centre and squashed up at the sides. If co-incident cardioids are to be used as a main pickup, they should be mounted at an angle of about 135° as this produces a useful L-R stereo recording angle and the least angular and level distortion. However, cardioids mounted at 90° can be useful for placing on a single, smaller instrument for use as an ancillary pickup. They will produce a very phase-coherent, centrally focussed image. Figure 3.2 shows the mounting and approximate stereo recording angles of cardioids at 90° and 135°.

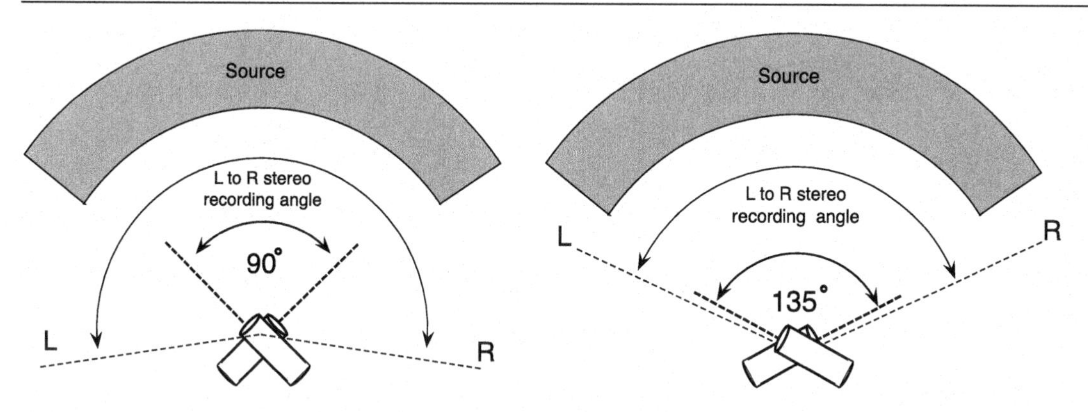

Figure 3.2a–b Cardioids mounted at 90° and 135°, respectively

Co-incident pairs create an image that is stable and coherent but lacks a feeling of spaciousness, and because directional microphones are used, the fullest LF range is not captured.

3.2 Spaced omnis

Using spaced omnis as an overall pair has both positive and negative characteristics. In its favour, the tonality will be excellent, with full LF range being included, and there will be a spacious feeling to the recording partly due to collection of more room sound, and partly because of the phase differences between the left and right channels. The downside is that the image will not be sharply focussed or precisely located, and if the microphones are too far apart, the centre of the recorded image will be low in level and the outer edges will dominate (the 'hole in the middle' effect). The lack of image precision can work to the engineer's favour if spot microphones are going to be extensively used; it means that they can be panned in a wider range of positions without causing conflict with the spaced omni pair. (See Chapter 13 for recording the classical brass ensemble as an example.)

The microphones will need to be mounted on a wide stereo bar (see Chapter 2), and it is common practice to angle them outwards at about 30° to 45° to take advantage of any HF directionality in the microphones. The stereo image when using spaced microphones as an overall pickup is predominantly created by timing differences between the two signals, although there will be some level differences at HF due to the HF directionality of omni microphones if they are angled outwards. The resultant source image width is therefore dependent on the spacing between the microphones, and this must be adjusted to take into account the size of the instrument or ensemble, its distance away, and the desired recorded image width. (See also section 5.5.2.) Spacing between fully panned omnis reaches a natural limit at about 1 m (3′4″), and beyond this, the timing differences start to become too large and there is not enough correlation between the microphone signals to create a stereo image. Reproduction of a large orchestra using spaced omnis is best addressed by use of the three-, four-, or five-microphone Decca Tree (see Chapter 8).

A small table of typical spacings used in practice to produce a conventional image width is included in Figure 3.3. These should be taken as a good starting point from which to adjust

Instrument/ensemble	*Guide to image width as percentage of whole*	*Microphone spacing*
Solo piano	50%–75%	25–33 cm (10" to 13")
String Quartet	50%–75%	60 cm (2')
Chamber choir (20)	90%–100%	80–90 cm (2'7" to 3')
Chamber orchestra	90%–100%	80–90 cm (2'7" to 3')

Figure 3.3 Table of typical omnidirectional microphone spacings

position and spacing for image width and amount of direct sound versus reverb. To make the sound less reverberant, the microphones should be moved closer to the source, bearing in mind that this might also increase the image width by virtue of being closer to the players, so some microphone spacing adjustment might also be needed.

As noted in Chapter 5, spacing greater than about 33 cm (13") on piano starts to produce an over-wide image that has poor localisation. Conversely, if a 30 cm (12") spaced pair is used on an orchestra and fully panned, the image will still be too narrow to fill the space between the loudspeakers.

The microphones should be panned fully left and right (or very close to it). Panning a main pickup inwards to reduce the image width of the source has the effect of narrowing the sense of space and reverb around the instruments as well, and this produces a rather mono-sounding recording. Additionally, because this technique involves the creation of timing differences (and therefore phase differences) between the two microphones, panning them inwards to any significant degree (less than about 70% left and right) will start to produce some colouration due to partial summation of signals of different phase. When using spaced omnis as a main pickup (rather than for a 'stereo spot' as discussed in Chapters 6 and 7), it is best to space the microphones a little less if the image width needs to be reduced and to pan them fully.

3.3 Spaced and angled cardioids

A very useful technique that combines some aspects of co-incident pairs and spaced omnis is the use of cardioids that are spaced a small distance apart and angled away from each other. This general technique produces a lot of possible combinations of angling and spacing that might prove useful (in that they produce a useful L-R stereo recording angle and no significant angular distortion); these were quantified by Michael Williams[2] in 1984. The stereo image is created by a combination of level differences (due to the directionality of the microphones) and timing differences (due to the spacing between the microphones). The resultant image is more focussed than the spaced omnis, but with a more spacious feel than a co-incident pair. Where the extended LF range of omnis is not required (such as for a choir), they make a very useful overall pickup. As with the spaced omnis, the pair should be panned fully left and right as the default, and altering the angle or spacing would be preferable to panning them inwards beyond about 75% left and right.

Of all the possible spaced and angled pairs, two have passed into common use: the ORTF (Office de Radiodiffusion Télévision Française) and NOS (Nederlandsche Omroep Stichting) pairs, as shown in Figure 3.4.

It can be seen that the NOS pair has wider spacing and narrower angling than the ORTF pair, indicative of the trade-off between the two aspects of this sort of pair. As we saw in section 3.1

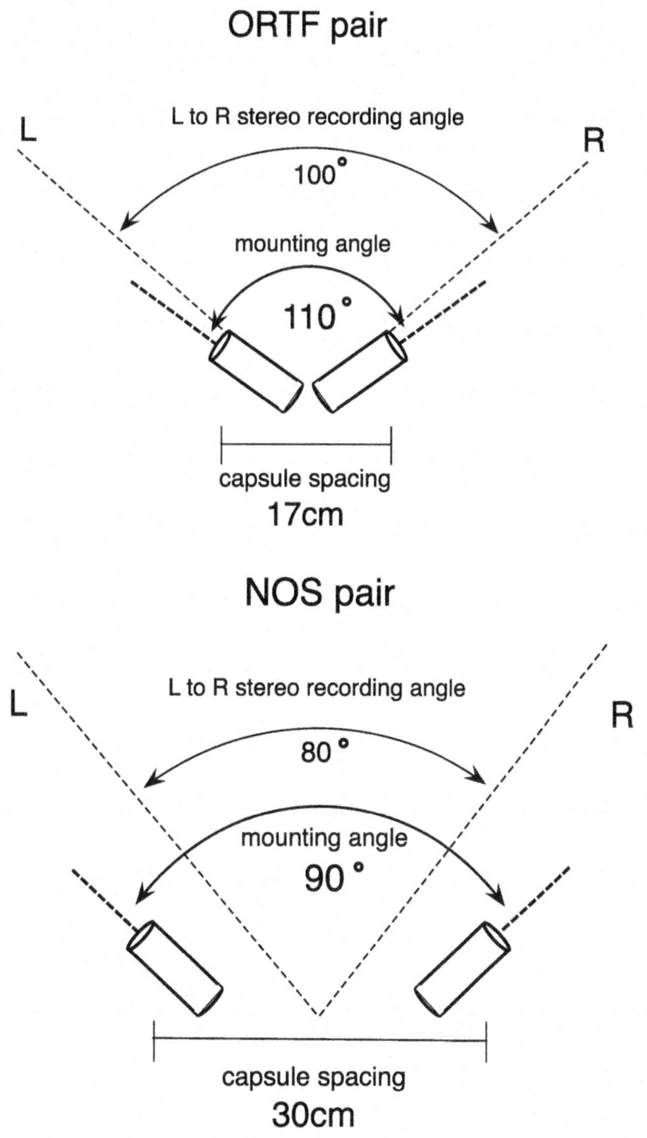

Figure 3.4a–b ORTF and NOS pairs

(co-incident pairs), a wider mounting angle means a wider image, and from section 3.2 (spaced omnis), wider spacing produces a wider image. The ORTF-type technique can be used flexibly in practice, and the angling or spacing can be altered a little to change the image width if changing the distance from the source is not an option. This might arise because the balance between direct and reverberant sound works well at a particular distance, or because the rig cannot be moved elsewhere because of space restrictions.

Elsewhere in this book, parallel-mounted pairs of spaced cardioids have been used extensively as spot microphones on soloists. This technique could be seen as part of the family of spaced and angled pairs as they have a mounting angle of 0° and spacing in the 20–30 cm (8″ to 12″) range. These 'stereo spot microphones' produce an image of the soloist that is wider than a single mono spot but is still appropriately restricted in width when fully panned.

Notes

1 See *Rumsey & McCormick: Sound And Recording*
(Focal Press ISBN-13: 978–0240521633)
7th Edition p498 for a useful summary of the research that underpins this finding.

2 *The Stereophonic Zoom: A Practical Approach to Determining the Characteristics of a Spaced Pair of Directional Microphones*
Author: Williams, Michael
AES Convention: 75 (March 1984) Paper Number: 2072
Publication Date: March 1, 1984
Subject: Studio Technology
Permalink: www.aes.org/e-lib/browse.cfm?elib=11692
There are interactive displays of stereo recording angles for various stereo techniques at www.sengpiel audio.com/Fragen08.htm

Chapter 4

Solo instruments

This chapter aims to cover those instruments for which there is a large body of unaccompanied classical repertoire. This includes both polyphonic instruments that provide their own harmonies (such as the guitar, harp, harpsichord, and other early keyboard instruments) and instruments that primarily produce a single line (such as violin, cello, and woodwinds). For instruments that are almost always employed as part of an ensemble, it is more practical and useful to consider how to record the whole ensemble. Therefore, discussion of brass instruments has been left to Chapters 7 and 13.

Throughout this chapter, the discussion is concerned with placement of spot microphones on a single player. It is assumed that in addition to the spot microphones, there will be some sort of overall room pickup to create a good backdrop of stereo, reverberant sound into which the spot microphones can be blended. Unless you are lucky enough to be working in an excellent acoustic environment, using two pairs will be needed to produce a sound that works well in all aspects. The overall room pair could be an ORTF-type, or omnis spaced at around 60 cm (2′), at about 2.5–3 m (8′ to 10′) high and set back by 2.5–4 m (8′ to 13′), depending on the acoustic.

When recording a solo instrument, the width of the recorded image of the instrument is a primary consideration. The sense of space and reverb around the instrument should fill the image width between the loudspeakers completely, but the width of the instrument within this should not fill the stereo picture if it is to give a realistic illusion of a live performance. A single microphone placed appropriately will be able to capture a good tonal overview of an instrument as long as it is not too close, but the result can be very one-dimensional unless there is also some sort of stereo pickup contributing significantly to the instrument's sound. This aspect of imaging as it relates to a solo singer is discussed in Chapter 6, and the same principles can be applied to solo instruments. Even with additional stereo artificial reverb, or stereo ambient pickup microphones, a single spot microphone can dominate the mix, and the instruments' image may become too narrowly localised, as if listening 'down a tunnel'. Therefore, the use of two spot microphones for solo instruments is something to be seriously considered, to create a sense of width and 'bloom' around the image. The desired lateral image width of a solo harpsichord will be a greater than that of a classical guitar, which will be a little greater than that of a solo violin or oboe, but none of them should be expanded to fill the whole width of the stereo field or collapsed to a mono point. There is, of course, a range of image widths that will produce a great-sounding illusion of a solo instrument playing in a real space, and the final judgement is for the engineer to make. A narrow image can make the listener feel further away, even if the microphones give a sense of being closer. For the sense of distance to work, there should be no conflict between the width of the image, the close detail of the sound, and the amount of reverb.

4.1 Classical guitar and flamenco guitar

These instruments are very similar to one another, both being nylon-strung (unlike the steel-strung acoustic guitar), and the recording approach can be the same.

Guitars, lutes, and theorbos are all quiet instruments with a restricted dynamic range, so the recording venue should have a very low level of background acoustic noise, and the microphones should have low levels of self-noise. Occasionally, classical guitarists will use a small amplifier with a freestanding microphone if performing solo or a concerto in a very large venue that is better suited to orchestral music. This amplifier is not an integral part of the sound; it is there to make it louder as transparently as possible.

The sustained sound from the guitar comes primarily from the resonating front body; the string vibrations are transmitted to the body which acts as an acoustic amplifier. The sound hole is not the main source of the sound, although it is a tempting target at which to aim microphones. From the point of view of the recording engineer, the next most prominent features of guitar playing are plucking sounds from the right hand, and noises and squeaks from the fingerboard. Players always try to keep these noises from left-hand position changes to a minimum, but will vary in their ability to do so. They are usually very concerned with noises not being prominent in the recording, and you are likely to have to spend some time trying to minimise them at the time of recording (the best option) or painstakingly removing them during post production (soul-destroying, time-consuming, and therefore expensive). In flamenco playing, striking the front of the guitar (*golpe*) is part of the technique, and this can produce a large LF resonance from the whole body of the instrument as well as a sharp transient at the point of impact. Use of this important percussive technique will increase the overall acoustic level when compared with a purely classical guitar, and might also be used in more contemporary classical repertoire. Other player noises such as stomach rumbles can become obtrusive when amplified alongside a low level guitar signal.

4.1.1 Microphone placement and image

There is a lot of material written about pop microphone placement for acoustic guitar, but they are generally placed too close to the instrument for a classical-style solo recording. The emphasis is often on two separate spot microphones placed around 20–30 cm (8″ to 12″) away from the instrument, each capturing a certain aspect of the sound such as the main body and the fretboard. These microphones are usually too far apart to form a coherent enough stereo pair, and so they form two pools of mono signal that are combined to create an overall tonal representation of the instrument, if not a good spatial impression. They may be panned almost to the same place, but the aim in a pop song is not to create the sense of a guitar playing in a real space but to capture the closer sound of the instrument that can be placed into a mix.

For classical guitar, a simple narrowly spaced stereo pair can be placed at about the height of the centre of the guitar, perpendicular to its front surface. If the guitar is angled upwards slightly when played, move the microphones higher to look down a little. Fine tuning the placement of the microphones can be helped by sitting on the floor with your head at microphone height, and moving around the instrument to see how the sound changes. As a starting position, aim the left microphone at the bottom of the body and the right microphone at the bottom of the fretboard. This will place the fret noise a little to the right, but this seems to work acceptably when listening to a guitar recording. Lateral movement around an arc will help to reduce or increase the noise from the fretboard if it is too intrusive. (See Figure 4.1.)

The pair should be spaced around 25 cm (10″), pointing forwards at a distance of 70–100 cm (2′4″ to 3′4″) from the guitar (if the room is a nice sounding one, a little closer if not) with a suitable amount of natural, smooth reverb. The image width that works best for classical guitar is to fill around 30% of the total stereo field. If it were too wide, it would give the listener an unreal sense of the instrument's size, and the creation of a real sense of perspective and space would not be successful. A guitar would only fill the listener's field of view if they were seated about 30 cm (1′) away, but the microphone placement and amount of reverb used give cues to the listener that place the player further back than this.

Finding an appropriate balance between the reverberant and direct sound by altering the distance at which the microphones are placed can be difficult to get right with classical guitar. There is a conflict between trying to pick up this quiet instrument in a reverberant environment by

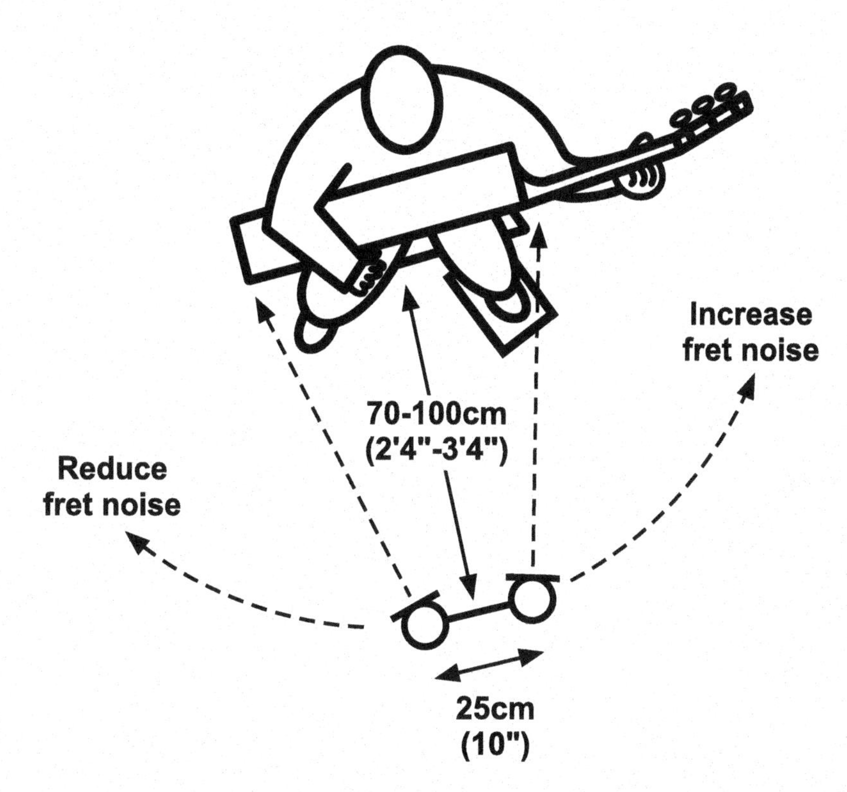

Figure 4.1a Classical guitar microphones – plan view showing line of lateral adjustment

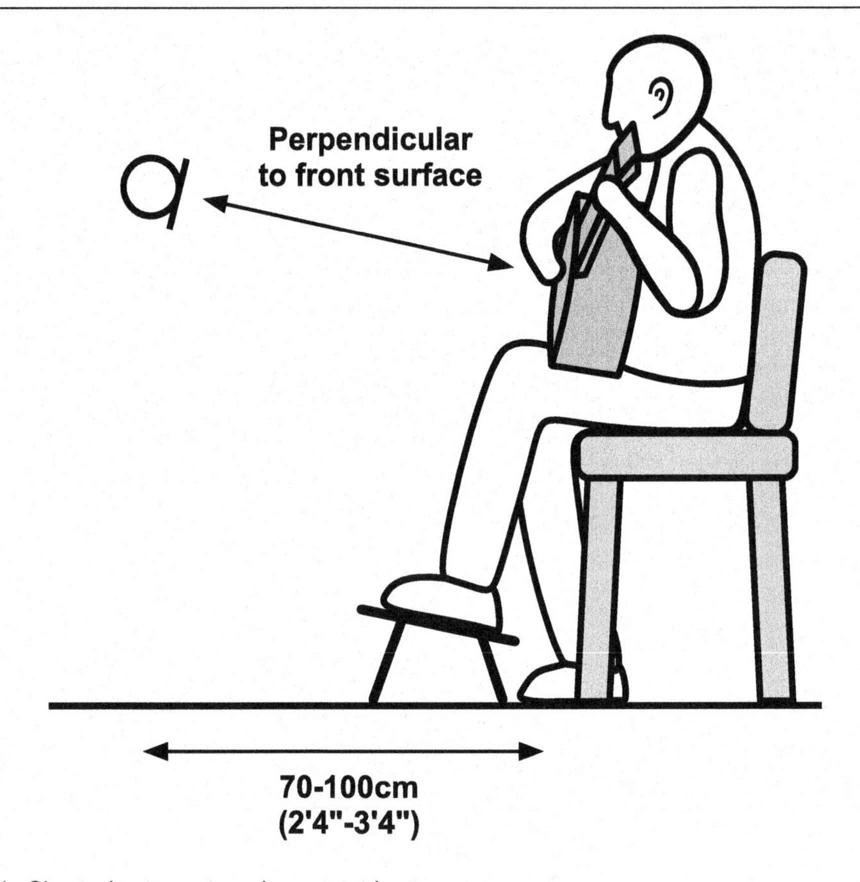

Figure 4.1b Classical guitar microphones – side view

moving in closer, and avoiding the close microphone placement that produces intrusive fret noise or a very localised tone quality. If you have problematic fret noise, move the microphones to at least 1 m (3′4″) away, and adjust their position around the arc by ear (see Figure 4.1). If you find that omnidirectional microphones at this distance are now picking up too much reverb, switch to directional microphones. This will cut out some of the reverb and give you a dryer but not too close sound to work with. Ribbon microphones (Royer R-121 or Coles 4038) in particular can be useful because their HF roll-off will soften fret noise, and an additional, more ambient pair of omnis can be used as a room pair to collect some reverb and warmth of tone.

A room with a high level of background noise will present similar, related difficulties. Microphones need to be close to record sufficient level from the instrument over the background noise, but this means that you might struggle with prominent finger squeaks and lack of reverb. Additional room microphones will be needed for the room reverb, and using ribbon microphones might help with the squeaks.

If you have to record a guitar, lute or theorbo in an orchestral or large ensemble context, you will need to be closer to pick up sufficient level and detail. This will of course be in addition to an overall orchestral pickup, so will be primarily used to make sure the transient details are not lost, and some contact with the finger noise is actually desirable to help the instrument be heard. A distance in the order of 15–25 cm (6″ to 10″) would be recommended, but you will need to beware of the possibility of proximity effect (an increase in the low frequency components) on any directional microphones at this distance, and correct for any excessive LF colouration if it is audible once mixed with the orchestra. If you have to use a single microphone in this context, start with it about halfway along the total length of the guitar (neck and body combined) and move it laterally until the balance between fretboard and main body of the sound is satisfactory. (See also Chapter 11 for concerto recording.)

4.1.2 Using artificial reverb on guitar

If the space in which you are recording is a dead room, or has poor sounding reverb, then artificial reverb can be added afterwards instead. However, artificial reverb can inadvertently emphasise finger noise that was relatively unobtrusive when it was recorded dry. A squeak is much more troublesome to the listener if it has 2–3 seconds of bright sounding reverb tail attached. In a real space, the HF content of the squeaks does not travel too far into the room because high frequencies are partly absorbed by the air. Hence, the ratio of 'squeak to music' picked up by the microphones gets higher the closer to the instrument the microphones are. The squeaks are not as present in the real room reverb as the other parts of the sound because they are at a reduced level by the time they reach a room boundary for reflection. If you are adding artificial reverb, this is something that you can try and control as part of setting the parameters. If you are using a programme that does not enable you to control the HF behaviour of the reverb algorithm in a sophisticated way, you can use a gentle HF roll-off on the auxiliary send that you are using to drive the reverb, and a similar roll-off on the reverb returns. Adding a small amount of EQ to both send and return will often give better results than EQ'ing only one of them more drastically.

4.1.3 Microphone choice

The lowest note on the classical guitar is E_2 at approximately 82 Hz, and so some good cardioids (e.g. Neumann KM84) will adequately cover the frequency range without the loss of any fundamental frequencies. Omnis (such as the KM83) might be preferred for the main guitar microphones if the location is good, and a natural balance between the room and the instrument can be found. Because they have a more extended bass end, you are likely to find that they will pick up some lower resonances from the body, but the preference for the sound is a personal one; the use of directional microphones should not affect important frequency content unless they have a roll-off that affects the 50–100 Hz range. As noted earlier, ribbon microphones will work very well, and they have the advantage that the HF roll-off will reduce the prominence of any fret noises and squeaks which can be particularly useful if you have to get in a bit closer. With fig of 8 ribbon microphones in particular, you should be aware of the possibility of the proximity effect starting to cause some artificial bass colouration if you are closer than 60 cm (2′) or so. This can be attractive in small amounts, but LF is seductive; you should be careful of being seduced by a sound that is large, warm and boomy but not actually like a classical guitar. As discussed earlier, using omnidirectional microphones for an ambient room pair, and ribbon microphones or cardioids for a more detailed pickup of the guitar can be a very successful combination. Avoid microphones that have any sort of HF lift above about 7 kHz that can make the instrument sound thin. If electrical noise (hiss) is proving a problem and you have optimised the rest of your signal path and gain structures, it is worth remembering that large diaphragm condenser microphones have a lower self-noise and higher sensitivity than those with small diaphragms, so require less mic gain.

4.1.4 *Floor and surfaces*

Floor reflections from a wooden floor can really enhance the sound of a classical guitar. Stone floors are too bright and immediate, as are some types of modern varnish on a wooden floor. If you have to record in a carpeted space, try placing some sheets of wood, plywood or medium-density fibreboard (MDF) under the player's chair to improve the sound of the instrument.

4.2 Harp

The orchestral concert harp has a wide dynamic range, and it is a large and powerful instrument when compared with its folk music cousins. Its sound consists of a distinctive plucked transient followed by a sustained resonating tone that comes both from the strings and the soundboard (although this is more of a resonator like a guitar body than a soundboard like that of a piano.) The range of its strings' fundamental frequencies extends from C_1 at approximately 32 Hz to G_7 at 3.1 kHz, so it approaches the piano in terms of its range once overtones are included. Therefore, in order to pick up the full effect of the lower frequencies in a solo recording, using omnidirectional microphones to form at least part of the sound is to be recommended.

The concert harp also has pedals which are manipulated to raise and lower the pitch of a given set of strings, creating the ability to play sharps and flats as required, and making the instrument chromatic in nature. These pedals will make some noise when they are being engaged or disengaged, but as long as they are not too obtrusive in the recording, they can be considered as part of the instrument's natural sound in the same way as woodwind key noise and guitar fretboard sounds. The opinions of engineers, producers, and performers concerning how much instrumental noise is tolerable can vary widely; it is unrealistic to try to eliminate these sources of noise completely, and accepting them as part of a real human performance on a real instrument will avoid the recording being processed into a state of unnatural and clinical quietness.

4.2.1 *Solo harp image*

The temptation when recording a solo harp might be to create an over-wide, rather 'technicolour' harp so that a full glissando travels from one loudspeaker and into the other. While this might have been fun in the early days of stereo and film scoring, or if used as a magical sound effect, we are trying to create the illusion of the sound of a real instrument in a real space, and therefore we need to constrain the image to something more central. In accordance with the techniques outlined for soloists so far, a forward facing pair spaced at 25–30 cm (10″ to 12″) apart can be used to create an image of the harp that is not overwide even when fully panned, although angling the microphones apart a little will increase the image width if this is desired. Spacing the microphone more widely will result in two less correlated sources, one picking up lower strings, and one higher. This will create some dramatically wide glissandi when panned left and right, but little sense of a real instrument or a real space. (See Chapter 9 for dealing with the harp in an orchestral context.)

4.2.2 Microphone position and choice

The player sits with the harp resting on their right shoulder with the soundboard close to their body. The player can then look along the rows of vertical strings, with the highest notes nearest to their right ear and right hand. This allows the player to reach forward with the left hand to the lower strings, and to have a clear view of the music stand which is placed on the left-hand side of the instrument (from the player's perspective).

Figure 4.2a and 4.2b show the harpist and some microphones positions to be discussed below. Positions A and B are useful; position C is to be avoided.

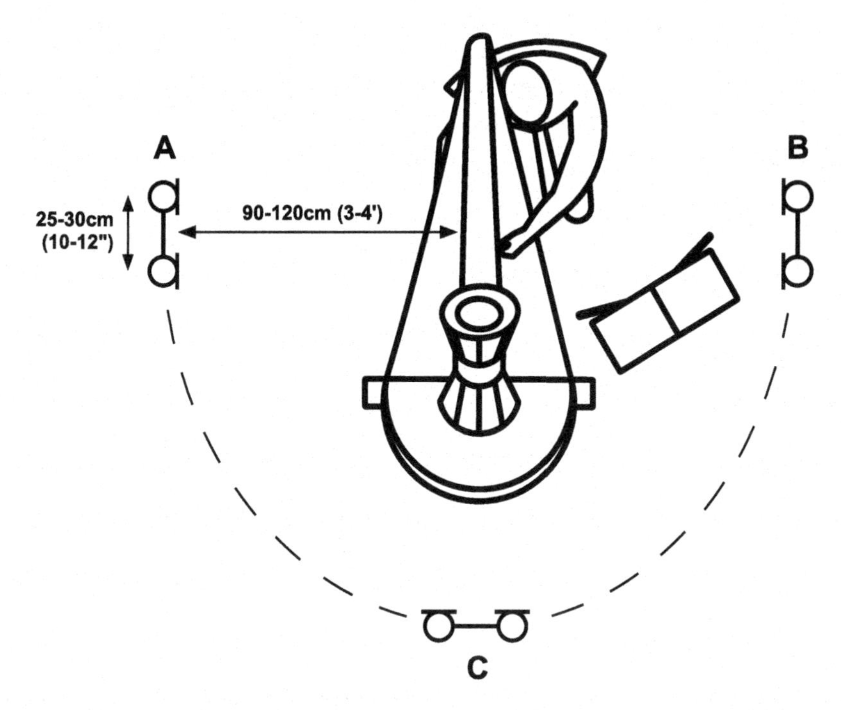

Figure 4.2a Harp – plan view showing alternative microphone positions A and B. Position C is not recommended

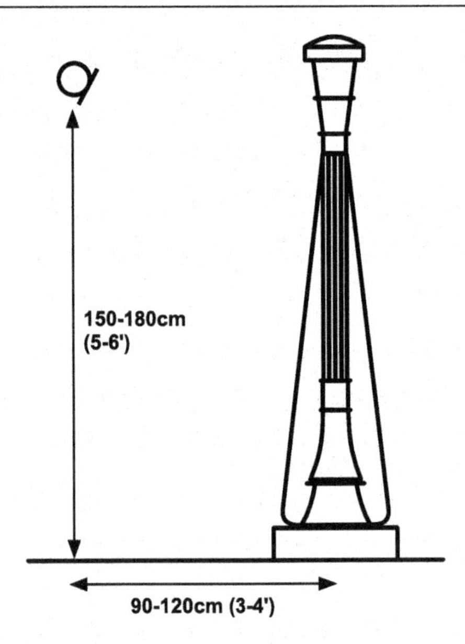

Figure 4.2b Harp – end view showing height for positions A and B

The microphones for positions A and B are placed sideways on to the instrument, aimed at a position midway down the strings from a height of around 150–180 cm (5′ to 6′) and about 90–120 cm (3′ to 4′) from the harp. In position A, the microphones are on the player's right-hand side, and this can be preferable to position B, as the finger noise is a little reduced from this side. In all other respects, position B will also work well, as long as the microphones can be placed to look above the music stand. Omnidirectional microphones would be the preferred choice if the room is sufficiently nice sounding, because of the extended frequency range of the instrument. The distance of 90–120 cm (3′ to 4′) from the harp will be close enough to get some good detail and pick up the quietest passages from the top strings, but not so close that the sound becomes localised to one part of the instrument.

If the engineer concentrates on the soundboard as the source of all the sound and moves the microphones around in a 90° arc to look straight at the soundboard from the front end of the harp (position C), odd phase relationships and cancellations can result. A more extreme version of this approach is to put one microphone on either side of the harp; this will result in two rather uncorrelated microphone signals which will not create a convincingly located instrument in the recording image. As noted at the start of the chapter, an overall room pair should be added to complete the sound.

4.2.3 Floor and surfaces and noises

As with the guitar, a nice wooden floor will enhance the resonance and sound of the harp. However, the pedals will create some noise acoustically and also have the potential for transmitting noise mechanically through a sprung floor. The best way to ameliorate this is to use some good isolation mounts for the microphones to avoid pedal noise transmitting itself in this way. If you are still picking up too much pedal noise, have a listen in the room with the player and see if this is how the instrument sounds anyway. If it is noisy in real life, talk to the player about it to see if anything can be done. If it cannot be improved by the player adjusting the instrument, it can be worth adding a small piece of carpet under the pedals as you might for piano, where the back of the player's heel is on the floor. Avoid using a large rug as it will be better for the overall sound for most of the interaction with a wooden floor to be included. Remember that some pedal noise is fine as it is part of the instrument; along with finger noise, it is only a problem if it starts to sound very close and out of context with the rest of the sound. This arises because we usually place microphones closer than we would sit to a performer in a concert, and if misjudged, this can lead to instrumental noises becoming too obtrusive.

4.3 Violin

The violin (along with the guitar and orchestral stringed instruments) consists of resonating strings that are coupled to a body that acts as an acoustic amplifier. The body produces the majority of the sound from its own vibrations and that of the air inside it. It has a very extended frequency spectrum containing high levels of HF overtones that give the instrument its character. Both the higher frequencies (for the sense of contact and connection with the instrument) and the lower frequencies (for a feeling of warmth and resonance) are needed for a beautiful and well-balanced solo violin sound.

The radiation pattern is complex, but taken as a simplified overview, the lowest frequencies are radiated more or less omnidirectionally, and the higher frequencies are beamed upwards perpendicularly to the front face of the body.[1] The higher the frequency, the narrower the beam of radiation. Consequently, placing microphones above the player will capture all the frequencies produced but may not be very flattering if the player is less skilful or harsh in tone. In a concert situation in an appropriate environment, the room will integrate the whole sound before it reaches the listener, in a similar way as happens with a classical voice. When recording a live concert, the choice of where to put microphones is restricted because of the resulting visual intrusion; thus microphones are often placed much lower down, and hence they do not capture the upper overtones. The engineer is then reliant on capturing some of this HF on other microphones within the room, usually on some sort of overall pickup placed higher and further back.

4.3.1 Microphone placement

Violinists are like singers in that they will move around quite a significant amount during a performance, and any microphone technique needs to be able to manage this situation adequately, without the image moving in and out of focus or from side to side. The majority of soloists will perform standing up, so all distances given in this section are in relation to a standing player of middling adult height, perhaps 1.7 m (5′7″). If the player is seated, you will need to lower the microphones accordingly.

With a professional player with a good tone, a reasonable starting point would be to place a pair of microphones about 90–100 cm (3′ to 3′4″) back from the player, looking down at the instrument from a height of 2.4–2.85 m (8′ to 9′6″) (see Figure 4.3a and 4.3b). This should produce a well-balanced overview of the whole sound, the aim being to achieve a good balance between a sense of space around the instrument and giving the listener a feeling of immediate emotional contact with the player. The balance between contact and spaciousness can be altered with the height of the pair, and it must be judged in a given context; the height that works will depend on how reverberant the room is and the directivity pattern of the microphones. The distance between the microphones and the violin also serves other purposes; it will reduce any uncomfortable closeness and intrusive bowing noise by means of the loss of HF with distance, and it will reduce the impact of player's movements on both the tonality of the sound and the stability

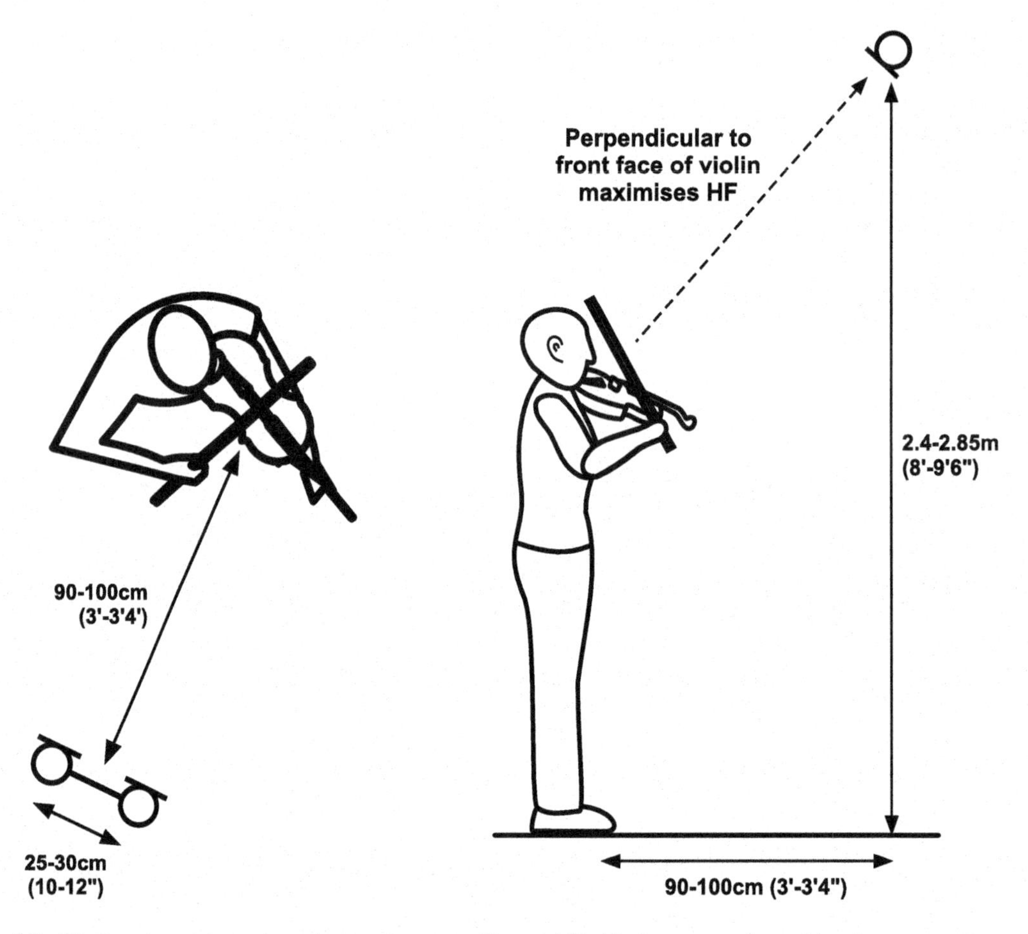

Figure 4.3a Violin spot pair – plan view

Figure 4.3b Violin spot pair – side view

of the image. Microphones placed lower down in front of the player at around the instrument's height of perhaps 1.5 m (4′11″) will lose the 'presence' of the bowing and HF components, and will need to be further back than 1 m (3′4″) to recover an appropriate balance between direct and reverberant sound within the room. Overall, it is harder to achieve a really pleasing balance between spaciousness and detail with the microphones placed lower down.

Player movement is something that can cause problems, particularly when using a pair of spot microphones. Initially, the microphone pair should be 25–30 cm (10–12″) apart, parallel to one another, and panned fully left and right to produce a central image with some 'bloom' around it. If the image swings around a lot when the player moves, these can be panned in up to around 75% (following the same principles outlined in Chapter 6) to reduce image instability. If this isn't sufficient to tame the movement, you can reduce the spacing of the pair to the lowest end of the suggested range, and physically angle them inwards slightly. Another technique that works extremely well is to add a central ribbon microphone (Royer R-121 or Coles 4038) to the violin pair and use this at a lower level in the mix to stabilise the image, as discussed in section 4.3.2.

Figure 4.3 shows a good starting position for placing the close pair on a violin.

The most common problem will be too bright a tone, or too harsh a tone with a less good player; this can be reduced by changing the microphones' horizontal angle and pointing them a little over and beyond the instrument (Figure 4.4a). This will move the instrument away from the front axes of the microphones and thus reduce the HF content. If this is not sufficient, you could reduce the height of the microphones in order to move them out of the way of the highest frequency projection that is perpendicular to the instrument's bridge and front face (Figure 4.4b). However, the microphone height does give a good feeling of space, and it can help avoid floor reflections, so rather than sacrificing height, you can also try moving the microphones around in an arc as shown in Figure 4.4c. The further around towards the left (on the diagram) that you go, the more the HF will be reduced as it will be effectively blocked by the player's head. Microphones behind the player might be useful in a concert scenario; however, they will also result in a loss of HF as the player holds the instrument to project forwards and upwards – not behind him or her. The best choice of technique for altering the HF content might depend on how the player characteristically moves; spend some time watching them play and observe where they tend to move and how much they move. This should enable you to get a good microphone position according to their 'average' playing position. If movement is extreme, microphone placement off-axis from the instrument's bridge will reduce the degree of HF change that occurs as the instrument swings around. (Figure 4.4a through 4.4c show different ways to alter the amount of HF on a violin.)

As before, an overall room pair should be added to complete the sound.

4.3.2 Microphone choice

The lowest fundamental frequency of the violin is G_3, at 196 Hz, and so a good quality condenser cardioid or omnidirectional microphone will be sufficient to pick up the lower end and warmth of the tone. As with all acoustic recording in a live space, the off-axis response is important when collecting room reverb, but in the case of a solo performer, other off-axis instruments do not need to be considered. Cardioids will pick up less of the room reverb at the suggested height and would be a good choice in a very live space. They will enable you to remain at a comfortable distance from the instrument while keeping the amount of reverb under control.

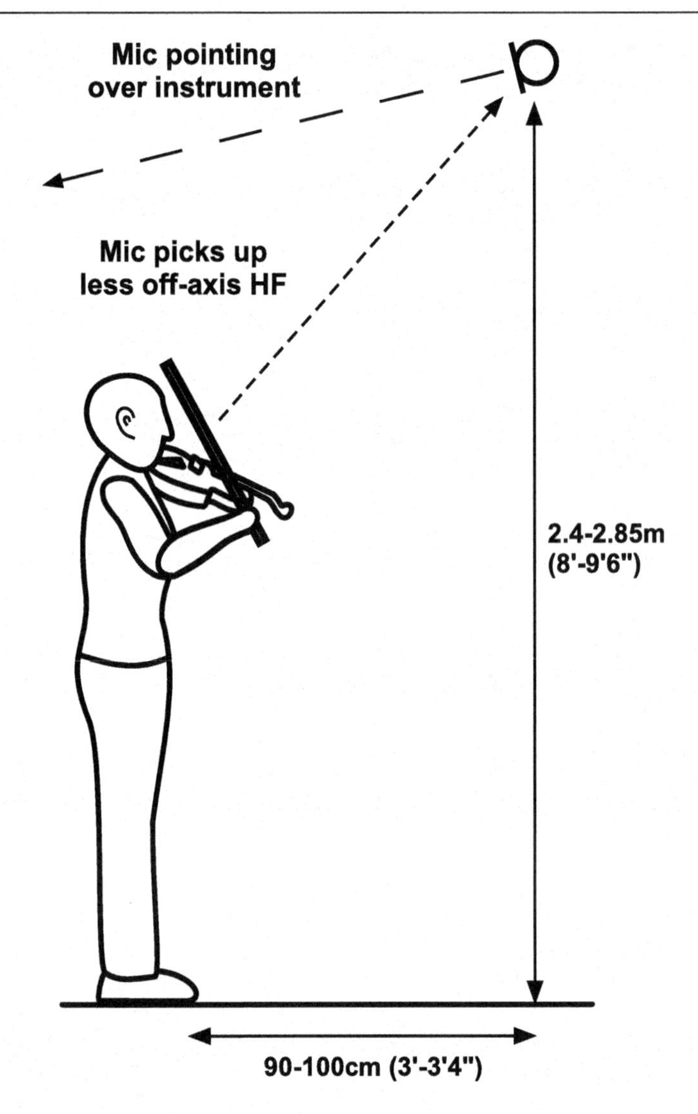

Figure 4.4a Change microphone angle to look over the instrument

In addition to the suggestions in Figure 4.4 for reducing the amount of extreme HF picked up, ribbon microphones can also be considered, either for the pair or as an additional central microphone (see Figure 4.5). The smooth and gentle HF roll-off will ameliorate any harshness in the tone but will still pick up some room reverb on the rear lobe. A useful technique for a recording session, as previously noted, is to combine a pair of condenser microphones mounted a little higher, at around 3 m (10′) (this increased height will reduce the amount of HF slightly as it is absorbed by the air), and a single ribbon microphone mounted about 30 cm (1′) lower.

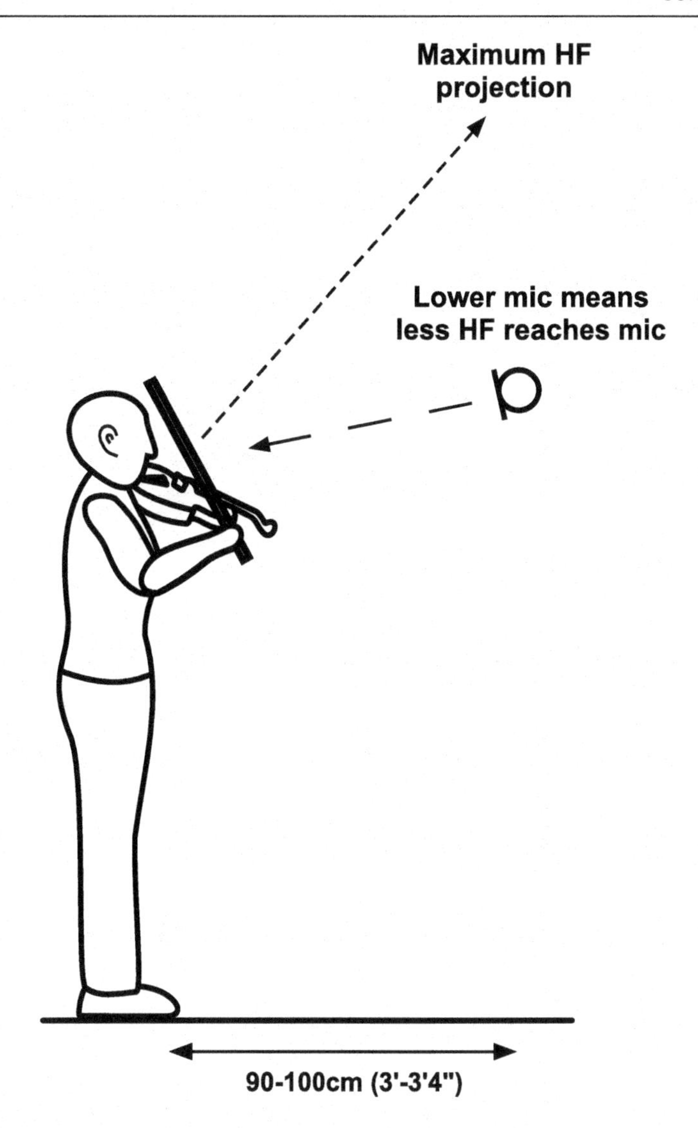

Figure 4.4b Lower the microphones

The ribbon can be placed closer to the violin because of its HF roll-off, and this might be a useful asset where being closer is necessary, perhaps because of the acoustic or some other constraint on microphone placement. This combination of a pair of condensers plus a central ribbon microphone allows you to use the ribbon microphone to fix the image more securely in the centre if it exhibits a lot of lateral movement on the condenser pair when the player swings around. The ribbon microphone will usually be mixed in at about 6–8 dB lower than the condensers, so it plays a secondary role in forming the sound, but it can be used to have a beneficial effect on the

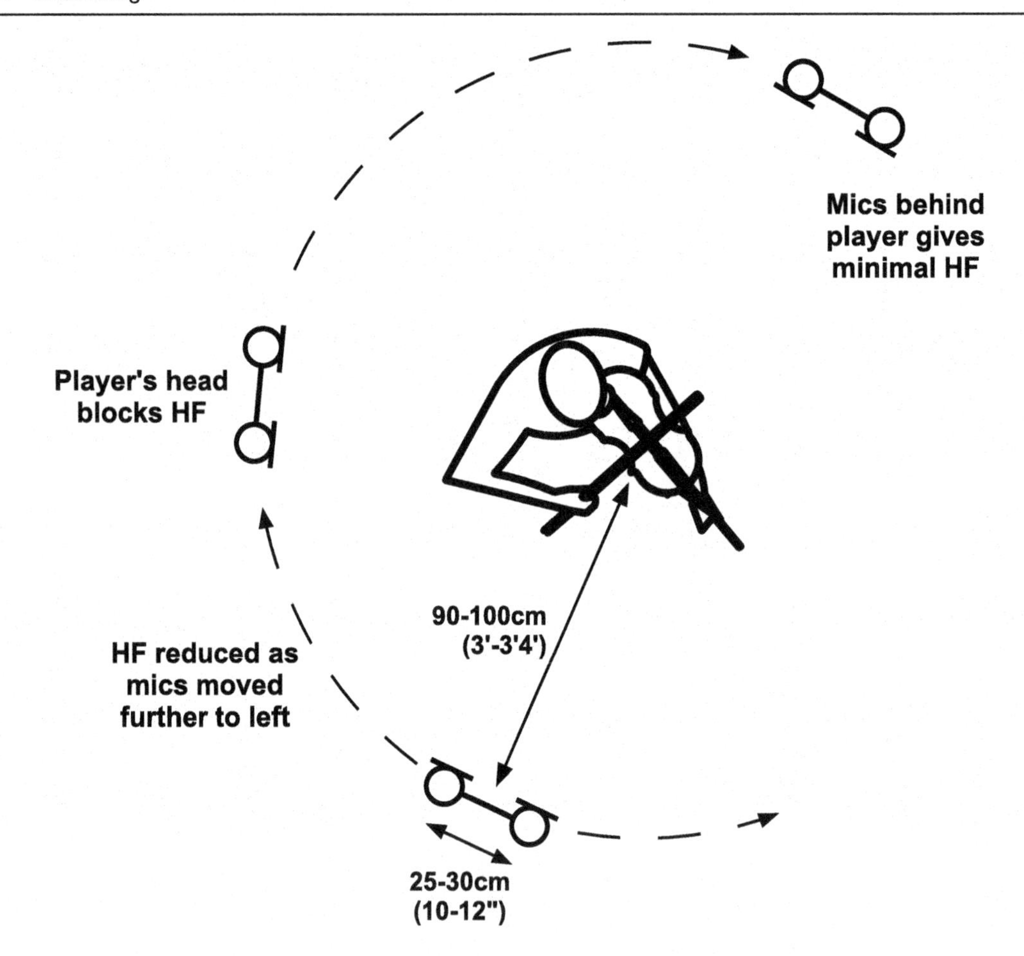

Figure 4.4c Move them round in an arc maintaining height

tonality as well as on the image stability. It will give a warmth and sweetness to the tone, while the condensers pick up more of the attack, bowing noise, and other HF components. The violinist Itzhak Perlman particularly likes ribbon microphones such as the RCA 44BX, and Coles 4038s were used while recording him in 1998 at the Saratoga Performing Arts Centre in New York during his collaboration with the pianist Martha Argerich.[2]

4.4 Cello

The cello radiates from its body in a similar way to the violin, in that the very highest frequencies are radiated perpendicular to the front surface and the bridge, and the overall pattern is quite complex. However, in comparison to the violin, it has been found that microphone placement is a little less critical, and moving on- and off-axis from the bridge will affect only the highest HF

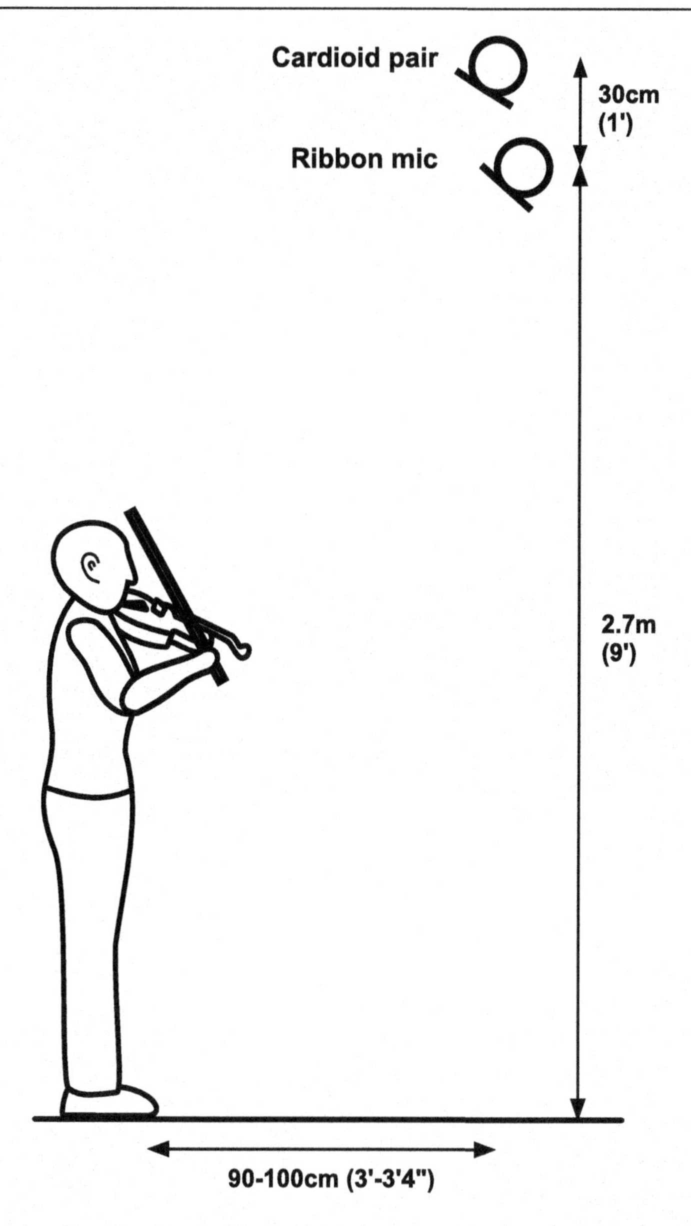

Figure 4.5 High condenser pair plus a ribbon microphone on a violin

content. Its lowest fundamental is C_3 at 130 Hz, which suggests that a good cardioid will extend low enough in the frequency range, although an omni might give a better tone.

The instrument's playing position is very different to that of a violin in that it rests on and is close to the floor. The reflections from the floor are important to the sound, and so a solid wooden surface would be the preferred choice when making a recording. A marble or stone floor is not flattering because of its bright reflections, so if you have to record using such a floor in a church for example, then put a small amount of carpet down underneath the player's chair. Carpet will absorb HF, so bear this in mind when choosing to use it in other circumstances, and avoid using a large area of it.

Some solo players like to sit on a podium to play because it can act as a further resonator and enhance the sound, but a podium's performance will depend on how well it is constructed. Some will enhance the sound, but others will absorb some of the LF, or produce an unpleasant resonance, or simply be creaky and noisy as the player moves. If the podium is working against you, you will have to negotiate with the player to find another solution.

4.4.1 Microphone placement and choice

The technique outline for the violin can be adapted for the cello, using a narrowly spaced pair of microphones (25–30 cm (10″ to 12″)) to give the solo image more substance and width, and aiming to capture a good balance of bowing detail, full tone, and sense of space. Because the playing position is different to the violin, the microphones do not need to be as high to capture the HF, but they can be placed in the region of 2.25 m (7′6″) high and 1.8–2.1 m (6′ to 7′) away, depending on the room acoustic and the microphones used. The microphones should be pointed towards the bridge, perpendicular to the instrument's front surface, and as you move the microphones around in an arc L to R, you will find that the HF is maximised when they are centrally placed in front of the instrument. Figure 4.6 illustrates the microphone position suggested as a starting point. In order to get a good feeling of a real instrument in a real space, moving the microphones a little closer or further away will help adjust the balance between direct and indirect sound.

In a concert scenario, any microphone will have to be placed lower, but this is less critical than it would be for the violin; there is a wider acceptable range because the instrument is less directional in its radiation when floor reflections are taken into account, and it is orientated differently with respect to the microphone. A microphone low enough to be looking straight at the bridge will be visually unobtrusive for live work, and a fig of 8 ribbon (Royer R-121 or Coles 4038) could be used to usefully discriminate against the orchestra.

Condenser omnis or cardioids will work well, remembering that it is not necessary to have a large-diaphragm microphone in order to capture a good extended LF range. Any pencil omnidirectional microphone will have a sufficiently extended frequency response.

The cello is also well suited to ribbon microphones using the HF roll-off to produce a mellow sound, and it enables closer placement if necessary. At the microphone distances suggested earlier, the proximity effect from the ribbon microphone should not be at all evident, but it will begin to show at the lowest frequencies at around 0.7–1 m (2′ to 3′4″) away – more likely in a difficult live scenario. As with the guitar (which has a similar frequency range), care should be taken not to make the instrument sound unnaturally bass heavy and larger than life. Its tone needs both the warmth and the more astringent upper frequencies to feel well balanced and real. Although the ribbon microphones will miss some of the upper HF that would add some bite, this part of the tone colour will be picked up on the overall room pair of condensers.

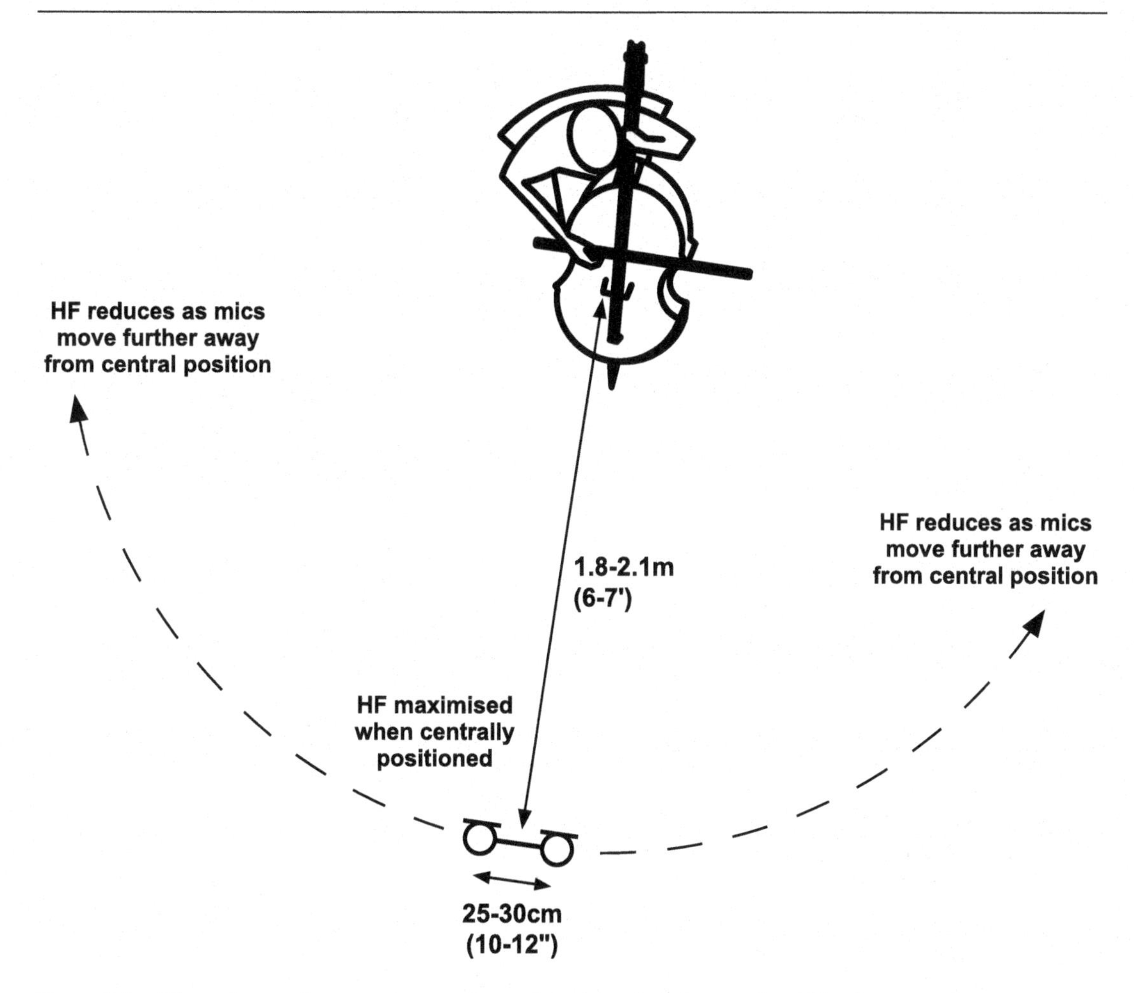

Figure 4.6a Microphone placement on cello – plan view showing arc for HF adjustment

4.5 Woodwinds

In radiation terms, the woodwinds can be divided into the oboe, clarinet, and bassoon, which all have reeds and are essentially closed at the mouthpiece end and open at the bell; and the flute, which operates using a stream of air striking the mouthpiece edge, and is open to the air at both the mouthpiece and far end. This has an impact on the radiation patterns, and thus which microphone placement options are most effective when recording them.

4.5.1 Oboe, clarinet, and bassoon

For the first group of instruments, radiation comes from both the open holes and the bell. The highest overtones pass right down the bore of the instrument, radiating partly from the open

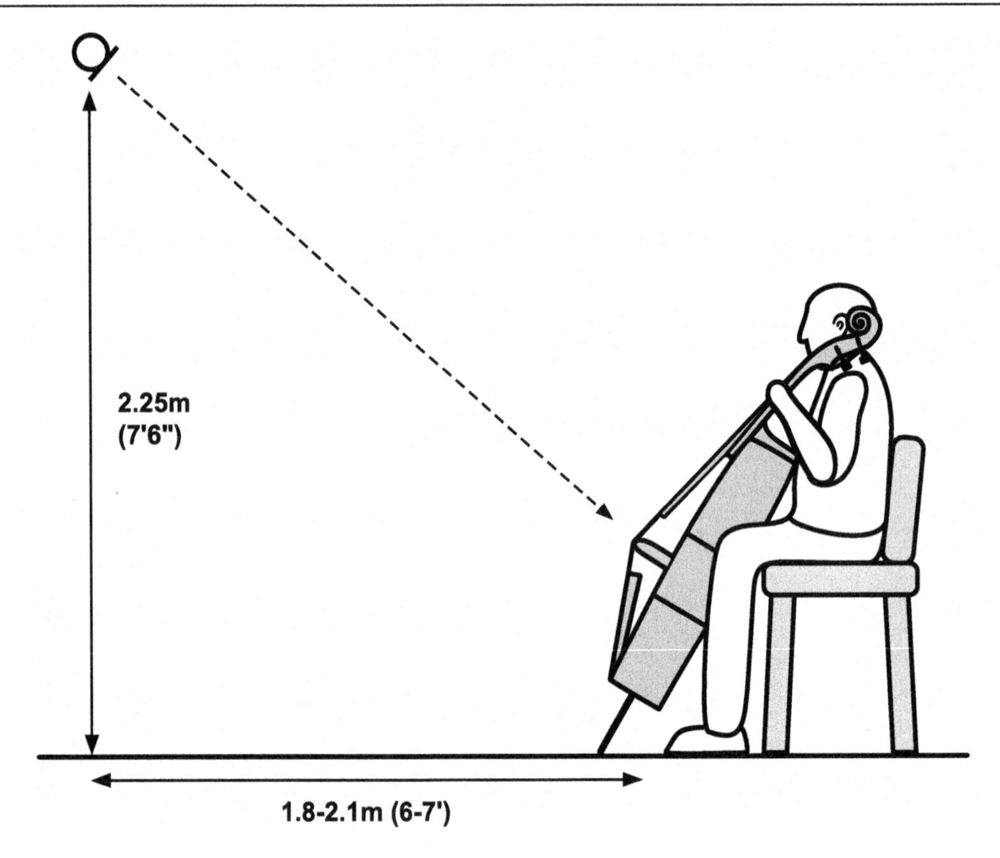

Figure 4.6b Microphone placement on cello – side view

tone holes but primarily from the bell, forming a cone-shaped beam aligned with the axis of the instrument. At the lower end of the frequency range (i.e. the fundamental frequencies of the lower register), the radiation becomes stronger from the first open holes than from the bell, and the overall radiation pattern becomes more perpendicular to the instrument's axis.[3] The higher frequencies emerging from the bell will be reflected from the floor (for the oboe and clarinet whose bells point in that direction) and projected diagonally upwards from the bassoon. The techniques used for the violin can be adapted for the forward-facing woodwinds, although HF content will be best adjusted with microphone height and angling rather than moving side to side as there is no head-shadowing. You should avoid placing microphones directly on-axis to the bell of the woodwind instruments, as you will collect more HF and wind noise and miss out on obtaining a good overall balanced tone, which is the goal of classical recording.

Microphones should be positioned in front of the player, but at a height of around 2.25–2.6 m (7′6″ to 8′6″) and looking down (assuming the player is standing; a bassoonist might prefer to sit and the microphones can be lowered accordingly). Height is less critical for the clarinet and oboe than it is for the violin because of the different radiation characteristics of the instruments

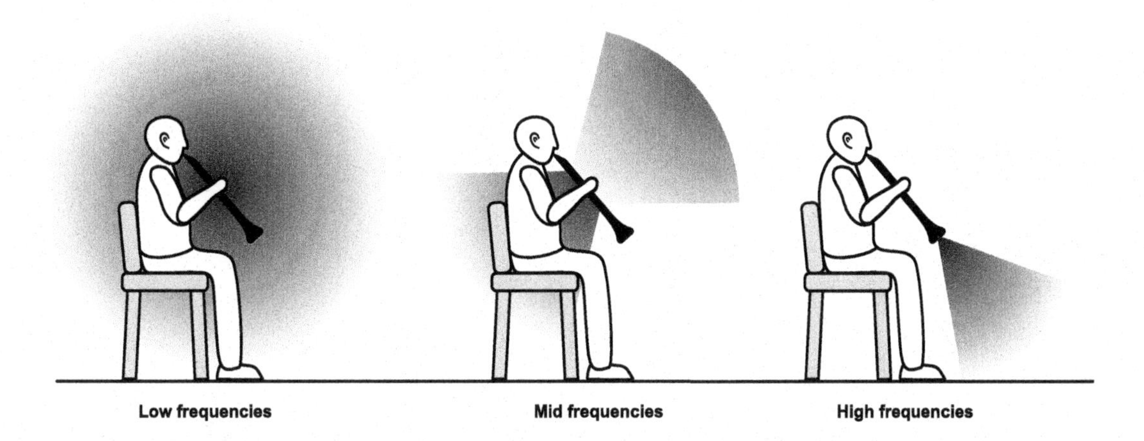

Figure 4.7 Generic end-blown woodwind radiation pattern at low, mid, and high frequencies. This applies to overtones as well as fundamental frequencies.

in conjunction with the floor reflections of the bell radiation, although key noise needs to be avoided. As outlined previously, to obtain an image that has some width and bloom around it, a pair of microphones instead of a single spot microphone is used.

Key noise is the aspect of woodwind recording that will cause the most difficulty, although bubbling and wheezing noises from worn or leaky reeds can also be a nuisance, and it would be best to politely talk to the player and see if they can change their reed where this sort of noise becomes intrusive. Key noise is part of the instrumental sound, so it has to be accepted to a certain extent, but it should not dominate. Keeping the microphones at a good distance (above about 2.25 m (7′6″)) will help with this, as the intrusiveness of transient-filled key noise is reduced by HF absorption into the air. If the microphones are too close, all key noise, reed noise, and mechanical sounds will be artificially exaggerated and feel very unnatural to the listener, who does not usually hear the players close up. If the room is reverberant, it will be better to avoid getting in too close (and picking up too much key noise) by using directional microphones rather than omnis. Ribbon microphones (Royer R-121 or Coles 4038) will also ameliorate key noise by virtue of the HF roll-off inherent in the design.

4.5.2 Flute

The flute can be placed in a different category of radiation pattern because it is supported more or less horizontally in its playing position, and its radiation comes from both ends of the instrument as well as any open holes.[4] The result is that it radiates almost equally to the front and to the back of the player, which can be very useful in opening up microphone placement opportunities in less than ideal circumstances. Placing microphones fairly close and halfway down its length should be avoided, as some phase cancellations may result due to some frequencies radiating in opposite phase from each end. The breath noise is more prominent from the front, as it is formed by air

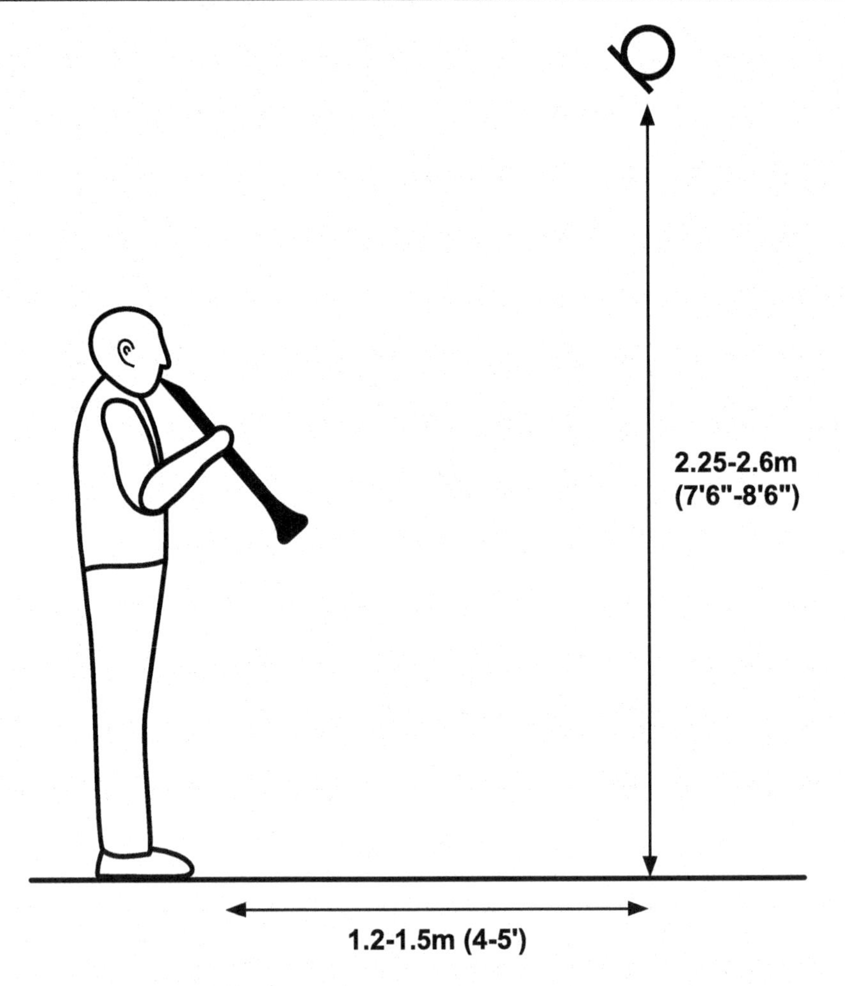

Figure 4.8 Woodwind player with microphones – side view

turbulence as the breath strikes the top of the mouthpiece. If too much breath noise is a problem, make sure the microphones are not in front of the player or at the height of the player's mouth; because of the equal front and rear radiation, you will be able to move the microphones behind the player to really reduce breath noise. Key noise, as with all the winds, will be reduced with distance, so if you are working in a reverberant space, keep the microphones at least a couple of metres (6′8″) away from the instrument, but use cardioids instead of omnis to reject a greater proportion of indirect sound.

Adding microphones from behind at approximately instrument height is a good, discreet option in a live situation, and unlike for the violin, this will not result in the loss of important HF content.

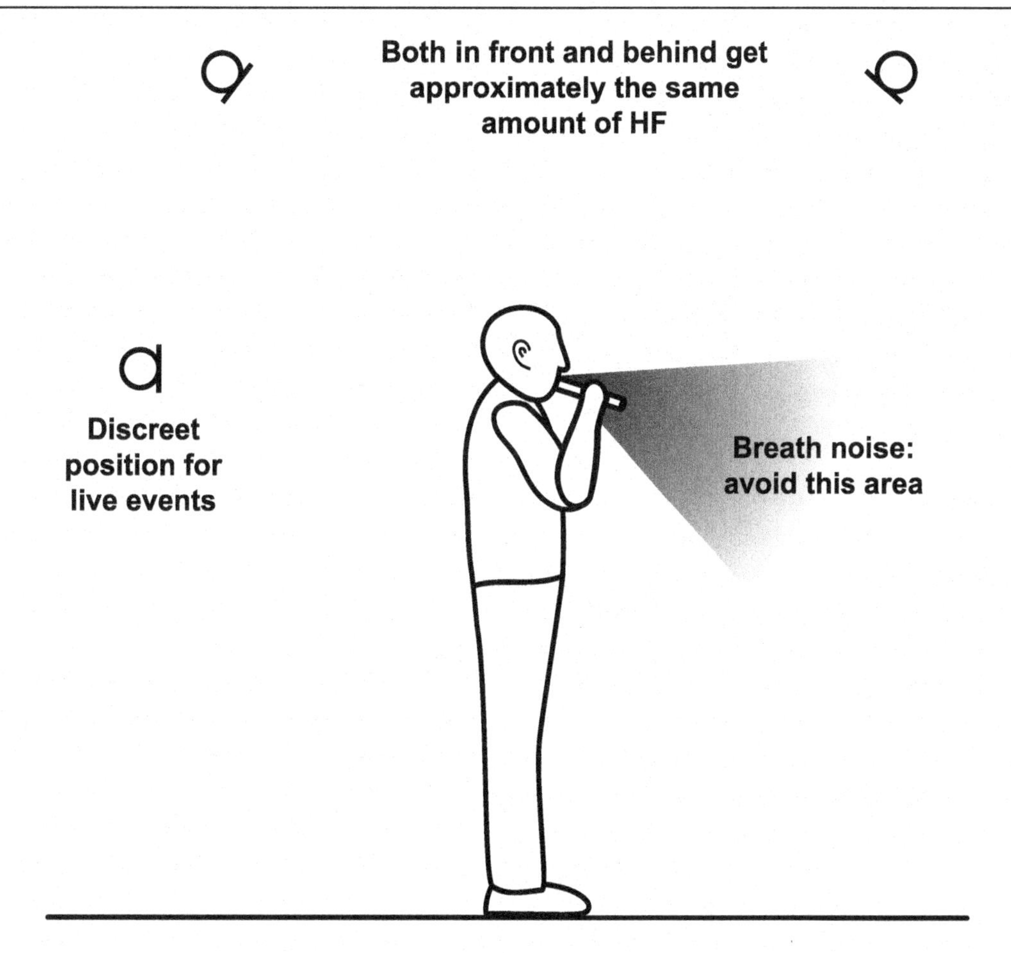
Both in front and behind get approximately the same amount of HF
Discreet position for live events
Breath noise: avoid this area

Figure 4.9 Flute – alternative microphone positions in front and behind

4.6 Harpsichord

Although the harpsichord is clearly an early relation of the grand piano, the technique of piano recording outlined in Chapter 5 cannot be simply transferred over to the harpsichord; its radiation pattern is different, and obtaining a balanced and characteristic sound needs a slightly different approach.

4.6.1 Harpsichord characteristics

The instrument is not as loud as a modern piano; the string tensions are lower, the soundboard technology is not as developed, and gradual changes in dynamics are not possible as the instrument's mechanism is not responsive to the velocity with which the keys are struck. To facilitate stepped changes in tone and dynamics, harpsichords have stops in a similar way to organs, which are a means of coupling together sets of strings in a variety of different octaves. Engaging and disengaging these will produce a significant amount of noise, as the mechanisms are relatively simple and primitive in design. The pedals can also make quite a lot of noise; placing a small piece of carpet on the floor under the pedals can reduce this.

The largest instruments have a range of five octaves, from F_1 at around 40 Hz to F_6 at around 1400 Hz. The exact pitch is dependent on the tuning being used; they are usually tuned to a lower pitch than A = 440 Hz, as would be normal for instruments of the baroque era, whether they are original instruments or reproductions. The upper frequency limit of the whole spectrum is much higher than 1400 Hz because of the strong presence of a multitude of higher partials. In order to fully capture the lower notes, omnidirectional microphones would be preferred.

4.6.2 Microphone placement

The majority of the harpsichord's useful sound is projected out to the side away from the lid and towards the audience, and the effect of the lid reflections is to add a layer of complexity and richness to the sound. When recording the piano, the practice of approaching it from the tail end is in part an attempt to reduce some of the complications that arise from the muddying effects of the lid reflections. With the harpsichord, these same lid reflections can be very useful in helping to integrate the sound and effectively glue it all together. Increasing the lid's angle so it is open more widely will affect the openness of the sound, and this is something you should experiment with. It is possible to remove the lid altogether, but the effect will be to lose some of the characteristic complexity.

If microphones are placed around the tail end, the mid and lower string resonances will be picked up, but not so much of the strong upper harmonic components of the strings and the characteristic plucking sound of the mechanism. All three are important aspects of the overall sound that needs to be captured to give a balanced view of the instrument.

Figure 4.10 shows the harpsichord from above and the side, showing the most useful arc in which microphones should be placed.

Given the lower overall level of output, it is usual to place the microphones somewhat closer than for a grand piano, and a good place to start will be in the region of 1.1–1.4 m (3′6″ to 4′6″) away from the well of the instrument at a height of about 1.5–1.65 m (5′ to 5′6″).

Increasing the distance of the microphones away from the instrument will reduce clarity and increase the amount of reverberation, but it will also reduce the amount of mechanism noise, which is potentially useful. However, the amount of mechanism noise can also be altered by the height of the microphones without the need to move them further away if this has a detrimental effect on the clarity of the sound. When the instrument lid is fully open, a lower microphone position will pick up more of the plucking, and a higher one will have more resonance and sustain. Microphone height adjustment will also affect the amount and quality of lid reflections collected, so it is worth spending some time altering this aspect of the placement to find the best place on any given harpsichord. This is also a good time to take a walk around the instrument and sit at different heights to give yourself an idea of just how much the sound changes.

The overall aim is to find a happy medium where there is enough clarity (so the recording is not dominated by room sound) whilst also ensuring that the mechanism sounds are not unduly intrusive. Making sure that the sustained part of the tone is well captured across the whole spectrum will help to make listening to the recording really enjoyable. Listening to a harpsichord recording where the transients are very harshly represented can become fatiguing.

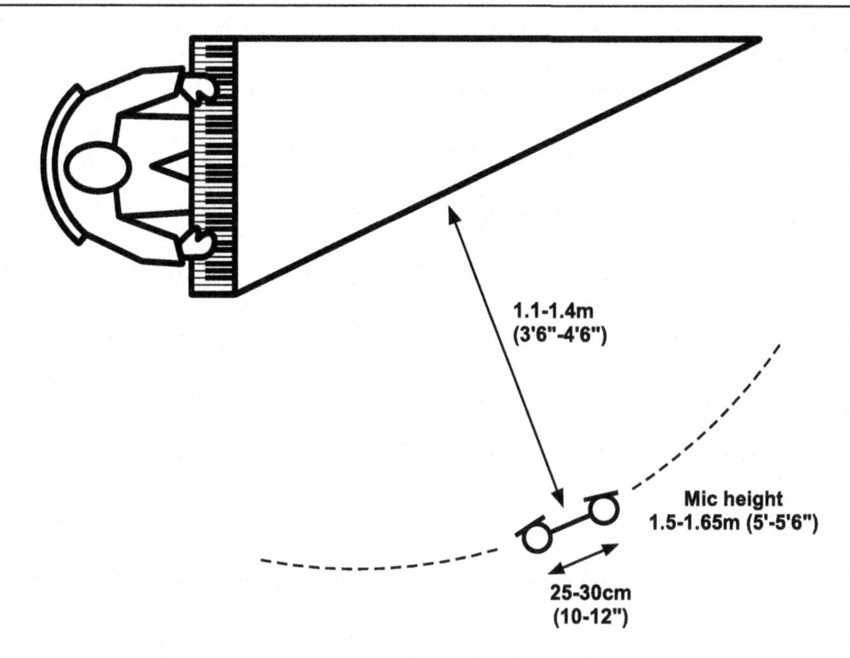

Figure 4.10 Harpsichord – plan view of pair of microphones

The microphones used in Figure 4.10 are a spaced pair at around 25–30 cm (10″ to 12″), angled slightly outwards, in addition to an overall room pair. The LCR three-microphone technique outlined in section 5.7.1 can also be used on harpsichord, although they will be closer together as the physical size of the instrument is not so great. Because of this, they could be panned almost fully LCR, but the final width of the image should inform the panning. In order to make the image and sense of space feel real, this technique will need more support from some sort of overall room pickup to act as a unifying element.

Notes

1 *Fundamentals of Musical Acoustics, sections 24.5*
Arthur H. Benade
Publishers: Dover
ISBN 0–486–26484-X
2 *MUSIC FROM SARATOGA/BEETHOVEN/FRANCK Itzhak Perlman/Martha Argerich/EMI CLASSICS (1999) UPC 717794908725*
3 *The Physics of Musical Instruments, section 15.5*
Neville H. Fletcher and Thomas D. Rossing
Publishers: Springer
ISBN 978-0-387-98374-5
4 *The Physics of Musical Instruments, sections 16.11.2, 17.7*
Neville H. Fletcher and Thomas D. Rossing
Publishers: Springer
ISBN 978-0-387-98374-5

Chapter 5

The piano

There is perhaps no other instrument that quite so divides opinion regarding the best way in which to record it, and there are, of course, many ways to go about it. This chapter aims to provide an overview of the features of the instrument and the principles involved in recording it, and then looks in some detail at the well-tried and trusted method originally developed by Decca engineers that is behind so many wonderful piano recordings. Many engineers with experience of other techniques are naturally sceptical of this approach and find it hard to believe that it could or would work. The best way to decide is to try it and make your own judgement.

The main focus of this chapter will be on grand pianos, and in particular full-size 2.8 m (9′) concert grand pianos such as the Steinway D. Although we are very aware that many readers will be working with smaller pianos than this, using the full-sized model offers the best way of demonstrating the key principles involved in piano recording. Once understood, they can be applied to any piano recording situation, whether on a baby grand or even an upright.

For recording piano with another soloist or as part of a chamber ensemble, see Chapters 6, 7, and 12, and for the piano concerto, see Chapter 11.

5.1 The nature of the sound of a piano

The complexity of the piano as a sound source can make it difficult to record successfully. When you strike most single keys on a piano, the hammer hits three strings tuned to that note, which all start vibrating at their fundamental frequency plus harmonics; these vibrations are then transmitted to the soundboard, which amplifies them in varying amounts according to its own physical properties and sends them out into the room. Along the way, different frequencies will be reflected off parts of the piano body and the lid in particular (see section 5.4), which directs some of the sound out towards the audience. In addition, some of the other 240 or so strings will experience sympathetic vibrations at multiples of their own fundamental frequencies if their dampers are raised at the time (the very highest notes have no dampers), and so the resulting patterns of vibration are very rich and varied.[1]

The complexity of all these strings causing and reacting to sympathetic vibrations in each other can easily be demonstrated. Sit at a piano, and rest your right arm across as many keys as possible from about middle C upwards. Push the keys down slowly to avoid making a sound; doing so will lift all the dampers for these notes off the strings, leaving them free to vibrate. With your left hand, play a short, loud note towards the lower end of the piano, and listen to the effect – you will hear a really complex mix of pitches ringing on after the loud low note has finished. These come from the strings of the notes you are holding down with your right arm, which have been set into motion by the transient of the low note, and by each other. You will also notice that they die away at different rates; the changing tone colour of piano die-away is a very complex characteristic of the sound that can also make editing more difficult. (See Chapter 18.)

Even though the interaction of vibrating strings is complex, the upper harmonics are not particularly strong. Figure 5.1 shows two spectrograms, one of a violin and one of a piano, recorded at about the same overall loudness as each other; the duration of each is about 20 seconds. Time is shown on the horizontal axis and frequency (0–21 kHz) on the vertical axis. The amplitude of each component of the sound is indicated by its brightness, with greater amplitude producing a brighter colour, or greater 'whiteness' in black and white.

The solo violin on the left clearly exhibits significant harmonic energy right up to the top of the human hearing range, with the harmonics appearing at integer multiples of the fundamental frequency. The solo piano on the right shows that the fundamental frequencies contain much more of the total energy (they are brighter) with the harmonics being much lower in level than those of the violin. This concentration of energy at the lower harmonics helps give the piano its warmth of sound, but also can mean that it is very easy for pianos to sound somewhat muddy and unclear, especially in dense passages when lots of notes are sounding.

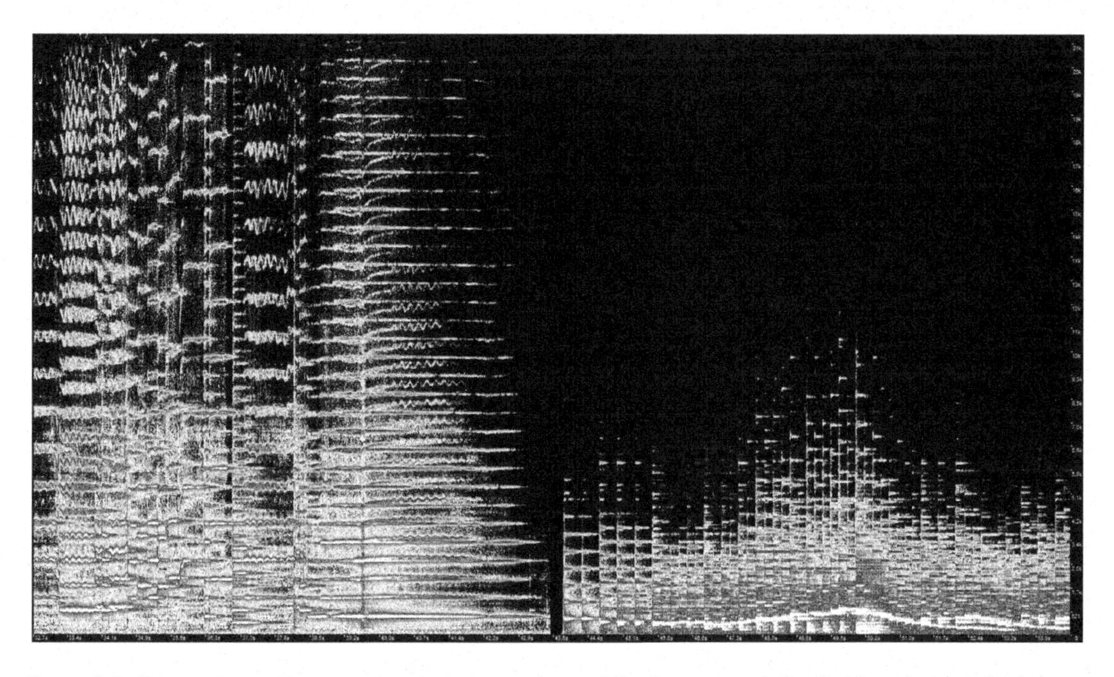

Figure 5.1 Comparison of energy levels in upper harmonics between violin (left) and piano (right)

5.2 The physical layout of a grand piano

Regardless of the manufacturer, the arrangement of strings inside a grand piano follows a similar pattern, with only the overall length of the instrument changing. The strings for the lowest notes are placed so that they cross over the strings for higher notes; this allows for greater string length within a given size of piano case. Reduced length can be traded against increased thickness when making bass strings, but the best tone comes from longer and thinner strings. For this reason, the bass end of a concert grand piano has a much more powerful tone than a baby grand or upright.

Figure 5.2 shows a plan view of a Fazioli F278, with the lid removed. The similarities in frame and string layout are obvious, and are much the same for other large pianos (Bechstein, Blüthner, Yamaha, etc.). The strings have been divided into zones A–D; the lowest bass notes only have one or two strings each whereas the middle and upper notes have three strings per note. All the strings in zones A–C have dampers to prevent them from resonating when they are not being played, and the small, very thin strings in zone D do not have dampers because they do not naturally ring on long enough to need them.

On some pianos, including older models of Fazioli F278, the frame layout is slightly different, meaning zones A–D are of different sizes. Figure 5.3 is a comparative table that brings together various parameters for three grand piano models, such as fundamental frequencies and note numbers. These are shown both as absolute values (from 1 for the lowest note to 88 for the highest) and MIDI note numbers. Note names are given in scientific pitch notation, where middle C = C_4.

5.3 The piano lid

A standard grand piano lid has two open positions, each propped up by one of two alternative hinged sticks:

- Full stick – raises the lid to an angle of about 35°
- Short stick – this is about an eighth of the length of the full stick; raises the lid to an angle of about 7°.

Some engineers also use custom-made longer props to raise the lid higher when recording; these should be purpose made, extremely strong, and used in accordance with instructions.

In the underside of the lid are two indentations to hold the end of the stick. It is absolutely essential that the correct indentation is used for each stick – the angle between lid and stick should be 90°. If the short stick indentation is used for the full length stick, it is likely that the weight of the lid will snap the stick, and the lid will come crashing down. Short stick is mainly used as a crude volume control, especially prevalent when accompanying soloists in concert, and its effect has been likened to putting a fire blanket over the piano! Good recorded results can be obtained with a short stick piano, but its use should be avoided if possible.

The sound from the vibrating strings and resonating soundboard projects predominantly upwards, especially at higher frequencies. For some recording purposes, removing the piano lid can be very useful (see Chapter 11 for double piano concertos or for those conducted by the soloist). Outside these unusual circumstances, the piano is designed to be heard from a position seated at some distance in front of it, and the purpose of the open, propped-up lid is to reflect the sound out towards the listeners. The reflections from the lid come together well at this listening

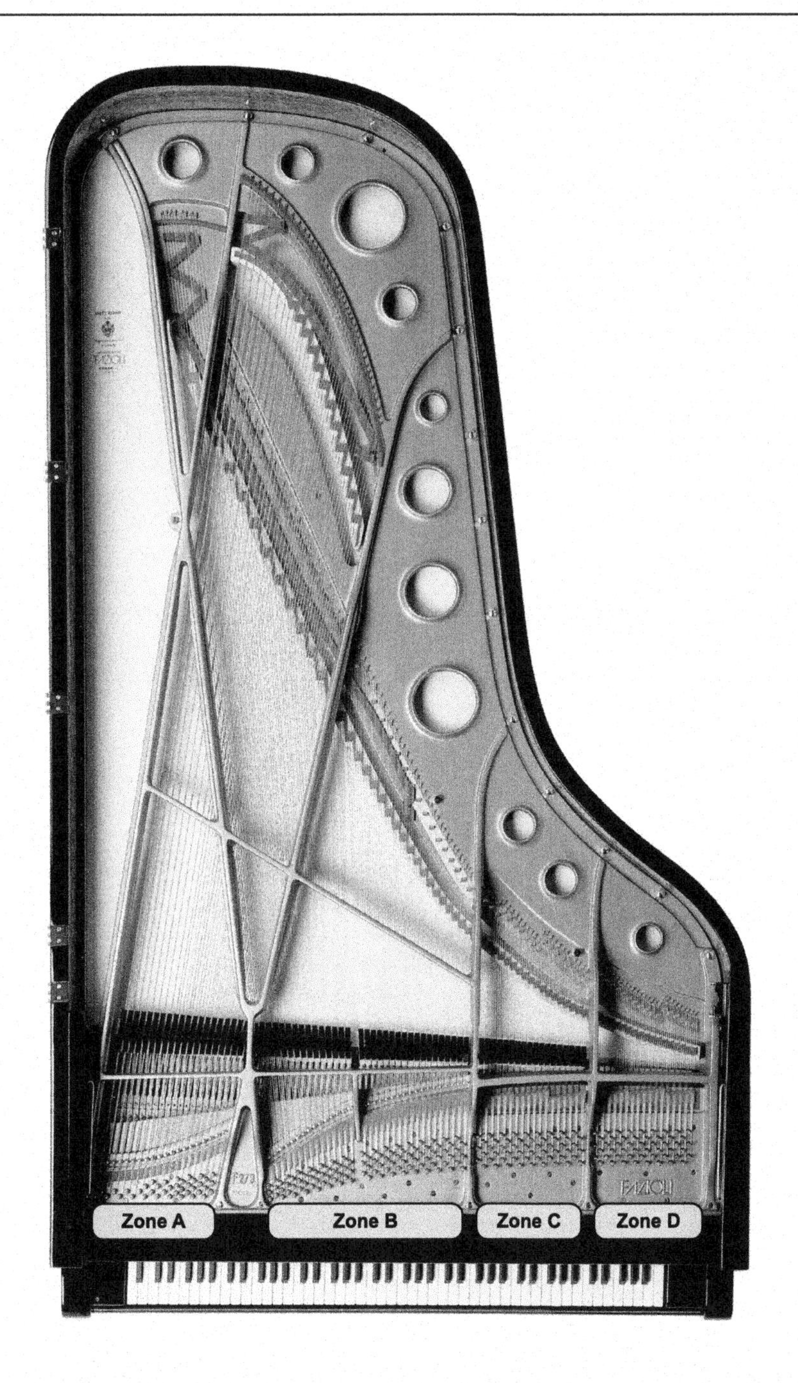

Figure 5.2 Fazioli F278 showing zones A–D

Image: © Fazioli Pianoforti SpA. Photo: Roberto Zava/Studio Step.

Note No.	*Steinway D New Fazioli F278*			*Old Fazioli F278*			*Note Name*		*Fundamental Frequency*	*MIDI*	*Black or White*
	Zone	*Strings*	*Damper*	*Zone*	*Strings*	*Damper*			*(Hz)*		
88		3	No		3	No	$\mathbf{C_8}$		4186.01	108	
87		3	No		3	No	B_7		3951.07	107	
86		3	No		3	No	$B\flat_7$	$A\sharp_7$	3729.31	106	
85		3	No		3	No	A_7		3520.00	105	
84		3	No		3	No	$A\flat_7$	$G\sharp_7$	3322.44	104	
83		3	No		3	No	G_7		3135.96	103	
82		3	No		3	No	$G\flat_7$	$F\sharp_7$	2959.96	102	
81		3	No		3	No	F_7		2793.83	101	
80	**D**	3	No		3	No	E_7		2637.02	100	
79		3	No		3	No	$E\flat_7$	$D\sharp_7$	2489.02	99	
78		3	No		3	No	D_7		2349.32	98	
77		3	No	**D**	3	No	$D\flat_7$	$C\sharp_7$	2217.46	97	
76		3	No		3	No	$\mathbf{C_7}$		2093.00	96	
75		3	No		3	No	B_6		1975.53	95	
74		3	No		3	No	$B\flat_6$	$A\sharp_6$	1864.66	94	
73		3	No		3	No	A_6		1760.00	93	
72		3	No		3	No	$A\flat_6$	$G\sharp_6$	1661.22	92	
71		3	Yes		3	No	G_6		1567.98	91	
70		3	Yes		3	No	$G\flat_6$	$F\sharp_6$	1479.98	90	
69		3	Yes		3	Yes	F_6		1396.91	89	
68		3	Yes		3	Yes	E_6		1318.51	88	
67		3	Yes		3	Yes	$E\flat_6$	$D\sharp_6$	1244.51	87	
66		3	Yes		3	Yes	D_6		1174.66	86	
65		3	Yes		3	Yes	$D\flat_6$	$C\sharp_6$	1108.73	85	
64		3	Yes		3	Yes	$\mathbf{C_6}$		1046.50	84	
63		3	Yes		3	Yes	B_5		987.77	83	
62	**C**	3	Yes		3	Yes	$B\flat_5$	$A\sharp_5$	932.33	82	
61		3	Yes		3	Yes	A_5		880.00	81	
60		3	Yes		3	Yes	$A\flat_5$	$G\sharp_5$	830.61	80	
59		3	Yes		3	Yes	G_5		783.99	79	
58		3	Yes		3	Yes	$G\flat_5$	F♯	739.99	78	
57		3	Yes		3	Yes	F_5		698.46	77	
56		3	Yes		3	Yes	E_5		659.26	76	
55		3	Yes	**C**	3	Yes	$E\flat_5$	$D\sharp_5$	622.25	75	
54		3	Yes		3	Yes	D_5		587.33	74	
53		3	Yes		3	Yes	$D\flat_5$	$C\sharp_5$	554.37	73	
52		3	Yes		3	Yes	$\mathbf{C_5}$		523.25	72	
51		3	Yes		3	Yes	B_4		493.88	71	
50		3	Yes		3	Yes	$B\flat_4$	$A\sharp_4$	466.16	70	
49	**B**	3	Yes		3	Yes	A_4		440.00	69	
48		3	Yes		3	Yes	$A\flat_4$	$G\sharp_4$	415.30	68	
47		3	Yes		3	Yes	G_4		392.00	67	
46		3	Yes		3	Yes	$G\flat_4$	$F\sharp_4$	369.99	66	
45		3	Yes		3	Yes	F_4		349.23	65	

Figure 5.3 Comparison of grand piano parameters

Note No.	Steinway D New Fazioli F278			Old Fazioli F278			Note Name		Fundamental Frequency	MIDI	Black or White
	Zone	Strings	Damper	Zone	Strings	Damper			(Hz)		
44		3	Yes		3	Yes	E_4		329.63	64	
43		3	Yes		3	Yes	$E\flat_4$	$D\sharp_4$	311.13	63	
42		3	Yes		3	Yes	D_4		293.66	62	
41		3	Yes		3	Yes	$D\flat_4$	$C\sharp_4$	277.18	61	
40		3	Yes		3	Yes	**C_4**		261.63	60	
39		3	Yes		3	Yes	B_3		246.94	59	
38		3	Yes		3	Yes	$B\flat_3$	$A\sharp_3$	233.08	58	
37		3	Yes		3	Yes	A_3		220.00	57	
36		3	Yes		3	Yes	$A\flat_3$	$G\sharp_3$	207.65	56	
35		3	Yes		3	Yes	G_3		196.00	55	
34		3	Yes		3	Yes	$G\flat_3$	$F\sharp_3$	185.00	54	
33		3	Yes		3	Yes	F_3		174.61	53	
32		3	Yes	**B**	3	Yes	E_3		164.81	52	
31		3	Yes		3	Yes	$E\flat_3$	$D\sharp_3$	155.56	51	
30		3	Yes		3	Yes	D_3		146.83	50	
29		3	Yes		3	Yes	$D\flat_3$	$C\sharp_3$	138.59	49	
28		3	Yes		3	Yes	**C_3**		130.81	48	
27		3	Yes		3	Yes	B_2		123.47	47	
26		3	Yes		3	Yes	$B\flat_2$	$A\sharp_2$	116.54	46	
25		3	Yes		3	Yes	A_2		110.00	45	
24		3	Yes		3	Yes	$A\flat_2$	$G\sharp_2$	103.83	44	
23		3	Yes		3	Yes	G_2		98.00	43	
22		3	Yes		3	Yes	$G\flat_2$	$F\sharp_2$	92.50	42	
21		3	Yes		3	Yes	F_2		87.31	41	
20		3	Yes		3	Yes	E_2		82.41	40	
19		3	Yes		3	Yes	$E\flat_2$	$D\sharp_2$	77.78	39	
18		3	Yes		3	Yes	D_2		73.42	38	
17		3	Yes		3	Yes	$D\flat_2$	$C\sharp_2$	69.30	37	
16		3	Yes		3	Yes	**C_2**		65.41	36	
15		3	Yes		3	Yes	B_1		61.74	35	
14		3	Yes		3	Yes	$B\flat_1$	$A\sharp_1$	58.27	34	
13		2	Yes		2	Yes	A_1		55.00	33	
12		2	Yes		2	Yes	$A\flat_1$	$G\sharp_1$	51.91	32	
11		2	Yes		2	Yes	G_1		49.00	31	
10	**A**	2	Yes	**A**	2	Yes	$G\flat_1$	$F\sharp_1$	46.25	30	
9		2	Yes		2	Yes	F_1		43.65	29	
8		1	Yes		1	Yes	E_1		41.20	28	
7		1	Yes		1	Yes	$E\flat_1$	$D\sharp_1$	38.89	27	
6		1	Yes		1	Yes	D_1		36.71	26	
5		1	Yes		1	Yes	$D\flat_1$	$C\sharp_1$	34.65	25	
4		1	Yes		1	Yes	**C_1**		32.70	24	
3		1	Yes		1	Yes	B_0		30.87	23	
2		1	Yes		1	Yes	$B\flat_0$	$A\sharp_0$	29.14	22	
1		1	Yes		1	Yes	A_0		27.50	21	

Figure 5.3 (Continued)

position; it can help to think of the lid as a lens that focusses the sound at this distance. However, microphones placed this far away will sound too reverberant and lacking in detail when heard without the visual cues that come from being able to see the instrument. When the microphones are moved in closer to the instrument, the lid reflections can combine with the direct radiation from the strings and soundboard to produce a muddy sound with a perceptible loss of clarity. (See 5.4.1 for the essential balance of 'clarity' and 'warmth' in a good piano sound.)

The positive side of the lid's action is that it acts as a binding agent, bringing the sound of individual strings together into a coherent and warm-sounding whole. This action is especially effective at higher frequencies, where the geometry of lid, soundboard, and strings allows for multiple resonances and reflections for these shorter wavelengths. The balance between positive and negative effects of the lid will depend on exactly where the microphones are placed, and much of the challenge of getting the desired balance of 'clarity' and 'warmth' in the recorded sound comes down to controlling how much influence the lid will have on the recording. The effect is not easily predictable from one piano to another; many engineers have experienced setting up microphones on a piano in the firing line of the lid, fine-tuning the set-up to get a good sound, and then subsequently trying to recreate that good sound on another occasion without success, despite the microphones being in the same position. This kind of problem is primarily caused by the lid, and any reliable, re-creatable microphone technique seeks to ameliorate these effects.

5.4 Recording aims

The attributes of a good piano recording are really the same as for any other instrument although not always easy to achieve: a pleasing tonal balance, a recorded image of believable width and stability, equal sense of perspective and clarity across the whole range of notes, and an enjoyable blend of the instrument with the sound of the room.

5.4.1 Tonality

The fundamental qualitative trade-off, which is a good idea to keep at the forefront of your mind when recording a piano, is between *clarity* and *warmth*. The aim is always to keep an appropriate balance between the two, the ideal being to get exactly the right amount of both. However, when you take steps to increase clarity, it is possible that these will have the side effect of reducing warmth and vice versa. Note that there is not one absolute right answer on this balance; it very much depends on the situation, such as the nature of the repertoire being recorded and the other instrumentation, and it is ultimately a subjective decision. But concentrating on getting this balance right (whatever 'right' means in any particular context) is the key to getting the best possible recorded sound.

Clarity
• Individual notes are distinct • Complex, fast passages can be heard clearly • The sound grabs the attention of the listener • Details don't get lost in the dense texture of sound • Quiet notes are clearly audible, even when other notes are sounding

Warmth
• Multiple notes sound part of a coherent whole • The sound isn't too strident or fatiguing • The complex richness of interacting strings is audible • The room reverb blends pleasingly with the piano

5.4.2 Stereo imaging and width

When looking at a piano from the audience's perspective, you might be tempted to think that you would hear a clear stereo spread because of the string positions within the frame; that is the high notes would appear to be on the left and the low notes on the right. In reality, the complex interaction between the vibrating strings, soundboard, frame, and lid mean that although the instrument sounds wide, there is no obviously clear directional source for any range of notes once you stand more than about a metre away. From the player's point of view, there is a clearer left-right differentiation (low notes on the left, high notes on the right), but for an audience member, this layout is not apparent. This aspect of the instrument needs to be considered when deciding what kind of recorded sound one hopes to achieve. In pop and jazz piano, it is often the convention to present the instrument with the notes laid out as if from the player's perspective, but for classical/acoustic recording the goal is a more indistinct image as heard from the audience. The Decca technique described in section 5.6 does just this.

A common pitfall when recording classical piano is to give the instrument excessive stereo width and allow it to fill the whole space between left and right loudspeakers. This gives the recording an unnatural feeling, as if it has assumed gargantuan proportions and taken on the size of the room. Because all classical recording tries to set the instrument into some sort of realistic space, it is important that the sense of perspective is maintained. For perspective to be convincing, several aspects of the sound have to reinforce one another: the image width, the proportion of direct to reverberant sound, and the amount of 'close' instrumental noises. A grand piano will only fill your whole field of view if you stand right next to it with your head almost inside it. But from this distance, the sound would be dominated by direct sound and a lot of hammer noise. If a more distantly recorded piano is allowed to occupy the whole width, it stops feeling realistic and starts to feel overblown. Personal preference comes into this to some extent, but as a guide, the width of the instrument on a classical piano recording should occupy around 50%–75% of the whole width, with the reverb and sense of the room occupying 100% of the width.

The use of a phase scope as an aid to making judgements about image width is discussed in Chapter 2. Figure 5.4 shows the kind of display you would expect to get from a good piano recording. There is a nice amount of width, but it doesn't extend all the way between left and right, and almost all of the components of the sound are in phase. This 'fat cigar' shape is a reliable indicator that all is well.

Figure 5.5, on the other hand, is too narrow, with the width getting very close to pure mono. Going the other way, Figure 5.6 is really too wide, with the sound spread out almost to the extremes of the loudspeakers. Going even further is Figure 5.7, where the 'blob' is becoming squashed horizontally instead of vertically, and you can clearly see some very horizontal elements – all of which indicate a lot of out-of-phase components. Out-of-phase recordings can sound very disorientating and unsettling, with no clear sense of where sounds are coming from, and can make some listeners feel quite unwell. The most important factor in controlling the amount of out-of-phase components when recording a piano is the spacing between microphones; this is discussed in section 5.5.2.

5.5 Recording a solo piano: the spaced pair

The most common method of recording a classical piano is to use a spaced pair, and this section will look at both microphone spacing and position around the piano. This will lead us on to the very particular use of the spaced pair that is the Decca piano technique in section 5.6.

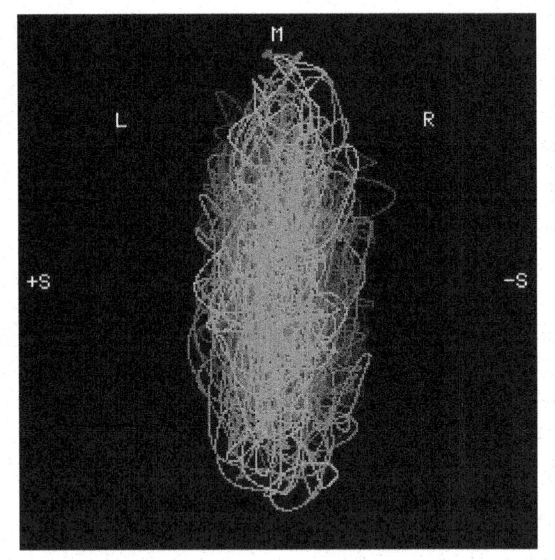

Figure 5.4 Ideal width for a solo piano recording

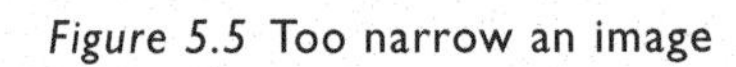

Figure 5.5 Too narrow an image

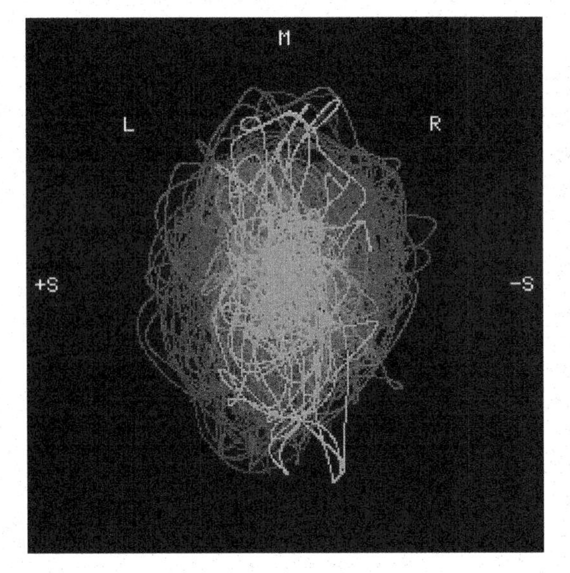

Figure 5.6 Too wide an image

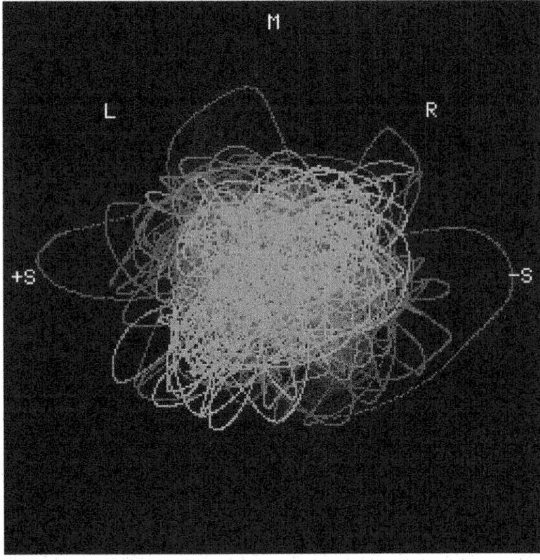

Figure 5.7 Far too wide, with out-of-phase notes clearly visible

5.5.1 Microphone positioning

Figure 5.8 shows an arc of positions for a spaced pair, from the keyboard to the tail. As a guide to the distance away from the edge of the piano, in position D at the tail end, the microphones are in the range of 1.2–1.65 m (4′ to 5′6″) away. Position B, opposite the well of the piano, is a little further away when measured in a straight line into the curve of the well. The microphones can be omnidirectional, cardioid, or even fig of 8 ribbons; the stereo image depends on the spacing of the pair and not the microphone directivity as the microphones are parallel to one another. Ribbon microphones (such as the Royer R-121 or Coles 4038) usually sound very good on piano; their sound is often described as warm, sweet, and lush, and the typically attenuated HF response of a ribbon doesn't matter on an instrument that is dominated by its lower frequencies. However, omnidirectional mics are usually preferred because of their extended LF response; this is an important consideration for classical repertoire that encompasses the whole instrument (the bottom A on a piano has a fundamental frequency of 27.5 Hz).

Different positions along this arc will produce different effects, and walking around the piano to have a listen while someone is playing can give you an idea of where to start. It is sometimes

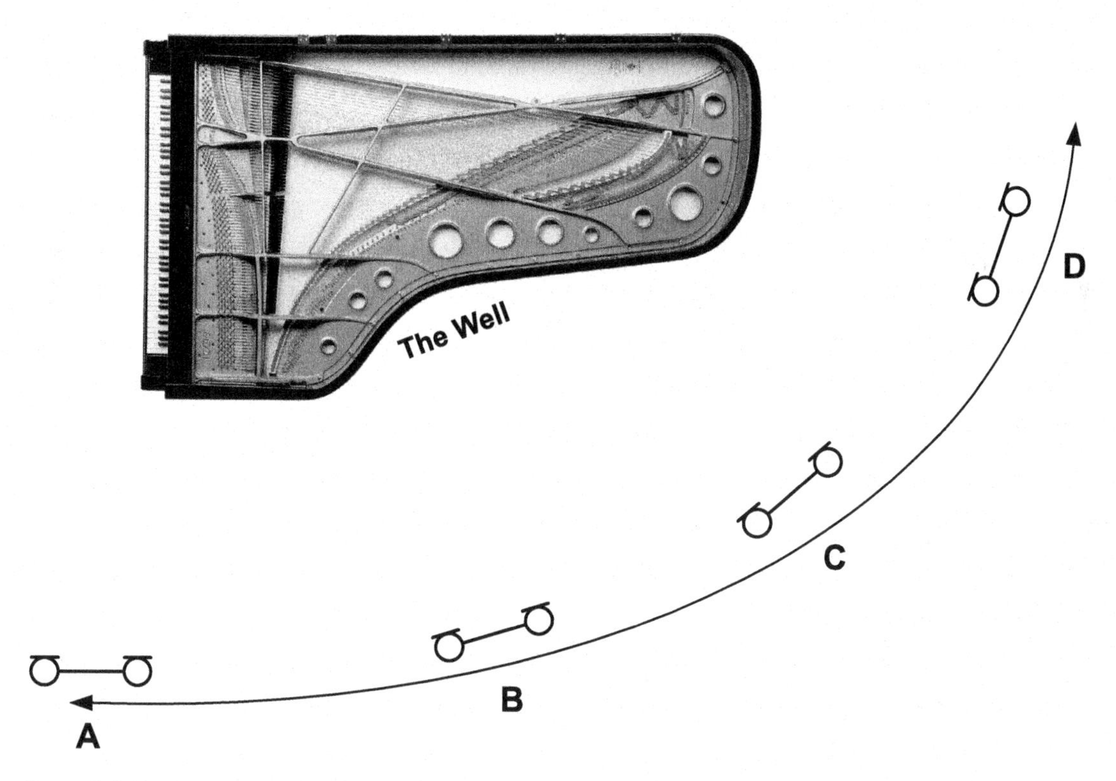

Figure 5.8 Positioning a spaced pair of microphones around a piano

Piano image: © Fazioli Pianoforti SpA. Photo: Roberto Zava/Studio Step.

thought that you will get more HF towards position A, and more LF towards position D, but it is often not quite like this; LF stays reasonably consistent around the arc, and there is a general trend towards reduced HF at the tail end.

As suggested in section 5.3, the lid and the reflections it produces are the main reason for the differences in sound along the arc. Because of its projection towards the audience, the muddying factor is heightened around position B; clarity improves and lid effects reduce as you move towards position D. Placing microphones between C and D allows the clarity/warmth balance to be controlled as required – left towards C to get more warmth and integration of the sound from the lid; right towards D to get more clarity. As you get closer to position D, the stereo image narrows (for a constant microphone spacing), and there is a slight drop in HF, which is often compensated for by using omnidirectional mics with a slight HF boost, such as the Schoeps MK2S.

Adjusting height also provides another way of controlling the effects of the lid; raising the microphone pair so that it approaches the height of the plane of the lid can reduce muddiness and give a similar clarity/warmth control to the left-right movement discussed earlier.

After lateral and vertical positioning, the next most important factor to consider is how close to the piano the pair should be. Closer microphone positions have two main effects: one is to increase clarity and instrumental mechanical sounds such as hammer noise, and the other is to make the image of the notes' layout more distinct, with the high notes on the left and the low notes on the right. Both these attributes are more sought after for pop or jazz piano; a brighter sound is one that will cut through a mix more easily, and using cardioids and a high-pass filter (HPF) can be useful to stop LF content cluttering a busy mix. However, for classical recording, the greater distance brings a feeling of homogeneity, room ambience, and warmth, and we want to avoid the sound that feels as if we are standing right next to the instrument.

5.5.2 Microphone spacing

When using a spaced pair of microphones with them both pointing forwards, the spacing distance is the most important factor in controlling the stereo width, regardless of their directivity. If the microphones are too close together, the sound will be too narrow and mono, and if they are too far apart, the image becomes too wide; a 'hole in the middle' can develop, and there can be a lot of unpleasant out-of-phase effects between left and right channels, making the source hard to locate in the stereo image. (The closely related topic of using spaced omni microphones as an overall pair is discussed in Chapter 3.)

For pianos, when recorded somewhere around position D in Figure 5.8, we have found that the optimum separation distance is about 30 cm (12″), measured from capsule to capsule, with full panning L-R. At distances of 25 cm (10″) or less, the sound becomes noticeably more mono, and at 15 cm (6″) the image is really far too narrow. At separations much greater than 33 cm (13″), the image becomes wider and starts to lose focus at the centre. By 45 cm (18″) it is starting to be too wide, and by 60 cm (24″) it is definitely too wide.

Figure 5.10 shows a useful chart derived from data collected by Michael Williams.[2] It shows the expected image width in degrees that will be produced for a given microphone spacing and physical source width, also in degrees. The full image width between loudspeakers set up in an equilateral triangle is assumed to be 60° (±30°) (Figure 5.9b). The bottom axis shows time difference between sound arriving at each microphone; this depends on both microphone spacing

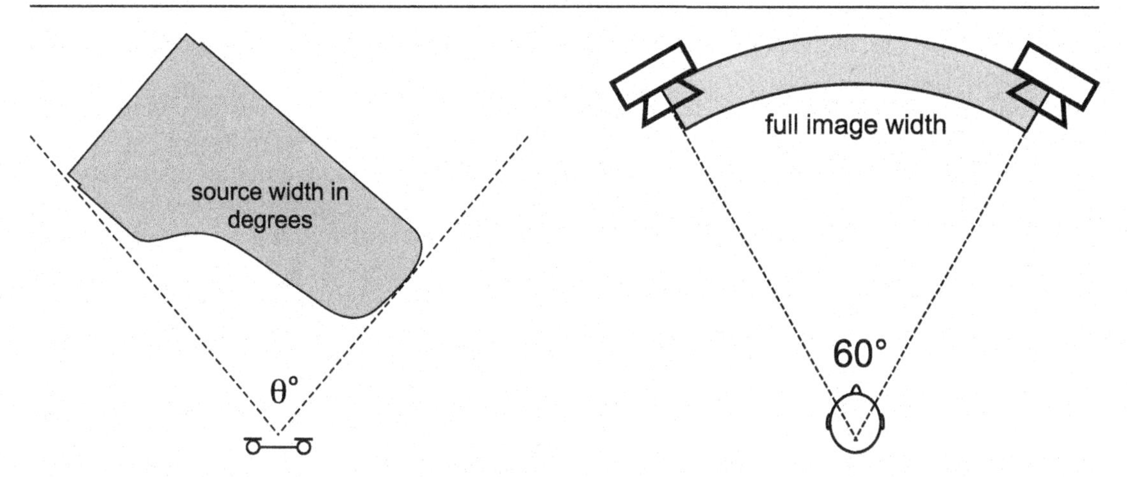

Figure 5.9a Source width at the microphones

Figure 5.9b Full image width on loudspeakers

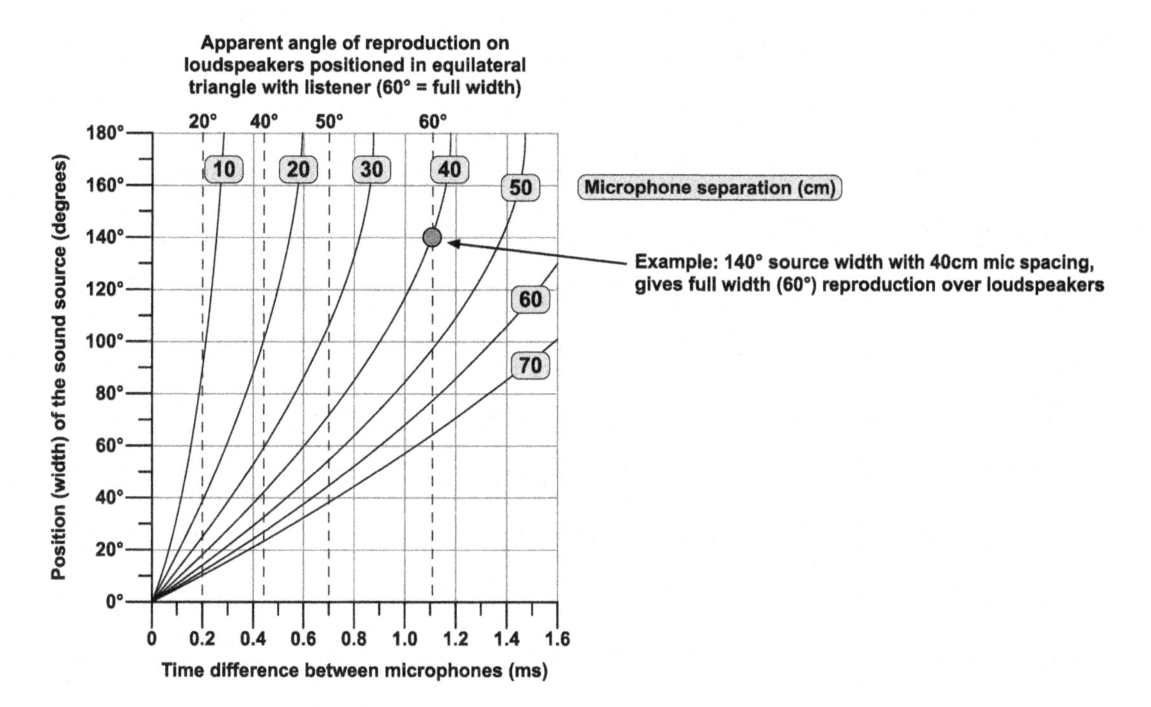

Figure 5.10 The effect of microphone spacing on perceived stereo image width. Data is for omni microphones but also applies to directional microphones provided they are both pointing forwards.

Source: Michael Williams.[3]

and the location of an individual source (such as a single player) in degrees away from the central line (Figure 5.9a).

Each of the bold diagonal lines corresponds to a particular separation distance between the microphones, as labelled (in centimetres). The image width (angle of reproduction) is directly correlated to the time difference between the microphones, and is shown by the vertical dotted lines labelled in degrees. The circle on the chart indicates that for a source width of 140°, a spacing of 40 cm (16″) between microphones will produce an image that is full width between the loudspeakers (60°).

To use the chart as a guide to microphone spacing for omnis or parallel directional microphones, first stand at the sort of distance from the piano (or other source) that your microphones are going to be. Point your left arm to one extreme of the piano and your right arm to the other extreme, and have a look at the angle between your arms; this is your source width at the microphone position. Next, decide on your final desired image width; for piano, we earlier suggested about 50%–75% of the full width, which is about 30°–45°. You can then use the chart to work out a suggested microphone spacing. Microphone distance from the source will alter the source width; this distance will usually be chosen according to the balance of direct and reverberant sound, unless your microphone placement is constrained by other factors.

It goes without saying that this is a guide only; the ultimate decision about microphone distance and spacing rests with the engineer on the basis of how it sounds.

5.5.3 Measurements of out-of-phase components

Another important aspect of the microphone spacing is the impact it has on the amount of out-of-phase components that are present in the sound, which is always a potential problem with spaced microphone techniques. The different path lengths taken by soundwaves to reach each microphone will result in some phase differences between the left and right signals, and their significance depends on the wavelengths involved. For example, the wavelength of a 1 kHz signal is approximately 34 cm (13″), and this is the kind of distance by which we are separating the microphones. The wavelength of 100 Hz is around 3.4 m, so a microphone spacing of 30 cm (12″) is less significant for these frequencies.

Phase differences for individual notes can be seen quite clearly if you get a pianist to play individual notes one at a time and look at the results on a phase scope. Some notes will be in phase (predominantly vertical display on the scope), some will experience an intermediate phase shift (generally circular display on the scope), and others will be distinctly out of phase (predominantly horizontal display on the scope). It is these latter ones which are potentially a problem. With spaced microphone techniques you can never get rid of them altogether, but if there are too many then it will start to affect the overall feel of the sound, and it is a good idea to reduce (or at least control) the number of notes that are distinctly out of phase.

In a test performed by Mark Rogers, all 88 notes on a Steinway D were played individually and recorded simultaneously at three different microphone spacings: 20 cm (8″), 40 cm (16″), and 60 cm (24″). The results are shown in Figure 5.11, where it can clearly be seen that the number of notes that are distinctly out of phase increases along with the distance between the microphones,

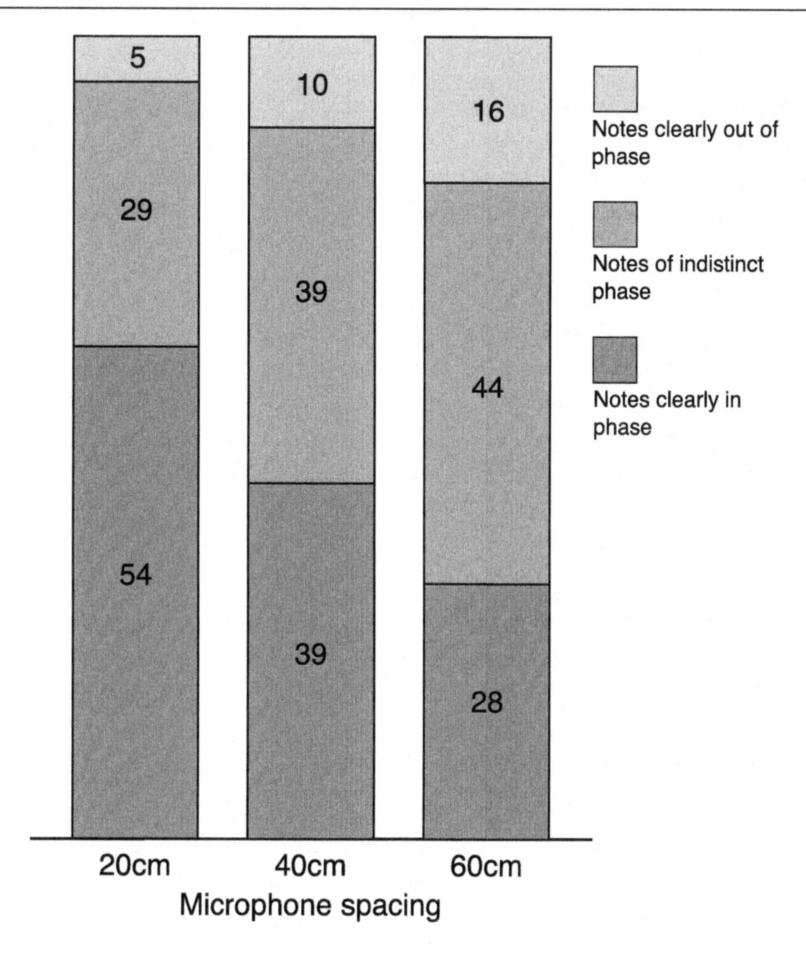

Figure 5.11 Effect of microphone separation distance on phase response of individual piano notes

while the number of notes that are clearly in phase decreases. The out-of-phase notes tend to be at the higher end of the frequency range, as the lower notes have wavelengths that are too long to produce this effect.

Having a few out-of-phase notes does not adversely affect the listening experience, and again this suggests that a spacing of around 12″ (30 cm) is most likely to give a satisfactory width without too much in the way of out-of-phase effects.

5.6 The Decca piano technique

All of the aforementioned factors led Decca engineers to develop a practical technique that gave consistently excellent results and allowed adjustment of the tonal palette with simple, small adjustments. Figure 5.12 shows a pair of omnidirectional microphones, spaced 30 cm (12″) apart and positioned 120–150 cm (4′ to 5′) from the end of the piano, measured from the closest piece of wood to the microphone capsules. Height is usually around 150–180 cm (5′ to 6′) – at greater heights, the mics are looking over the lid.

The default starting position is for the right microphone to look directly along the middle bass string, pointing at the top of the metal frame at the keyboard end (see Figure 5.13). If observed from the keyboard end, this will be approximately in line with the lowest C (see Figure 5.14).

The left microphone points at the highest note with a damper (see Figure 5.15). The pair is rotated so that a perpendicular centre line from midway between the microphones points at the middle of the keyboard (E/F above middle C). Microphones should be panned fully left and right.

The given positions are just starting points; there is nothing sacred about them, and it is likely that some adjustments will be necessary:

- *Forwards* (closer to the piano) *and backwards* – adjusts the balance of direct to reverberant sound.
- *Sideways movement* – adjusts the amount of the mid and high frequencies while leaving the LF more or less constant. Moving to the left will increase them, and moving to the right will reduce them. Therefore, to make the pianist's left hand (mainly LF) more prominent, move the microphones a little to the right. Often a small correction of a few centimetres or inches will make the required difference.
- *Height* – this can adjust the amount of lid colouration. With the pair below the lid, some of the 'glue' effect of the lid that integrates the sound together will be obtained. Microphones placed in the plane of the lid will remove much of the lid effect altogether. If the pair is up at 180 cm (6′), it could be looking right over the plane of the lid. Sometimes this doesn't really matter, but if the R microphone is above the plane of the lid and the L microphone is below it, a slightly phasey effect on the bottom end can result. The microphones can be lowered, or sometimes a small move to the left will bring both of them just under the lid. As mentioned

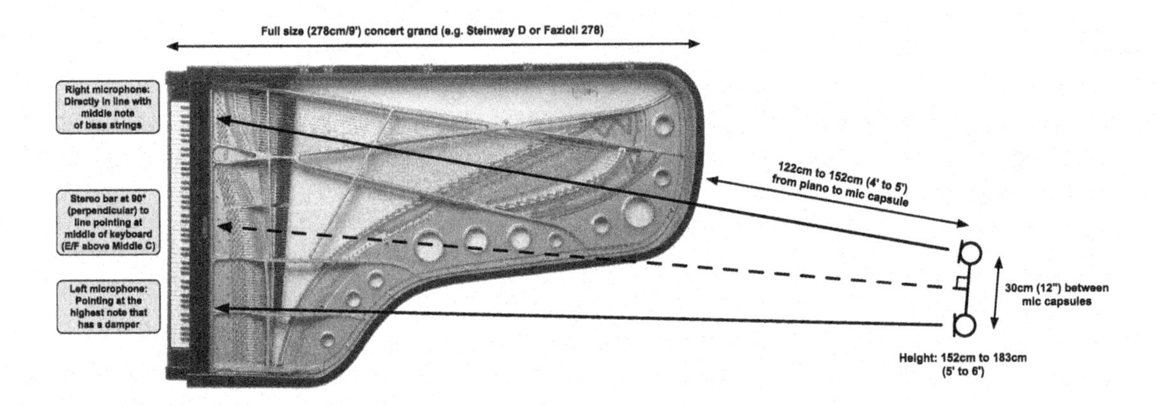

Figure 5.12 The Decca piano recording technique

Piano image: © Fazioli Pianoforti SpA. Photo: Roberto Zava/Studio Step.

Figure 5.13 Right microphone pointing down centre line of Zone A strings
Photo: Mark Rogers.

earlier, some engineers use custom lid props to raise the angle of the lid wider, which allows for higher microphone placement and changes the lid's effect.

- *Microphone spacing* – adjusts the image width for a given distance from the piano. Again, only a small change can give the desired result, and in accordance with earlier comments, wider spacing than about 35 cm (14″) will start to make the image too wide.

There may be an overall lack of HF at the tail end, so it is common to use mics with an HF boost to give a little more reach or focus. Before Schoeps introduced the MK2S, Decca often used Neumann KM83s, M50s, and KM53s.

A strong point in favour of the Decca technique is that it generally produces a consistent perceived distance from the sound source across the full keyboard range. Despite the distance of the

Figure 5.14 Looking from the keyboard, the centre line of Zone A matches up with lowest C key
Photo: Mark Rogers.

microphones from the upper strings, the high notes are very clear and do not feel like they are further away than the low notes. Moving the microphones around to the left (towards positions C and B in Figure 5.8) often seems to make the higher notes recede a little, despite closer proximity to the upper strings.

In some rooms, and on some pianos, the Decca technique (with a 30 cm (12″) spacing) can make notes between C_5 and C_6 prone to boosting and/or cancellation. It is worth doing a test by playing a chromatic scale of consistent loudness in the room, recording it, and seeing which

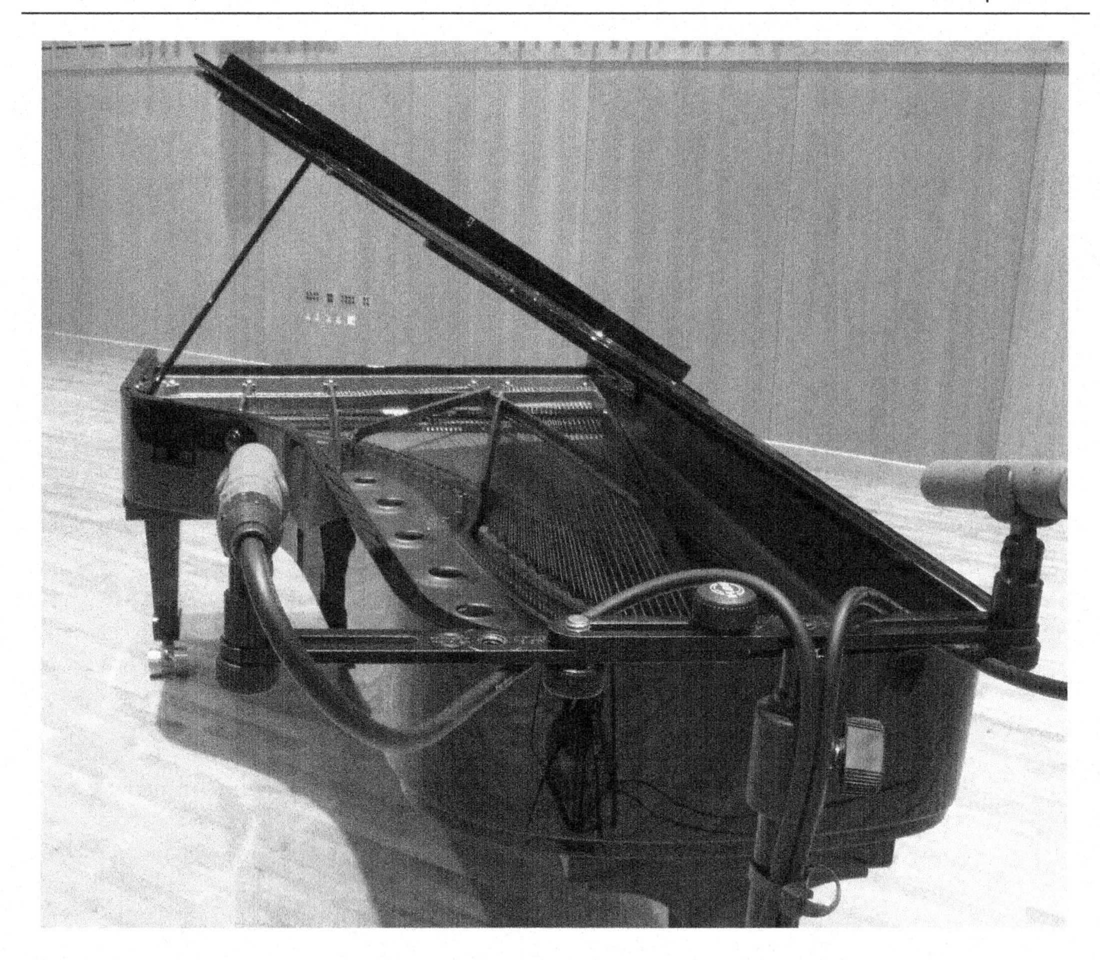

Figure 5.15 Left microphone pointing at transition from dampers (Zone C) to no dampers (Zone D)
Photo: Mark Rogers.

notes, if any, are louder or softer on playback. Very small movements of the pair can sometimes eliminate these problems.

5.7 Techniques for other scenarios

This section is intended to cover recording pianos in more difficult circumstances where it is not possible to use a spaced pair or the Decca technique.

5.7.1 The three-microphone LCR method

There are many occasions when it is necessary to get closer to the piano with microphones; this might be in a concert situation where space is tight or where a tail pair cannot be used for other

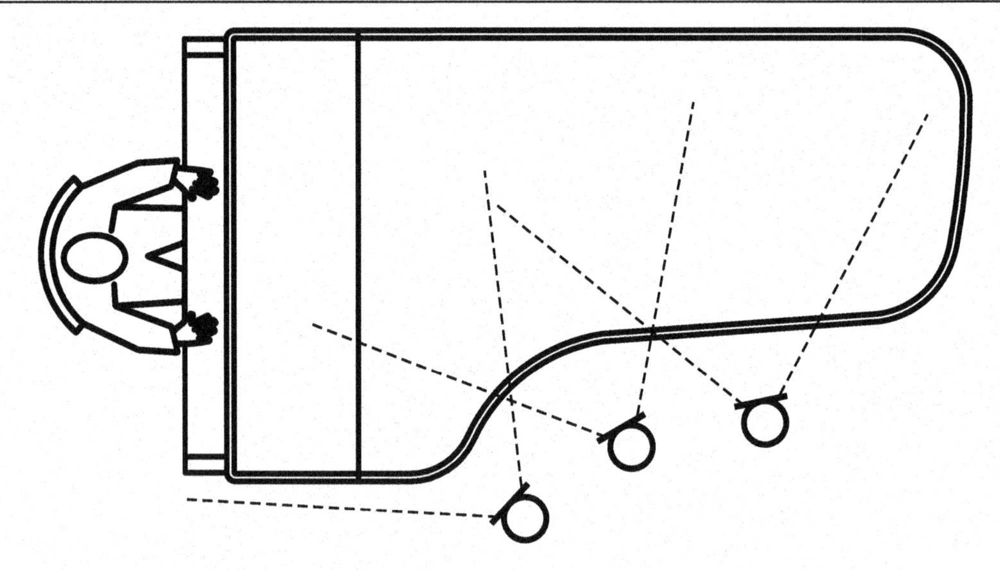

Figure 5.16 Plan view of three microphones covering a piano from a closer distance

reasons. It is more commonly used where the piano is performing as part of an ensemble rather than for solo recital. As with all close microphone placement, it is important to try and obtain even coverage of the instrument. The useful analogy with spotlighting when trying to obtain even coverage of a group of players is explored in Chapters 8, 9, and 15. The same principles apply when trying to pick up a large instrument evenly from a close distance. Given the sheer size of the grand piano, if you have to place spaced microphones quite close to it in the well of the piano, the chances are that two microphones will not be sufficient to cover all the strings evenly, and three microphones will have to be used. These can be placed at a height of about 1.2–1.5 m (4′ to 5′), looking down onto the strings with the capsules placed just back from the edge of the instrument. Avoid taking the microphones right under the lid and closer to the strings unless you are aiming for a more hammery, jazz-orientated sound.

Figure 5.16 shows a plan view of the microphones' locations and areas of coverage of the strings and soundboard. The microphones are aimed at the high, middle and low strings respectively and can be panned fully L, C, and fully R. The recording image does not come out as overwide because of the anchoring action of the centre microphone. The level of this microphone can be used to alter the middle voicing of the piano in a range of up to about ±4 dB with respect to the left and right microphones. Figure 5.17 shows a photograph of the technique from a piano trio session in Abbey Road Studio 2.

5.7.2 Recording a piano on short stick

Sometimes you have no choice but to record a piano whose lid is on short stick. Surprisingly, the Decca technique, with the height set so that the microphones point into the small gap between piano and lid, works quite well. Another good option is to use the three-microphone LCR

Figure 5.17 The LCR piano technique.

Photo: Carlos Lellis, Programme Director, Abbey Road Institute.

technique outlined in 5.7.1, again with the height adjusted to allow the microphones to point into the gap. The potential for a rather close and bright sound that over-emphasises the mechanics of the instrument can be ameliorated by use of ribbon microphones which soften any transients with their upper frequency roll-off. Wide cardioids (e.g. Schoeps MK21) are a good substitute for omnis if rear rejection is desired.

5.7.3 Microphones under the piano

Despite the temptation to hide microphones under the piano during a live event, this is almost never a good idea. The sound under the piano is highly coloured as it comes mainly from the resonating soundboard and not the strings. There is also potential for strong boundary effects with the floor and the setting up of standing waves between the floor and underside of the piano. The sound is often described as 'honky', and it needs a lot of EQ to make it usable.

5.7.4 *Upright pianos*

To get good results on an upright piano, the same principles we have discussed for grand pianos can be used, although there is no large lid to integrate the sound. One option is to open the top of the piano and point the microphones in from above, looking down the length of the strings. Alternatively, if the bottom front cover can be removed, they can be pointed upwards along the length of the strings from the bottom. Removing all the front covers and recording from the front means that the player is potentially in the way, unless you can place a pair high enough and looking downwards. Putting microphones behind an upright piano (behind the soundboard) is similar to putting microphones under a grand piano and can give a similarly coloured and unpleasant sound.

5.8 Practical issues when recording pianos

5.8.1 *Loudness and distance*

One of the reasons why the piano is such a good instrument both for soloing and accompaniment is that notes played harder do not just sound louder but also sound physically closer thanks to the increase in the transients. The tonal character changes such that if you shut your eyes and try to imagine how far away the sound source is, it will seem to get closer as it is played harder. Thus a skilled player can literally bring a melody forward whilst keeping other notes recessed. However, this feature of the piano can be a trap for the engineer: if you have spent time getting the microphone distance just right with the performer playing one piece, it can be disconcerting to hear it sound too close or too far away when the repertoire or performer changes. Therefore it is a good idea, whilst setting up, to get your performer to play a range of material at a range of dynamics, listen for the changing perceived distance, and settle on a happy middle position.

5.8.2 *Monitoring and image width*

As noted in Chapter 2, you need to be very careful if monitoring your recording on headphones. It is particularly easy to make a piano sound good on headphones only to discover that it sounds far less so on loudspeakers. Most typically, a relatively narrow and direct piano sound with a lot of additional full-width reverb will sound very impressive on headphones, but when played on loudspeakers the two elements seem quite disjointed and separate, with the reverb seeming washy and overbearing, making it hard to focus on the direct piano sound. However, what sounds good on loudspeakers will almost always sound good on headphones too.

5.8.3 *Pedal noise*

The pedals on a piano can be quite noisy. Some of the noise can be from the internal mechanism (that which lifts the dampers or shifts the mechanism for playing una corda), and some can be external (such as shoe scrapes on the floor). One of the great advantages of the Decca technique is that the microphones are a long way from the pedal mechanism, which helps to minimise the problem. However, it is always a good idea to take a small piece of carpet or a mat to a session

Figure 5.18 Piano recording session at the University of Surrey showing talk microphone for the pianist, Decca piano technique, mat under the pedals and stool, and screens to break up reflections.

Photo: Caroline Haigh.

that can be put under the performer's feet to reduce noises from their interaction with the floor (see Figure 5.18).

5.8.4 Screens and reflective walls

If the recording room has flat walls, reflections can be a problem, especially if the walls are parallel. It can be a good idea to avoid positioning the piano with its long hinged side parallel to the rear wall, and turning the piano to an angle allows the sound to bounce off the walls in different directions. Another very useful solution, if you have access to them, is to use acoustic screens around the piano. These should be placed at different angles and with gaps in between them rather than forming a continuous block. The aim is not to absorb sound waves but to break up reflective paths and make the internal surfaces of the room more complex. (See Figure 5.18 for use of screens.)

5.8.5 Floor effects

The underlying structure and surface material of the floor can have a big effect on a piano sound. Concert hall floors that are suspended or sprung can sometimes act as giant low-frequency resonators, which can have a beneficial effect in warming up and enhancing the low end of the piano. A solid floor has no spring at all and will not couple with the piano in this way. Very reflective surfaces will give a harder sound, and in venues such a churches with stone floors, placing absorptive material under the piano – such as carpet, kneeler cushions, or even the fabric piano cover – can help to counteract this.

5.8.6 Stereo balance and bias

The imprecise imagery of a piano – not really knowing where a sound is coming from – can lead you to set up your microphones so that the recording sounds good initially, but it becomes less pleasing to listen to over a long period. Often this is a result of some kind of left-right balance issue, which might be related to amplitude and/or timing.

It is essential that the two microphones have the same gain so that any imbalance between left and right can be corrected by adjusting only the timing (e.g. by rotating the pair so that one or other mic gets closer to the piano). Here is a useful set-up method:

1 Before mounting the microphones, hold them together so that their capsules are next to each other and talk directly on-axis into them whilst observing the signal on a phase scope. Any gain discrepancy will be seen as a bend to the left or right. Adjust gains so that the line is vertical.
2 Next, set up the microphones in position and ask the pianist to play whilst you listen on headphones and/or loudspeakers (headphones are particularly good for this test).
3 Whilst listening, repeatedly toggle a left-right reverse function on your monitoring system. This will switch between sending the left microphone to the left monitor and the right microphone to the right monitor, and vice versa.

The image will be different when you left-right reverse (the piano is not a symmetrical instrument), but if set correctly, you should not feel a major image shift from one side to the other. If you are experiencing movement, you will be able to work out if the pair is producing a signal that is left or right heavy (when panned the correct way around.) Given that the microphones have been electrically balanced in step 1, any left-right imbalance is due to microphone placement. To correct this, try these options:

1 Move the whole mic array left or right if it looks as if the piano is physically offset from the centre line of the pair.
2 Rotate the mic array (if fitted on a stereo bar) to adjust the time of arrival for the pair of mics. If the image is right heavy, then rotate the array clockwise (viewed from above) to bring the

left microphone closer and the right microphone further away. In the default Decca setup, a perpendicular line from the centre of the stereo bar points to the middle notes of the piano, but as the microphones are moved leftwards (towards C and B in Figure 5.8), you will find that the line points more towards the bottom end of the piano.

Middle keyboard sounds (two octaves around middle C) are particularly helpful for this adjustment.

5.8.7 And finally, talkback

As noted in Chapter 2, if your microphones are a long way from the player of the instrument, you cannot rely on them for communications as well. If you use the Decca technique, the microphones really are a long way from the pianist, and in a session environment it is essential to put up an extra microphone close to the player in order for them to be able to talk to you. (See Figure 5.18). This microphone should not be included in the recorded mix, but only in the monitoring. Inevitably, you will forget to mute it at some point, so if the current take sounds strange, your first thought should always be to wonder if you've left the talkback microphone on.

Notes

1 *The Musician's Guide to Acoustics, pp241–252 The Piano*
Author: Murray Campbell, Clive Greated
Publisher: OUP Oxford, 2011
ISBN: 978–0198165057
2 *Unified Theory of Microphone Systems for Stereophonic Sound Recording*
Author: Williams, Michael
Affiliation: Institut National d'Audiovisuel, Ecole Louis Lumiere, Paris, France
AES Convention: 82 (March 1987) Paper Number: 2466
Publication Date: March 1, 1987
Subject: Microphones and Recording
3 *The Stereophonic Zoom: A Practical Approach to Determining the Characteristics of a Spaced Pair of Directional Microphones*
Author: Williams, Michael
AES Convention: 75 (March 1984) Paper Number: 2072
Publication Date: March 1, 1984
Subject: Studio Technology
Permalink: www.aes.org/e-lib/browse.cfm?elib=11692

Chapter 6

Voice: solo and accompanied

This chapter is concerned with recording the classically trained voice as a soloist with piano. For other aspects of voice recording, see Chapters 15 and 16.

6.1 The singer in a recording session

Singers have the reputation of being more temperamental than most, and it is worth taking the time to prepare well for the session and try to pre-empt every possible source of difficulty, from temperature and draughts to refreshments and breaks. Because the singer's instrument is also part of their body, the need for them to feel comfortable during a recording session can be of even greater importance than for other musicians. A trained singer will be able to give of their best for about three hours in a day, and you should be aware of this; repeating difficult cadenzas multiple times will be inherently time-limited.

Make sure the artist feels comfortable and secure at the start of the session. Listen attentively to them singing live in the studio, and remember that the voice that they hear is not the same as the voice a listener hears; their voice is conducted to their ears through the bones of their head as well as via the surrounding air. You will have to convince them that what you are capturing is an honest representation of their voice, especially if they have little recording experience, so building up their trust in you is essential. It is also a good idea to talk to the accompanist, especially if they are experienced and tactful and know the singer well; where problems occur, the accompanist can often help to smooth things over by agreeing with your suggestions at the right time. If this all sounds like treading on eggshells, it can be exactly that.

6.2 The classical voice and microphone placement

The classically trained voice is designed to project over an orchestra in a big, live space without additional amplification, and as such, it has a huge dynamic range and power. This has implications for both recording and mixing; you will have to actively manage the dynamic range while mixing and leave sufficient headroom when recording. An experienced singer will save their voice to a certain extent during rehearsals, and so you should be prepared for an additional 10 dB or so in level once they are live in a session take or a concert. If you are in any doubt about your recorded levels, under-modulate; do not be overly concerned with recording at 10 dB below peak level, as in a practical sense it matters very little, and you can restore the level afterwards. Yes, the signal-to-noise ratio will be very slightly worse, but if you overshoot in the other direction, the recording can be unusable where it has clipped in the digital domain. The increased noise level will only start to be a real problem if you record a large multitrack at 16-bit, with all tracks under-modulated; if you are recording 24-bit and using relatively few microphones, some under-modulation should not be of any concern.

It is important to emphasise that in manipulating levels manually, you are not aiming to achieve the very restricted dynamic range of a pop vocal but more to support the voice when it is singing in a lower, weaker register by raising the level in the mix. While automatic compression is the norm when recording pop vocals, its use on a classical voice recording is not; it is important for the voice to retain the natural control, phrase shaping, and expression that the singer has worked so hard to achieve. A compressor is designed to act over a relatively short time frame and alters the envelope of the sound; skilful fader riding will be able to work with the musical phrase to achieve a really transparent and natural sounding result while supporting weaker passages and allowing crescendos and climaxes to achieve their full force. This sort of 'intelligent compression' (i.e. fader riding) will be discussed in section 6.9.

The distance at which a classical singer should be recorded is also different to that generally used for the pop singer. The voice is designed to be heard at a distance and relies on an acoustic space to integrate the whole sound together. The harmonic content of the voice changes with distance, and very close to the front of an operatic singer, the HF content will dominate the sound at the expense of warmth and richness, and the hidden gear changes from chest voice to head voice will be more apparent. The voice will have a 'presence' edge that is undesirable in classical singing because it feels too close to the listener, and the illusion of 'natural' perspective in the recording as a whole will not be sustainable. A microphone that is placed too close will only pick up the parts of the frequency content that are projected in the microphone's direction, and a much more complete blend of the overall voice is best achieved at a distance of at least 90 cm (3′) away. However, there will come a point in moving further away that the singer's microphones will start to pick up too much of the room and the accompanying piano. This will cause some loss of the desired clarity of sound, and the engineer needs to find the right distance to produce a good balance between clarity and picking up the full spectrum of the singer's voice. The exact distance will depend on the acoustic characteristics of the room, where the piano and singer are placed in relation to one another, and the microphones used.

Another factor that mitigates against closer microphone placement is the potential for the singer to move. This is part of the performance, and cannot be rigidly constrained if the singer is to perform expressively and musically. The change in tonality and level produced by movement of 15–30 cm (6″ to 12″) back and forth will be much more significant with a close microphone, but less significant when the microphone is placed in the 90–120 cm (3′ to 4′) range. A piece of carpet under the singer's feet will reduce noise from movement being transmitted acoustically and through the floor to the microphones, and to a certain extent, will act as a psychological barrier to prevent them moving off the carpet towards the microphone. Another barrier to movement would be to place a line of tape on the floor for the singer to keep behind; you should also mark the position of the music stand at the same time as you are marking microphone positions. (See Chapter 2.)

Microphone height above or below the mouth will affect the tone of the voice, although this effect is not as pronounced as it is when working at close pop microphone distances of 10–30 cm

(4″ to 12″). Microphone popping caused by plosives should not be a problem at greater distances either (even if using susceptible directional microphones), as the blasts of air from the mouth that produce this effect spread outwards and will have dissipated much of their energy by the time they have travelled 90 cm (3′).

Sibilance and HF content tend to be projected horizontally and slightly downwards, so a common choice is to place the microphone slightly above the mouth line and looking down (Figure 6.1a). This will produce a less harsh sound than directly on-axis if you find that on-axis is over-bright. The only potential drawback for this position is that if the singer has a music stand, there will be times when they glance down slightly to look at the music, and this slight downward movement of the head will result in a momentary loss of focus as the tone loses additional HF.

Figure 6.1b shows the microphone slightly lower than the mouth, but not so low that any stand is in the way. Singers will want to be able to see over the music stand if there is one, so it will not be positioned very high, and the microphone should be able to be positioned looking over the top. In this position, there is less of a change in tone when the singer looks down at the music or lowers their head for any other reason.

Figure 6.1c shows a microphone on-axis to the mouth, which in pop recording is likely to result in a problem with plosives because the microphone is so close, but it will be fine in classical recording as long as you are happy with the HF content of the voice at this angle.

Figure 6.1d shows the use of 'Pavarotti' microphones, such as the Schoeps RC Active Extension Tube arrangement, which consist of a normal Schoeps microphone capsule mounted on a long gooseneck extension with the microphone head amplifier mounted at the bottom of the stand. These are designed to be for discreet concert use, and their longest standard length (1200 mm) places the microphone below mouth height of an average singer, although this does not matter in practice. As discussed for Figure 6.1b, microphone placement below the mouth gives a wider safety margin for head movement as singers tend to dip their heads from horizontal rather than tipping them backwards.

To judge where your microphone is placed in relation to the singer's mouth, it is best to stand and look from the side while the singer is performing. It can be much harder to judge this placement looking on from the front, and the singer will be less distracted and be able to perform naturally if you are not in their sight line.

6.3 Using two microphones on the voice

This is a technique that is used for two different reasons. One is simply to have a backup microphone in a live situation, and it is only ever intended that one of them should be used in the final mix. For this reason, backup microphones and the main microphone will be placed very close together, within about 5 cm (2″), and Schoeps even make a double extension tube with this arrangement for live use (Schoeps R2C KC).

The second reason is to create some width around the singer's recorded image, as noted for solo instruments in Chapter 4. For this purpose, the microphones are spaced around 20–30 cm (8″ to 12″) and pointing forwards; both microphones will be used in this scenario. This is a very typical Decca technique that has been widely but not universally adopted, and will be discussed here as it is adaptable to most solo instruments (as noted in Chapter 4). The technique aims to give the illusion of size and space, or 'bloom' to the voice, but with stability of image as well; we

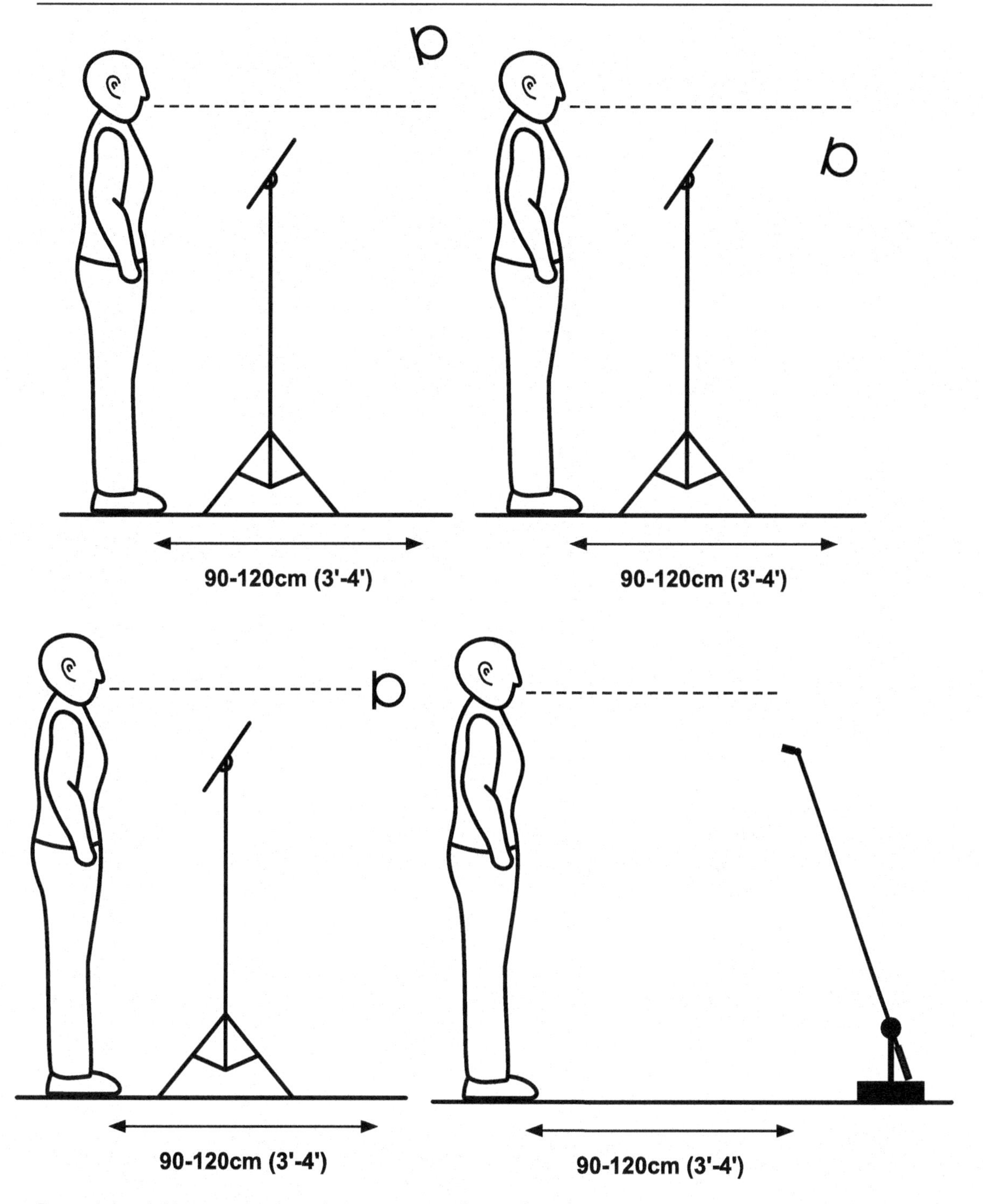

Figure 6.1a–d Alternative microphone positions for a classical singer

are attempting to trick the listener into perception of a stationary source that does not feel artificially narrowly located. The quality of 'bloom' around the singer is best described as a pleasant colouration or smudging that works for that particular voice and repertoire.

6.3.1 Single- versus double-spot microphone

In the context of a stereo recording, using a single microphone on a singer or other soloist will produce an image that is very laterally stable and locked in place, but it can also present as an artificially pinpointed source. As the microphone is faded up, the effect is like looking down an increasingly narrow tunnel at a very focussed, but small singer. This effect limits how much of the spot microphone can be used because as the singer is narrowed to a point source, the image starts to feel artificial. Two microphones, used appropriately, will give the singer a sense of width and space that feels more natural to the listener and blends much more easily with a stereo piano sound.

6.3.2 Suitable microphone spacing

As noted in Chapter 3, both microphone spacing and the mounting angle between a pair of directional microphones can be used to control the image width. With this 'double-spot' microphone technique, the microphones are placed parallel to one another facing forwards, which naturally constrains the width of the resulting image even if the microphones are panned fully left and right. As we shall see with this technique, it is common for the pair to be panned inwards a little, to anywhere between about 70%–90% left and right depending on the soloist's movements, and the engineer's preference for the quality of the sound.

The most successful spacing (arrived at empirically) is in the range of 20–30 cm (8–12″). Any wider than this and the singer's image will feel very spacious, but increasingly hard to locate and with a lateral position that partly depends on which note is being sung. The singer's movements will also become exaggerated because any widely spaced pair tends to widen the L-R lateral placement of the source location. Movement of the soloist while performing is a problem with both singers and musicians that perform standing up, such as violinists, flautists, and other wind players. Seated players, such as cellists, are less able to move their instrument around freely and present less of a problem in this regard.

The next consideration is panning of the pair, which is typically between 70%–90% left and right. Engineers engage in a certain amount of trade-off between microphone spacing and panning to give the right amount and quality of 'bloom' to the voice while keeping the performer's movements in check. Because there are phase differences at each microphone with a spaced pair, panning inwards involves partial summation of signals with different phase and therefore some small boosts and reductions in the level of some frequencies. If taken to the extreme (as when panning a pair of spaced microphones to mono), this could become more obvious comb-filtering, but the partial inward-panning used in this context produces an effect that is viewed positively as another tool to be used to shape the sound as desired. The small amounts of colouration involved are manipulated by the engineer as part of this trade-off between spacing and panning. The difference between spacing at 25 cm (10″) and panning 70% left and 70% right, and spacing at 20 cm (8″) and panning 90% left and 90% right, will be subtle in terms of colouration on the voice, but a preference can be chosen. It is worth taking the time to experiment at the time of recording so that you can choose your preferred combination of spacing and panning for a given voice.

6.4 Microphone choice

The next thing to consider is microphone choice. Singers usually love to have a big microphone to sing into, but it is important to debunk the notion that a large microphone is needed in order for the singer to sound 'large'. How 'large' a singer sounds has to do with the frequency response of the microphone, and in particular, the quality and extent of the LF range. The LF range of a microphone is not connected to the size of the microphone's casing or the size of its diaphragm; all that is required to reproduce LF is that the diaphragm is able to move back and forth at a reasonable amplitude at low frequencies, and it does not need to have a large diameter in order to be able to do this. Omnidirectional microphones are simple pressure transducers, and as such always have the potential for the most extended LF response regardless of the size of their diaphragm. Omni-condensers can extend down to 10–20 Hz (although this very low extension also collects acoustic rumble very well). Microphones such as the Neumann M49 and M50 appear large, but the actual capsule inside the case is quite small. The physical size of a microphone does have an effect on the off-axis HF, with larger microphones suffering more off-axis HF loss or colouration; in a classical recording context, this will affect the reverb that is picked up and the direct sound from other instruments.

The acoustic level that is produced by an operatic singer does mean that you may need to use the switchable attenuator that is built into the electronics of a condenser microphone. This reduces the level of the signal before it enters the head amplifier, and avoids the microphone distorting or overloading. Ribbon microphones will cope with high SPLs as long as they are not subject to moving air, but this will not be a problem with the microphone at these distances from the singer.

For pop vocals, the choice of microphone will depend on which gives the desired colouration on-axis that best suits the individual voice and also on whether the singer is to sing live in the room with the rest of the band (necessitating a microphone with excellent rear rejection) or is to be overdubbed later. For classical recording of singer and piano, a cardioid condenser microphone with a flatter frequency response is desirable. The off-axis response should also be as flat and uncoloured as possible as it will be picking up the room and the piano, and significant unflattering colouration on this microphone source will adversely affect the piano sound. As with all classical recording, this spill is not to be avoided but embraced; it is useful to the engineer in making sure the whole sound has a sense of cohesion so that the singer and piano do not feel artificially separated into different spaces. The unavoidable tendency for cardioids to be lighter at the bass end than omnis will be apparent on the piano and room spill. For this reason, it is best to use omnis for the piano microphones during a voice and piano recording to ensure that the overall piano and room sound does not suffer from a lack of extended LF. The proximity effect of the cardioids will not come into play when recording at these distances, but if you have to compromise by using closer microphone placement when recording a live concert, be aware of the possibility of LF colouration creeping in.

The Neumann U67 is a very well-liked classical vocal microphone for a couple of reasons. In cardioid mode, it has a little lift in response on the front axis between 5 and 10 kHz, and in the off-axis response at 180° there is also some lift between 3 and 8 kHz, which can be flattering for the room reverb. The Neumann KM84 or KM64 have a fairly flat frequency response, with a small HF rise, and some LF roll-off because they are cardioids. The KM84 in particular has a clean-sounding off-axis response. You should avoid microphones that roll off more quickly in the mid bass end because the loss of warmth at the bottom of the voice will be noticeable, even for a female voice, especially if a mezzo-soprano or alto. For the same reason, if you choose to put in an HPF to help with rumble, it should be no higher than about 75 Hz to avoid losing some body from the voice. It is probably wisest to do this during post-production and not at the time of recording so that you can be sure that you are not damaging the vocal sound.

Ribbon microphones also make a good choice for recording classical singers with piano, depending on the relative position of singer and piano. Figure 6.5 deals with a situation where fig of 8 ribbon microphones would be a good solution to the particular layout. Because the ribbon microphone (such as the Royer R-121 or Coles 4038) has a natural HF roll-off, it can be usefully used to tame very high harmonics in the voice and make it easier to blend in with the piano, which contains relatively lower levels of upper partials and thus sounds duller in tone than a bright soprano or tenor.

Using omnis or wide cardioids would undoubtedly produce a good vocal sound but they will pick up too much piano, and it will be difficult to use fader movements to keep the singer supported at all times without noticeably changing the piano sound.

6.4.1 *Managing with a poorer quality microphone*

Unfortunately, the solo classical singer has such a complexity of sound that it is very good at showing up deficiencies in microphones and other parts of the signal chain. In the days of vinyl (not counting the current revival in pop vinyl releases; classical recording has always tried to improve the transparency of recording technology), the poorer frequency response and increased tracing distortion that is present towards the end of each side of an LP became horribly apparent on soprano soloists in particular.

If you have limited microphone choice, or only a single microphone that is flawed and is producing colouration on the voice that you don't like, is there anything that you can do? One suggestion is to try placing the singer off-axis to the microphone by varying amounts, which can be done by raising or lowering the microphone as well as angling it. The frequency and directional response in the horizontal and vertical planes might not be even or symmetrical, and the colouration that is objectionable might only be present in a narrow acceptance angle centred on the front axis. A cardioid is only reduced in level by about 6 dB on its side axis, so going as far as presenting the side of the microphone to the singer will not result in a great deal of level loss but might help the frequency response.

6.5 Use of ambient pairs

An ambient pair is designed to pick up more indirect or reverberant sound from the room than direct sound from any of the instruments. It can be used to supply a clean source of reverberant sound for a particular instrument, to act as a 'glue' to support the spot microphones, and to anchor the instruments together in the same acoustic.

When placing a pair to obtain more reverberant sound, we tend to think of placing them horizontally further away from the players, but there is often very good reverberant signal to be picked up above and behind the players. One useful technique for live or studio work is to place an upwards facing ORTF-type or XY pair somewhere high above the performers. The rears of the cardioids are aimed at the performers, so they discriminate in favour of reverberant rather than direct sound. Placing these upstage behind the players can be very visually discreet in a concert recording, and if they are suspended, they will be barely visible. The reverb from this pair will be free of unwanted audience noise, although placing additional microphones in the audience to collect applause is a sensible practice.

In the studio context of recording voice and piano, an ambient pair can be usefully added 1.2–1.5 m (4′ to 5′) behind the piano, well above the lid at about 3 m (9′10″) high and pointing upwards. (See Figures 6.2–6.4 for inclusion of ambient pairs.) In the context of studio opera recordings, ambient pairs have been used high above the singers' heads to obtain some more 'air' around the voices when their solo microphones could not be placed further away. (See Chapter 16.)

6.6 Concert recording layout

Figure 6.2 shows a concert position layout with the singer facing outwards towards the audience. In this position, the solo voice microphones are in a good place to pick up the piano as well; the piano sound will be tonally good, if lacking some detail. The addition of piano microphones will help with this, and the options are to place a pair of omnis at the tail end or move the singer out towards the audience and put some piano microphones behind them. Piano microphones placed in the well of the piano will contain less room reverb and sound a little close, so care will have to be taken when blending them into the mix so that the piano is not inadvertently pulled in front of the singer in the recording perspective. Following the rule that the closer the microphones are placed, the more of them you will need to obtain a good tonal balance from a large instrument, using three microphones as L, C, R will give you more flexibility to adjust the middle voicing of the piano. The preferred solution in a nice sounding room would be the tail pair of omnis, but if the room has poor acoustics, the L, C, R microphones will give you better results (see section 5.7.1). In a concert situation, the use of short stick on the piano is a possibility outside the engineer's control, although this has a detrimental effect on the piano sound. Microphones can be arranged so that they peer into the small gap between the lid and the piano body in order to pick up some more piano detail. This can be done with either a tail pair or the L, C, R microphones in the well.

It is also possible to obtain a reasonable balance with a single pair if carefully placed, but the microphone position will be a compromise between piano and voice. If the piano lid is open at 35°, then for the piano sound, it can work well to have the microphones at a height that places them at 15°–20° above the horizontal level of the piano strings. Depending on the singer's height, this is likely to be higher than would be ideal for capturing the singer, so adjusting this until you have a good compromise would be the best place to start. The singer should be central on the microphones, and is best prioritised in terms of sound if you have to choose between the singer and piano (sincere apologies to all accompanists). Using a pair of spaced omnis for this approach will give you the full LF range and a good tone, but the image will not have the finesse and

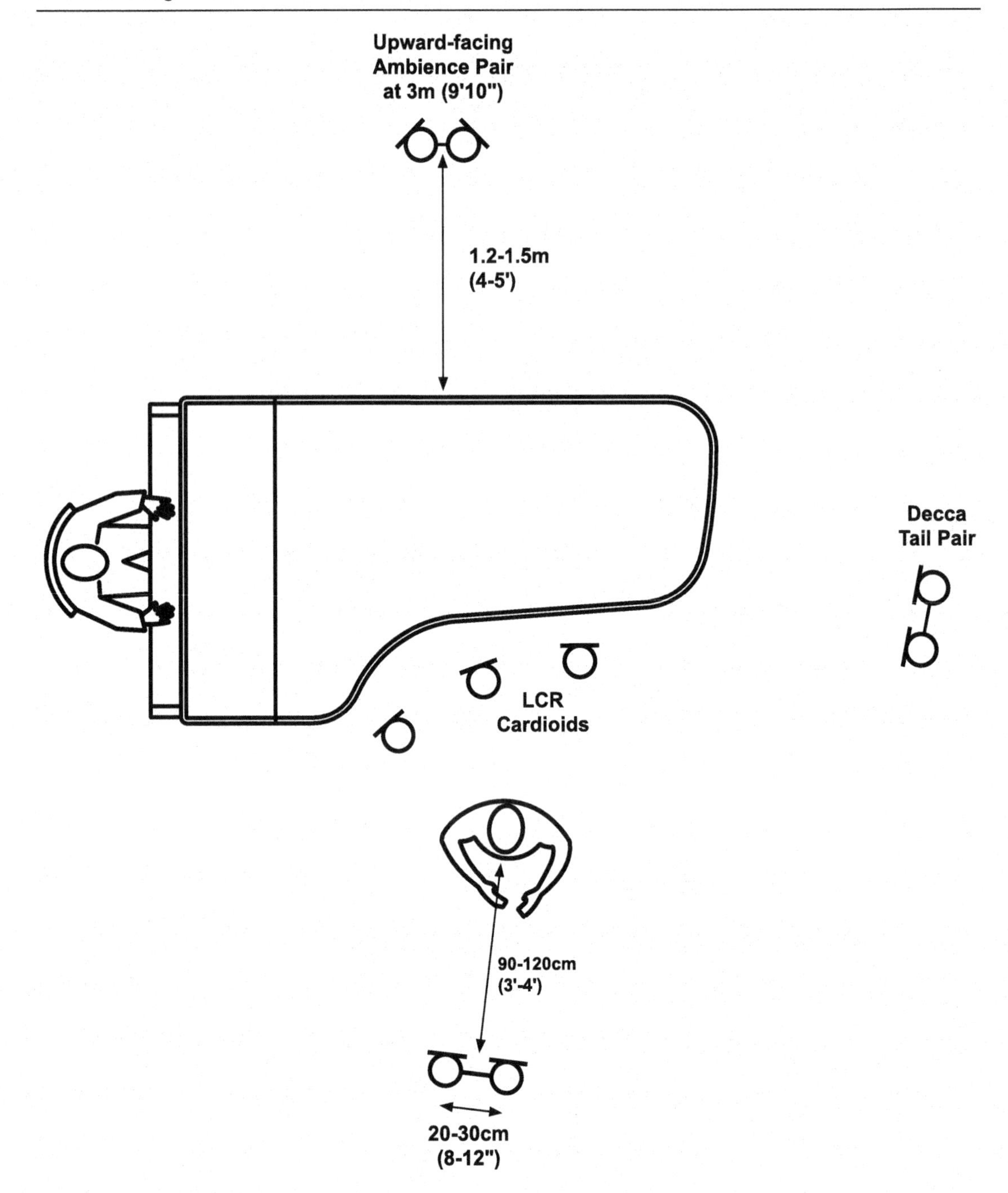

Figure 6.2 Concert layout for voice and piano

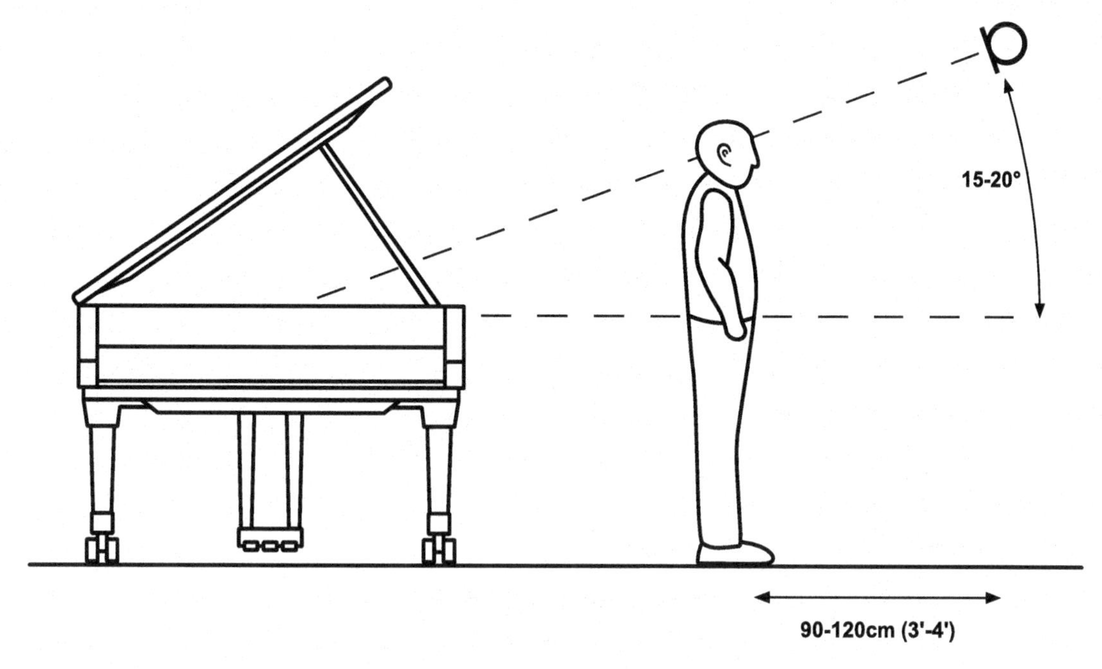

Figure 6.3 Concert layout for voice and piano using a single pair

focus that you will get from a skilfully balanced recording using a few more microphones. See Figure 6.3.

In many concert halls, you will find a spaced pair of omnis already rigged, slung in front of the stage, although usually too far back or high up to be useful as a single pickup for your voice and piano. They could be used as more overall pickup to give you some warmth and sense of space, in conjunction with a pair of cardioids on the piano and another on the voice to add focus and detail.

6.7 Studio recording and reverse concert positions

In a recording situation, there is no need to retain the concert position layout, and this opens up the potential for improving the communication between the singer and pianist by allowing them to maintain a good line of sight with each other. Some singers also like the opportunity to see the piano keyboard, and almost all of them dislike standing right in the well of the piano. Figure 6.4 shows a recording layout with an arc of possible singer positions, all of which face the piano. In addition, a 3 m (9′10″) high, upwards-facing ambient pair is included, either behind the piano, or just behind the singer.

Positions A and B are good for eye contact with the pianist, and position C means that the singer can also see the pianist's hands, but the pianist will have to turn his or her head a little to maintain eye contact. All these positions mean that the piano is being picked up on the back of the vocal cardioid microphones, hence the need for a clean off-axis response from these microphones. They need to be at least 90 cm (3′) away from the side of the piano (using a Steinway D model as an example). If they are closer, they will pull in too much of the piano onto the vocal microphones, and this will make it harder to keep the voice balanced against the piano in the mix.

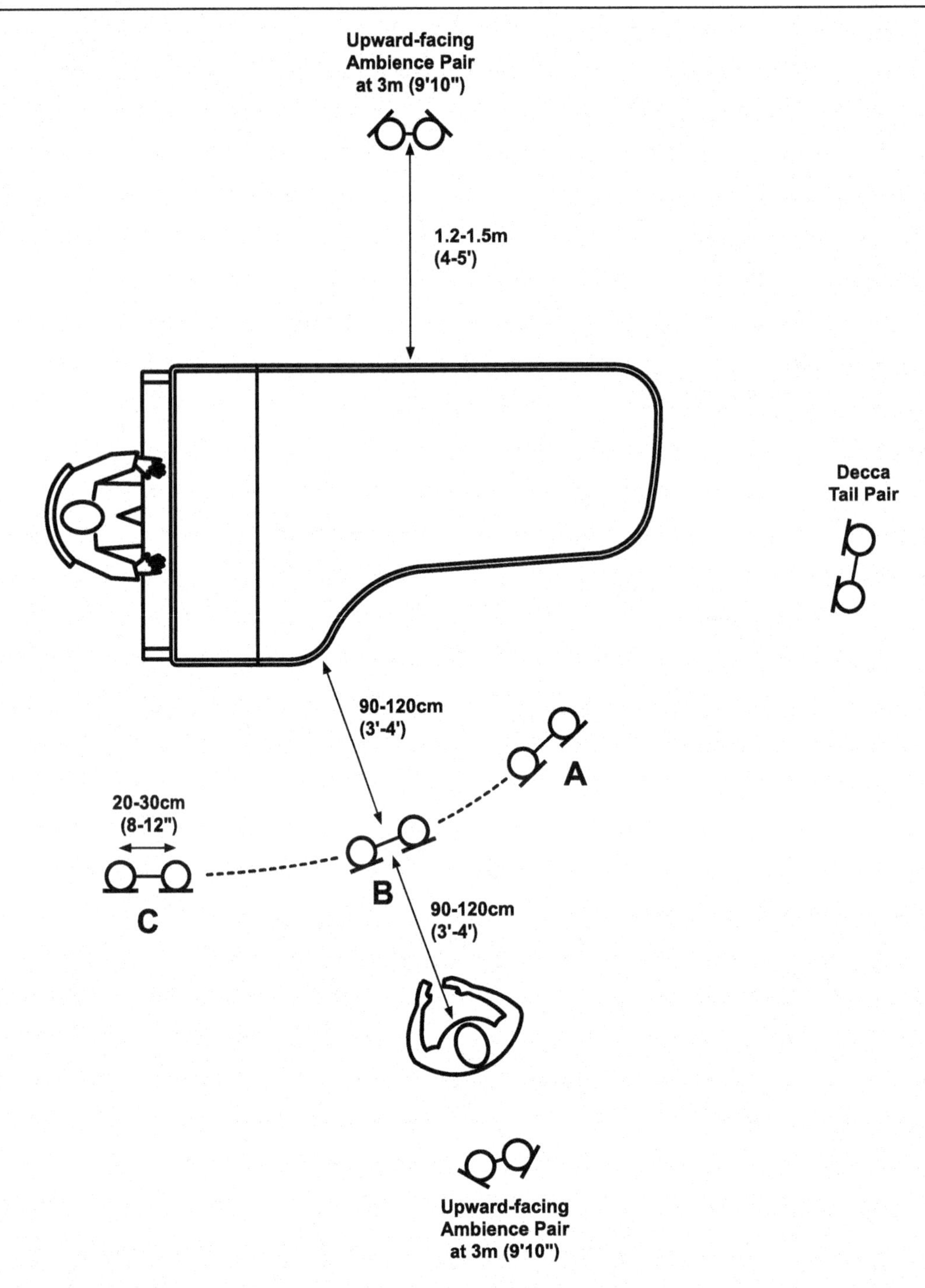

Figure 6.4 Studio layout with alternative singer positions A, B, C

Conversely, if you place the singer too far away, the piano spill will be too distant, the vocal spill onto the piano microphones will also be very distant, and the piano and vocal microphones may have such different sounds that they will not blend together well.

Reduction in level of spill is often achieved in other recording scenarios by choosing to point the back of the microphone at the source that is to be discriminated against. If you are working in a tight space with a dry reverb characteristic such as a theatre pit, the null point of a microphone's pickup can be used effectively; the acoustic has little reverb and you are close to individual sources. However, when working with piano and singer at microphone distances of greater than 1 m (3′4″) in a live space, the piano is not a point source at which the back of these singer's microphones can be aimed effectively. It is nearly 3 m (9′10″) long, and is present in the room reverb as well, so it will inevitably be picked up on the singer's microphones wherever they are placed. Therefore, the first priority should be making sure the singer is comfortable with their position relative to the piano and then to prioritise capture of a good singer sound when placing the vocal microphones. Because the spill is unavoidable and is to be utilised as part of the overall sound, its quality is of great importance. To assess the quality of the piano spill arriving off-axis on the vocal microphones, listen to the piano alone on these microphones, and you will be able to hear if there is a lot of colouration, usually in the form of rather muddy sounding middle bass. The tonal quality of this spill is all dependent on the characteristics of the vocal microphone that you are using, and if it is unsatisfactory, there is little to be done other than change the microphone.

As already mentioned, the piano microphones would ideally be omnis to make sure that the piano's full timbral range is captured. There will of course be singer spill onto these microphones, and this will be increased significantly if the piano microphones are placed in the arc of the piano rather than as a tail pair as shown in Figure 6.4. If for any reason, the piano microphones have to be placed in the piano arc, cardioids might yield an overall better recording, despite the lack of deeper LF.

Now that we have two stereo pairs of microphones placed on our singer and pianist ensemble, the next question that arises is whether this will cause confusion in the stereo imaging. After all, each pair has a very different 'view' of the same ensemble, so which is the most important one, and where do 'left' and 'right' lie? The answer is that there is enough de-correlation between the pairs for it to be possible to overlay the images from each with no ill effects. The piano pair (panned fully left and right) will provide a balanced piano sound, give a wide overall sense of space, and will also add some warmth and richness to the voice. This will form the basis of the sound, with the vocal pair blended in to give sufficient focus to the voice. Although there is a little more voice level on the left piano microphone, the voice appears only a little to the left on the image from the piano microphones, and this becomes unimportant once the vocal microphones are added and used to centre the voice placement in the image. The vocal microphones alone will not produce a wide image but will just add some focus around the voice, and as such, which way round they are panned when in positions B and C turns out to be not particularly critical. Try both ways (L-R and R-L) and see which you prefer. In position A, where the singer and piano pairs are closer to one another, it is likely to matter more which way around the vocal microphones are panned.

The final recording should not sound like a combination of spot microphones on two separate players, and so if each player feels quite clean and isolated, the microphones are too close and picking up too little of the room and of the other player. It is very important that the singer and piano feel as if they are inhabiting the same space by sharing the same reverb characteristics, and the piano microphones are the key to providing an overall sound of the room and the whole ensemble.

The balance in level and perspective between the piano and the voice is a finely judged one; it will sometimes depend on the musical content, but in general, the piano part is written to be more than a simple accompaniment, and it performs a real dialogue with the singer. However, it is natural that the singer should feel slightly closer than the piano rather than the other way around; this reflects the fact that the singer stands in front of the piano in a concert, and this is the image we have in our minds' eye when listening to a piano and voice recording. The danger of pulling the piano too close in the perspective is greatly reduced when you are not using close piano microphones in the well of the piano; the tail pair that is performing the role of overall ensemble pickup will be much easier to balance correctly. The singer should always be supported and audible and sufficiently focussed, but you are not aiming to replicate the levels of singer dominance that are found in, say, a Frank Sinatra–type recording where the soloist is very emphatically the central focus.

Figure 6.5 shows a slightly different layout with the singer standing more in the well of the piano and facing towards the pianist, thus placing the piano on his or her right-hand side, a scenario that might be arrived at because of singer preference.

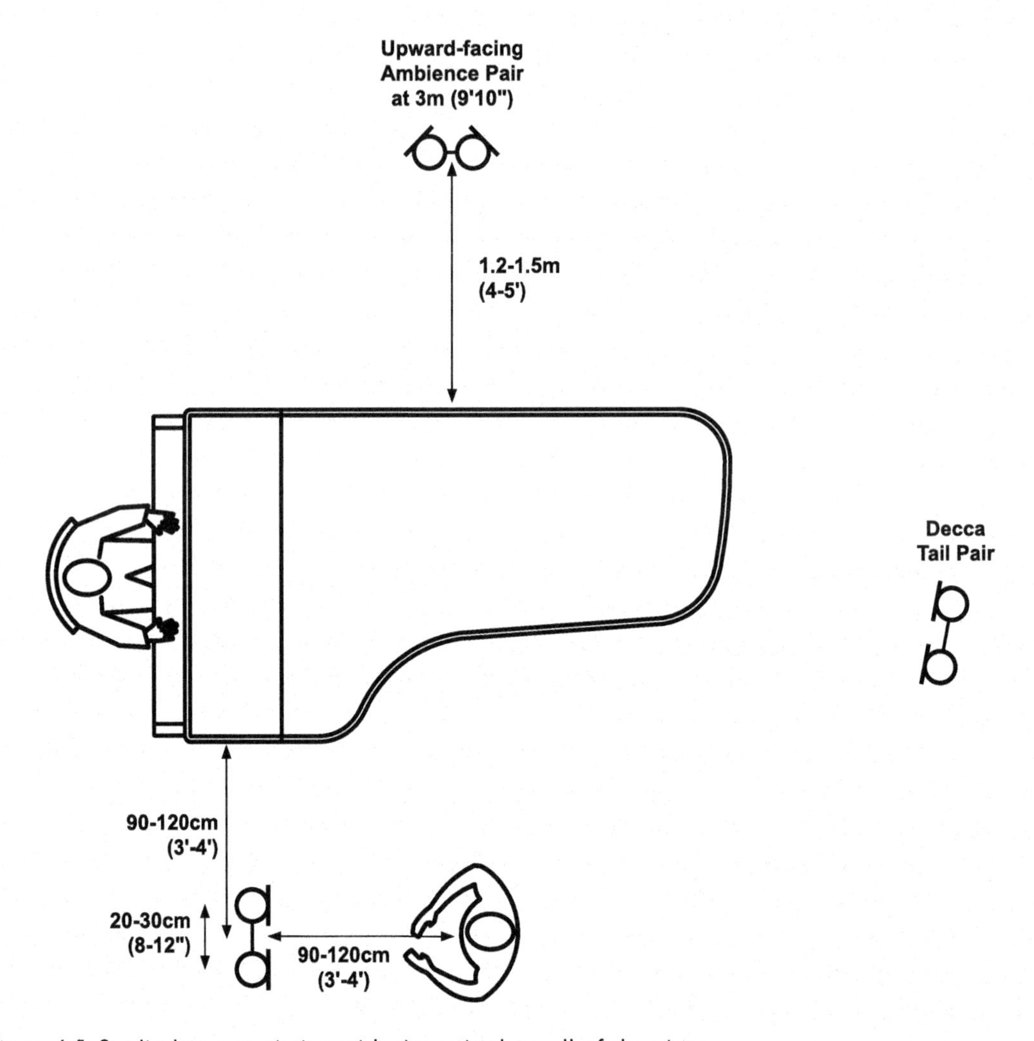

Figure 6.5 Studio layout variation with singer in the well of the piano

The most important difference between this and Figure 6.4 is that the piano will now be more present on the vocal microphones if cardioids are used because they are now picking up the piano on their side axes. As for positions B and C in Figure 6.4, the panning of the vocal microphones (L-R or R-L) is very much down to which works best when you try it, and it does not need to reflect the L-R panning of the piano pair.

To mitigate the presence of too much piano on the vocal microphones in this situation, ribbon fig of 8 microphones could be used to good effect, with the side axes pointing at the piano. The side axis null on a ribbon mic is more effective at source rejection than the rear null on a cardioid (especially on a high-end condenser cardioid, where a flat off-axis response is prioritised over off-axis rejection). Using ribbon microphones in this layout will enable the singer to be a little closer to the piano than in Figure 6.4 because of this increased rejection. The rear of the ribbon microphone in this situation will pick up a lot of room sound which will be useful for the overall mix if the room is good one, and the gentle HF roll-off of a ribbon microphone can be very flattering to a vocal sound and make it easier to blend with the piano. The spacing between the fig of 8s can be in the 20–30 cm (8″ to 12″) range used for the spaced cardioids previously. Small modern ribbon microphones (e.g. Royer R-121) are straightforward to rig this far apart, but be careful when rigging the larger and older Coles 4038 type ribbon microphones as a narrow pair. If they get too close together while you are attaching them to the stereo bar, they can be magnetically attracted to one another, and they can play havoc with your watch!

6.7.1 Vocal duet studio layout

Where there are two singers, they can be recorded with one microphone each, as the other singer's microphone will contain enough spill to add some sense of width to each singer's recorded image. Apart from this, the technique can be based on that for a solo singer (i.e. including a tail pair on the piano and an ambient pair in the room).

Figure 6.6a shows an adaptation of the solo singer layout from Figure 6.4 with both singers facing the pianist, side by side along a similar arc, but with only one microphone each. The singers are standing about 1.2 m (4′) apart – wherever is most comfortable for communication – and the microphone heights are the same as for a soloist. This layout is good for maintaining eyelines between both singers and the pianist, but care must be taken to avoid any singer reflections of the piano lid.

Figure 6.6b shows a sideways layout that will allow the use of ribbon microphones and hence reduced piano spill, but the sight lines for communication for the more distant singer are not as good, so the performers might not be as comfortable in this position.

6.8 Classical voice and lute/theorbo/guitar

Classical voice is paired with the lute in early music, such as Dowland's songs, and the classical guitar is often substituted for the lute. Everything that follows will assume that the guitar is being used, but the same principles apply to the lute or theorbo. All of these instruments are quiet, and the most pressing problem is the mismatch of acoustic power between the lute or guitar, which is very quiet, and the classically trained singer who has the ability to completely overpower the accompaniment if they move into a higher gear. This makes the final result very difficult to balance, and the priority has to be reducing the spill of the singer onto the guitar microphones as much as possible.

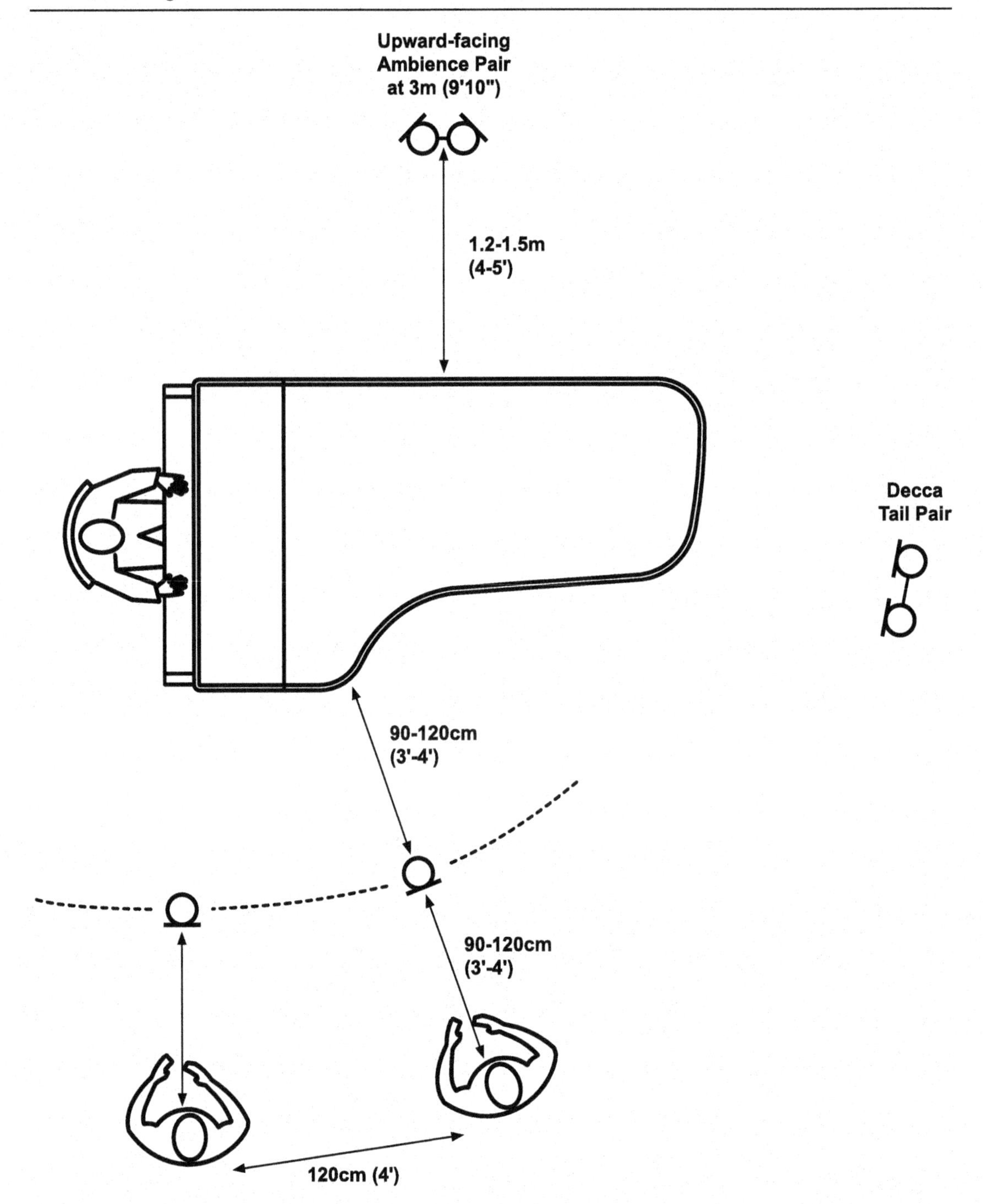

Figure 6.6a One layout for use with a vocal duet and piano

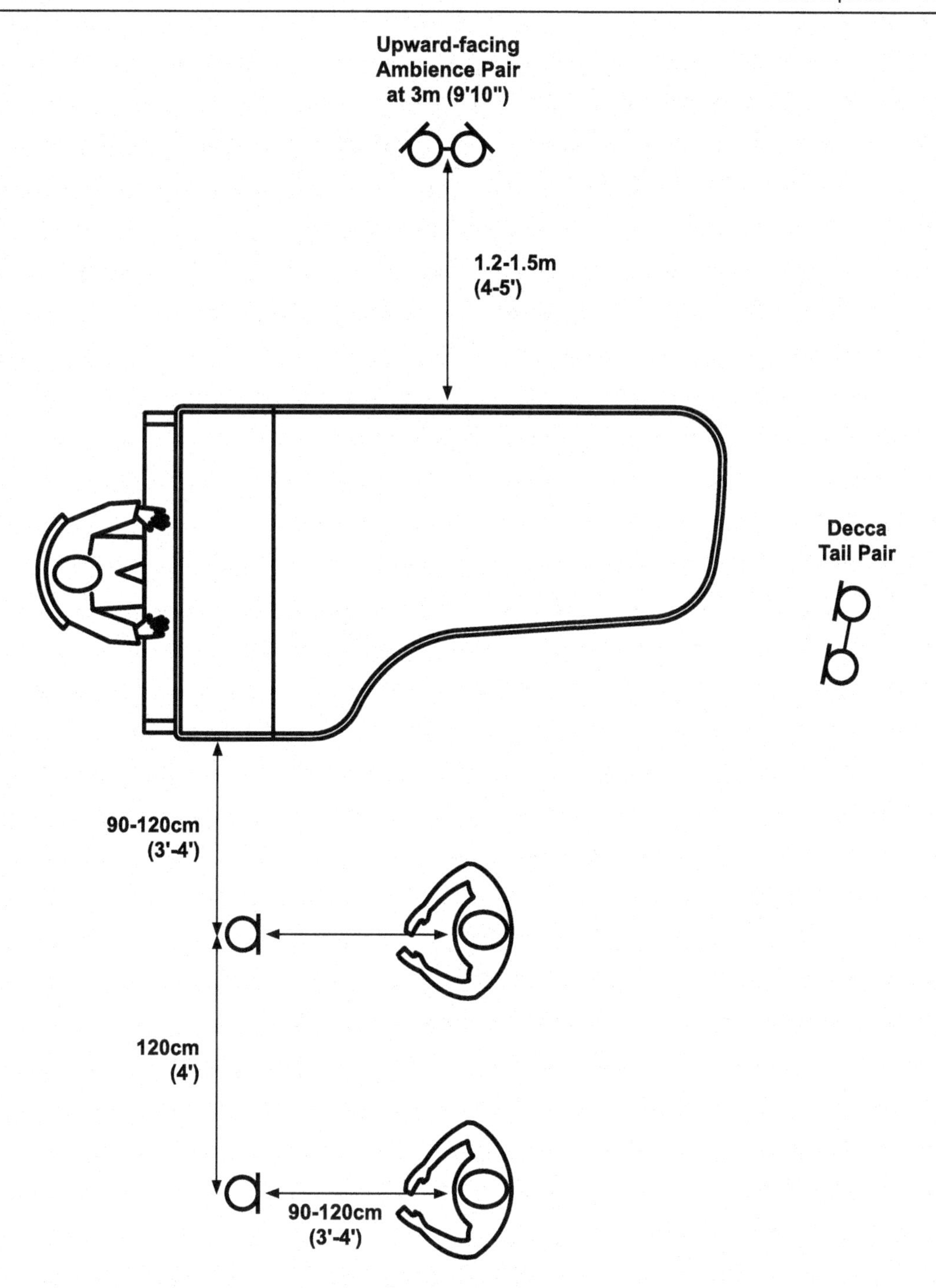

Figure 6.6b Another layout for use with a vocal duet and piano

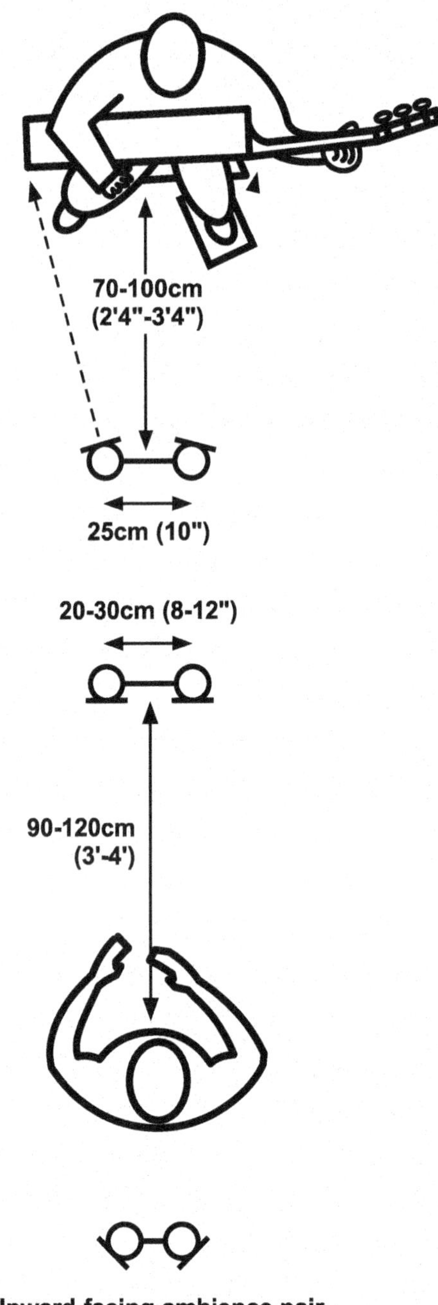

Figure 6.7a One position for singer and guitar on session

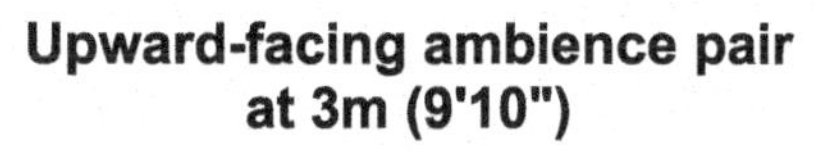

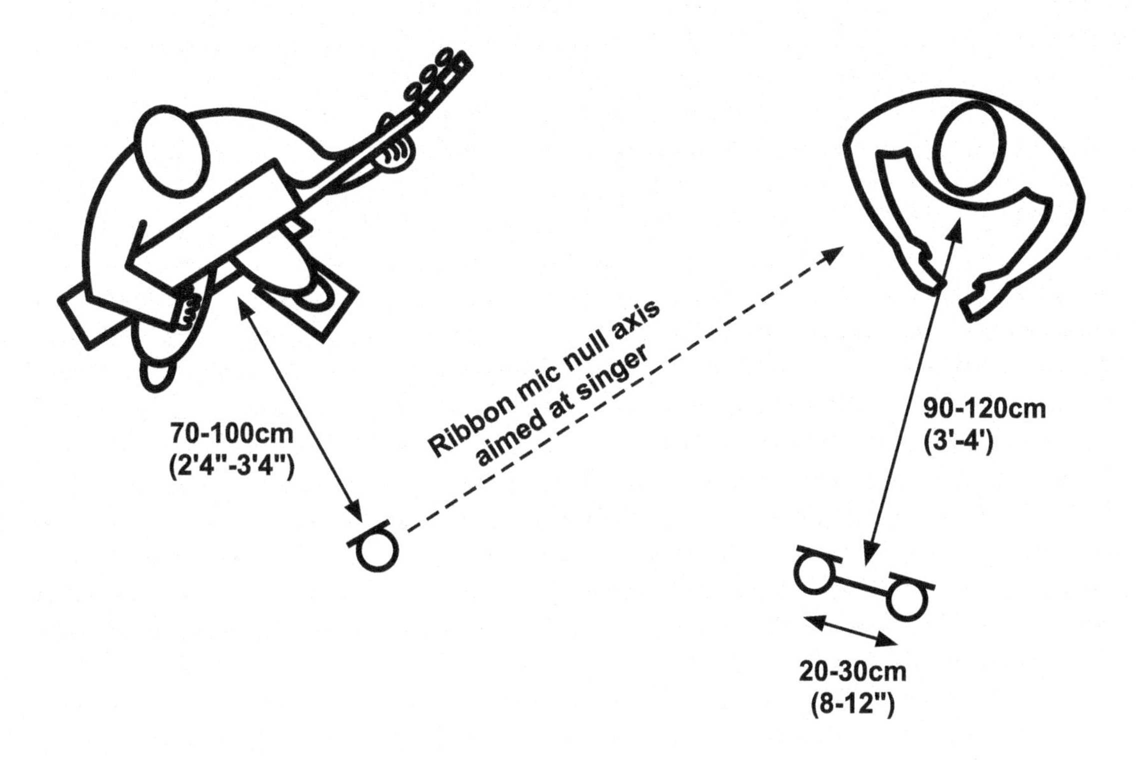

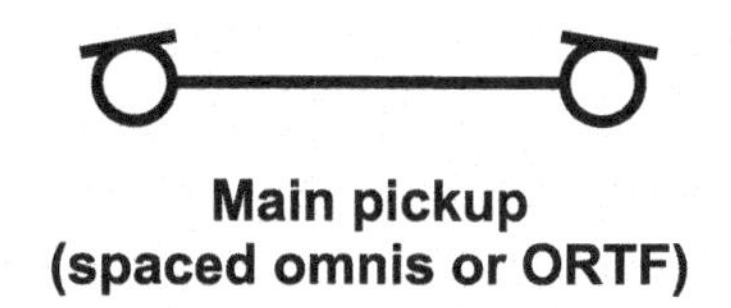

Figure 6.7b An alternative position for singer and guitar on session

The performers need to be comfortable with their positions, and in a recital would normally be side by side. In the studio, if you can persuade them to face each other, then you can use cardioids on both singer and guitarist, and you will have as good a separation as you can hope for. See Figure 6.7a and 6.7b. In addition to microphones on each soloist, you will need some overall pickup and a pair of ambient microphones if the room sounds good enough. If the performers prefer to be positioned more side by side or angled partly towards each other rather than facing each other completely, consider using ribbon microphones on the guitar using the null axis to discriminate against the singer. Chapter 4 gives more details on approaching recording the classical guitar as a solo instrument.

6.9 Mixing and fader riding for a singer

Classical singers have a very large dynamic range, and changes in level need to be anticipated when mixing. If there are going to be some lower, weaker notes that need supporting against the piano accompaniment, it will work better if the fader is already part way to where you need it to be; sudden large changes in level will be audible and disturbing. The scale of fader moves that can be used inaudibly will be anything up to about 4–5 dB.

Very large fader moves in the region of 10 dB on the vocal microphones will affect the piano sound, but smaller ones will be masked by the other microphone sources. (Remember that the main body of the sound will come from the piano microphones, and these faders will not be moved during the recording.) Fader moves will not necessarily solve every problem with level and you might need to encourage the performers to make changes as well. You can ask the pianist not to overpedal or play too loudly where the singer has weaker notes, and talk to the singer about leaning in towards the microphone very slightly when singing low and pulling back a little for high notes. Some singers cannot do this at all, and others will do it beautifully, but you cannot assume that a classical singer will have any microphone technique. It is not part of their traditional training as they are assumed to be performing without amplification; in contrast, pop or jazz singers who have been to a performing arts school or who have had lessons will have been taught microphone technique as part of their craft. Younger classical singers tend to have less well-developed dynamic control than those who are more experienced, and some good fader work can really help to support the performance where this is a problem.

Chapter 7

Solo instruments and piano

Although we looked at solo instruments in Chapter 4, it is more common to find that a soloist is accompanied by piano, and occasionally by guitar or harp. In order to tackle this combination of instruments, we can draw on the techniques and ideas outlined in Chapters 4 through 6. Section 6.3 details the technique of using a 'double-spot' microphone array on the soloist, and the same principles will apply to soloists in this chapter.

Because all these pairings involve the piano, it is worth mentioning some common features here. The piano microphones in all the following scenarios are a tail pair of either omnis or wide cardioids positioned with the aim of creating a good piano sound and without getting concerned over the more distant spill of the soloist. In addition, an upwards-facing ambient pair is included to capture some indirect sound from both instruments and can be used to set them both in the same space.

As we saw in Chapter 6, when recording more than one instrument in a classical context, acoustic spill is unavoidable because the players will be physically close to one another, and the acoustic space could be fairly live. Therefore, spill has to be embraced and can be positively utilised as long as we make sure that the spill is of good quality (meaning it has a good tonal balance and is not too reverberant), so that when it is included as part of the mix, it does not have a detrimental effect on the tone colour or perspective of the instrument. This is a very different approach to recording pop music, where spill is avoided as much as possible to enable closer-sounding instruments and very individual processing of parts. If you are used to pop recording, accepting spill and learning to use it to build the sound can take some getting used to.

Stereo imaging is another aspect of recording two players that can cause some difficulties, either in theory or in practice. As we will see, it is common to use two or three different pairs of microphones on the ensemble, each one of which will form some sort of stereo image. The question arises as to whether overlaying these stereo fields will produce a confusing overall image or whether this can be done without the different images affecting one another greatly. What is usually found in practice is that as long as the pairs of microphones are a certain distance apart (greater than about 2 m (6′7″)), there is enough reduction in correlation between the signals at the different pairs of microphones that their images can be superimposed without any problems occurring.

When mixing the two main components of the 'piano plus soloist' recording, it is important to remember that both sets of microphones are equally important, and neither is really a 'main pair' in terms of level (although the piano pair will give a wider overall view of the ensemble). Because neither is dominant, you should not use time delays on either pair. Time delaying microphones only works when one array of microphones forms the dominant part of the signal to which the other can be adjusted to be more time coherent. In the case of piano and soloist, the soloist's spill on the piano microphones will be later than the soloist on its own microphones, and the piano spill on the soloist microphones will be later than the piano on its own microphones. Therefore, introducing delays on either pair can cause more problems and should be left alone. (See Chapter 17 for appropriate use of time delays.) In the same way as for piano and voice, the piano microphones will be panned fully left/right, and the soloist microphones 70%–90% left/right depending on the amount of player movement and its effect on the stability of the image. The ambient pair should also be panned fully left/right. Start with the piano microphones, and then add in the soloist microphones to achieve the right perspective and detail on the solo player. The ambient microphones can be used at a low level to add a little more overall sense of space to both instruments.

It is worth taking the time on session to move the microphones around to get as close to the balance that you want as you possibly can. This is especially true for this sort of chamber music, where the balance of spill and direct sound is so much part of the microphone placement at the time, and all the microphones affect each other. If you assume you can sort out the balance and perspective between piano and soloist later when mixing, you might find that your set-up does not really work and a satisfactory balance is impossible.

Bearing all this in mind, let's take a look at the most common pairings of soloist and piano.

7.1 Violin and piano in concert

In a concert situation, the violinist will be standing slightly in the well of the piano, facing the audience, or slightly angled towards the pianist for eye contact. There might be restrictions on microphone position due to venue rules about audience sight lines, or the need for clear camera shots from certain angles.

7.1.1 Piano microphones

First of all, let's consider the piano. It will usually be possible to rig a tail pair as outlined in Chapter 5, which will not be in the audience's line of sight. It is common for a camera shot to be taken from the piano tail end looking at the pianist along the length of the strings, but this can usually be done whilst keeping the tail pair out of shot. However, you should be aware that the camera operator might get between your microphones and the piano, so take steps to avoid this! This pair should be placed with the aim of picking up a good piano sound without worrying about the inevitable slightly distant sounding violin spill. Given that this spill is onto the front of the piano tail pair, its tone quality will be good, and the distance will be useful in giving a sense of space around the violin. The violin's main tone quality and focus will come from a different set of microphones. The piano pair can be omnis, cardioids, or wide cardioids depending on whether the stage area is dry and how much reverb you want to collect.

7.1.2 Violin microphones

Placing microphones to pick up the violin in a concert presents a number of potential problems. When a stand is placed on the stage with performers, it is advisable to use a shock mount

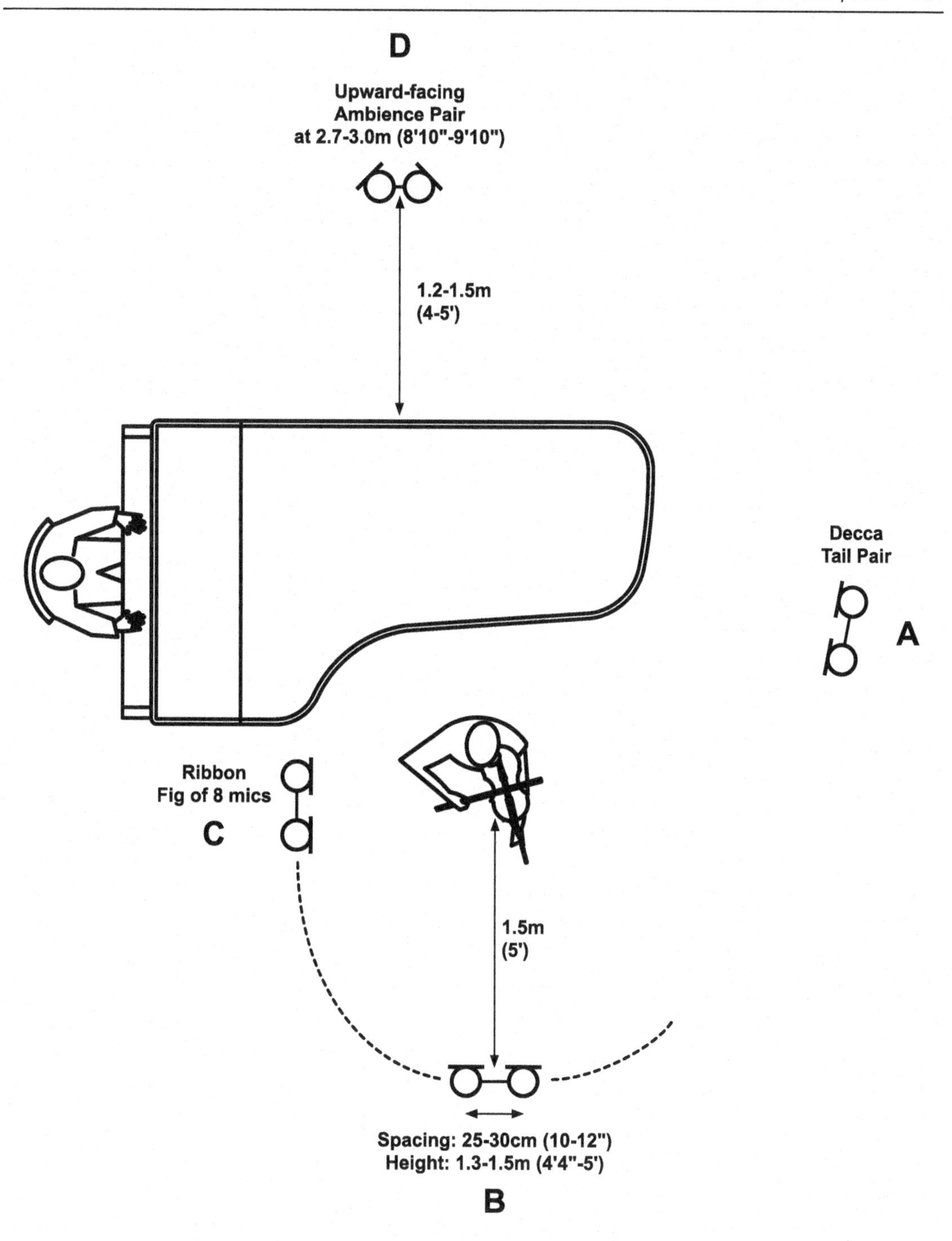

Figure 7.1 Concert layout for piano and violin

to prevent seismic transmission of vibrations through the floorboards and microphone stand. Anything placed in front of the violin will almost certainly be in the way of the audience's view, and it will also be in the firing line for direct piano reflections projected from the raised lid, thus making it harder to discriminate in the violin's favour. Microphones placed at a height of around 2.5 m (8′2″) on a stand in front of the player (as if recording a solo violinist) are particularly visually intrusive, and the options are to sling them if possible, or to lower the stand as far down as 1.3–1.5 m (4′4″ to 5′) (see pair B placed around the arc as indicated in Figure 7.1). To avoid sounding too close at this height, the pair will need to be placed at least 1.5 m (5′) back from the player. If the stand is placed on the stage, the distance available might be less than this, and the stand would then be better placed on the floor in front of the stage, or 'Pavarotti' microphones could be used (see Chapter 6) at a similar height. Although lower microphone positions will not pick up as much of the upper partials of the instrument, the tone should be acceptable. For all these positions, cardioids would be first preference in a live concert to reduce the pickup of audience noise.

The necessity for lower microphone placement might work to your advantage as it might be possible to place the microphones around the arc indicated in Figure 7.1 such that the violinist's body is blocking some of the piano sound. This will affect the mid to high frequencies more and help to discriminate in the violin's favour. Whichever solution is adopted, the amount and quality of piano spill present on the violin microphone(s) needs to be assessed to see whether it is detrimental to the overall piano sound. Cut the piano microphones to see what is being picked up, and also listen to the piano playing on its own. Where there is a problem, it usually takes the form of muddy piano bass, and some of the lower frequencies of the piano spill could be filtered off from the violin microphones without losing the essential richness of the violin. The violin's lowest frequency is around 190 Hz, so the filter should not extend this high to avoid hollowing out the violin sound; the violin sound contains many strong upper partial components, but without the warmth from the 190–400 Hz region, the instrument will sound very thin.

Another approach to discriminating against the piano with the violin microphones would be to use some ribbon fig of 8 microphones for their excellent side axis rejection. Their natural HF roll-off means that the microphones can be placed somewhat closer than brighter condenser microphones without sounding too close. They can then be placed as in position C, much further round the arc indicated in Figure 7.1, but at the same lower height. However, if there is going to be any filming or broadcast of the concert, ribbon microphones, even modern ones (e.g. Royer R-121), will usually be too big to be acceptable to the director. Fig of 8 microphones that are switchable condenser models do not have such good side rejection or the same characteristic HF roll-off as a ribbon microphone, and so will not perform in the same way in this context. Where there are practical objections to the use of a pair of microphones on the violin in a concert situation, a single microphone will have to be used instead, and care must be taken in mixing not to overuse the signal and make the violin sound too localised and mono.

7.1.3 Ambient microphones and mixing

The last component of the recording will come from a pair of microphones set up to collect more ambient and indirect sound, to give a good sense of space around both instruments. This will be used at a fairly low level in the mix to join together the sound from the piano and violin microphones. In this context, an upwards-facing XY pair of cardioids or wide cardioids placed behind the piano at a height of around 2.7–3 m (9′ to 10′) will be visually discreet and relatively free of audience noise. In any live situation, this will be an effective method of collecting clean, indirect sound and preferable to placing a distant pair of microphones further back into the auditorium. Viewed from the audience, they will be mainly out of sight, and might just be visible sticking up behind the piano's raised lid. Microphones that are susceptible to wind noise (those with more loosely tensioned diaphragms) should have windshields fitted if they are going to be used high in the air; air conditioning can cause gusts of air that are more noticeable higher up in the room.

When it comes to mixing the three pairs of microphones together, panning needs to be considered. The piano pair and the ambient pair should be panned fully, and depending on how much the violinist moves around, panning the violin microphones about 75% left and right should keep the player fairly stable in the image while still endowing it with a sense of width.

7.2 Violin and piano: studio layout

Once the players are freed from the concert hall scenario, alternative studio layouts can be adopted both to help with the recording and to improve communication between the players. The biggest alteration is that the soloist can turn around and face the pianist; this gives them an arc of possible positions in which to stand, some of which will work well and others which should be avoided.

Each scenario involves placing a tail pair on the piano (see Chapter 5), a pair of violin microphones, and an upwards-facing pair designed to pick up an overall blend of indirect sound from both. They all assume that the piano is on full stick; although the tail pair on the piano will pick up a surprisingly acceptable sound with the piano on short stick, this is not ideal, and from the engineer's point of view full stick is by far the best for recording.

7.2.1 Choice of violin position

Figure 7.2a–c shows three alternative positions for the violinist on session. Figure 7.2a shows the player in the well of the piano, Figure 7.2b shows the player moved out from the well of the piano towards the keyboard, and Figure 7.2c shows two positions, including one with the player behind the piano. The possibilities inherent in each position are discussed below.

Figure 7.2a is rather like the concert position, but the player has turned to face the pianist. In this position, the level of piano sound coming from a large Steinway on full stick can be quite overpowering, and the player might be used to playing live with the piano on short stick. Because the piano is best on full stick for recording, the soloist might not actually be comfortable standing in the piano well on a session. Another drawback of this position is the degree and quality of piano spill on the violin microphones. They are shown in front of the violin, looking down from a height of 2.5–3 m (8′ to 10′) in order to get a good distance away from the player. This enables the player to stand at a close enough distance to the pianist for communication to be good, but for the microphones still to be placed far enough away to get a well-balanced tone. In this position, they also discriminating to a small extent against the piano, but they will pick up a lot of

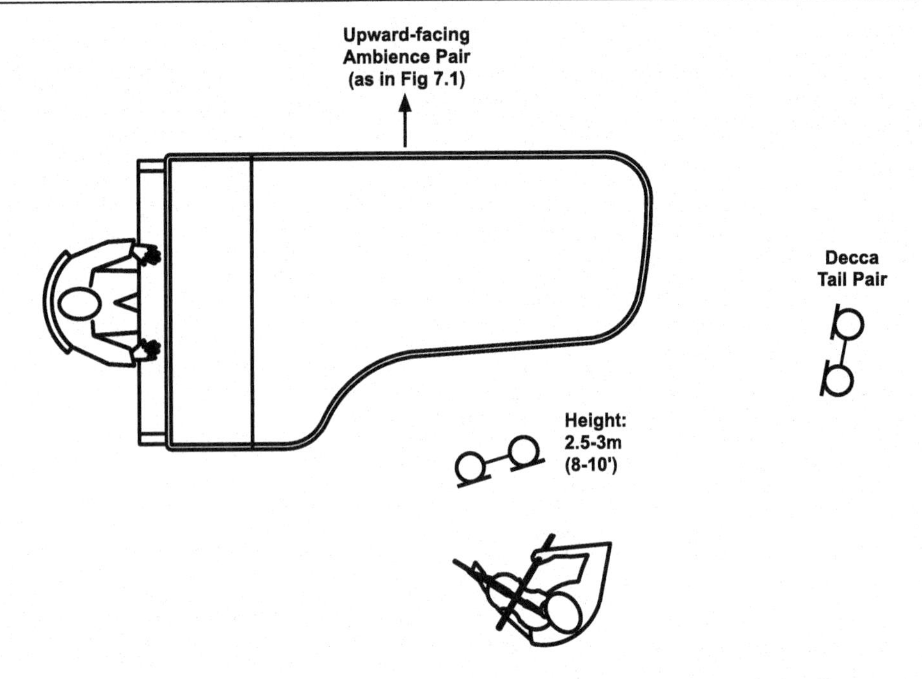

Figure 7.2a Studio layout 1 for violin and piano – reverse concert position in the well

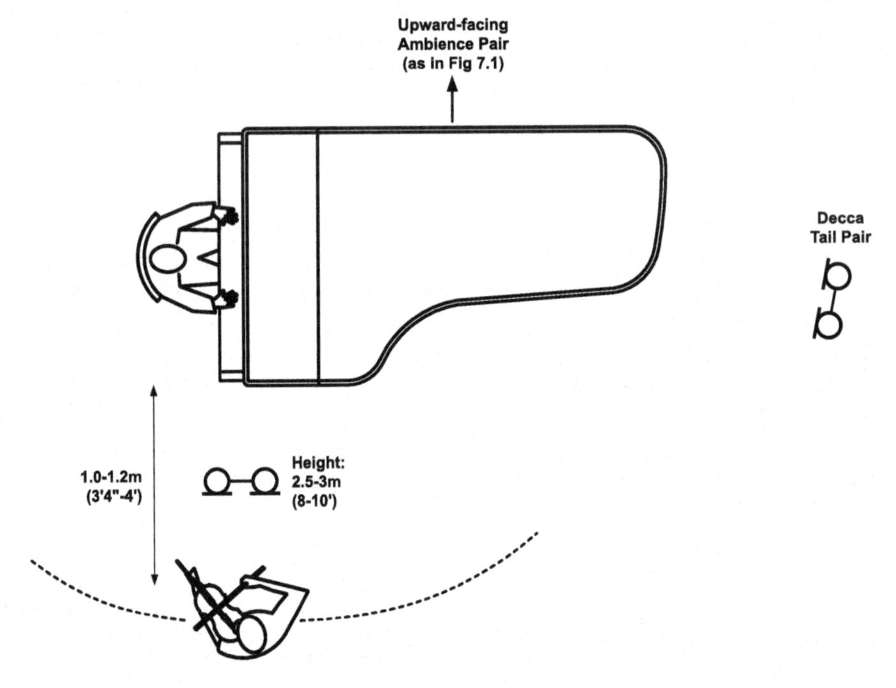

Figure 7.2b Studio layout 2 for violin and piano – out of the well near the keyboard

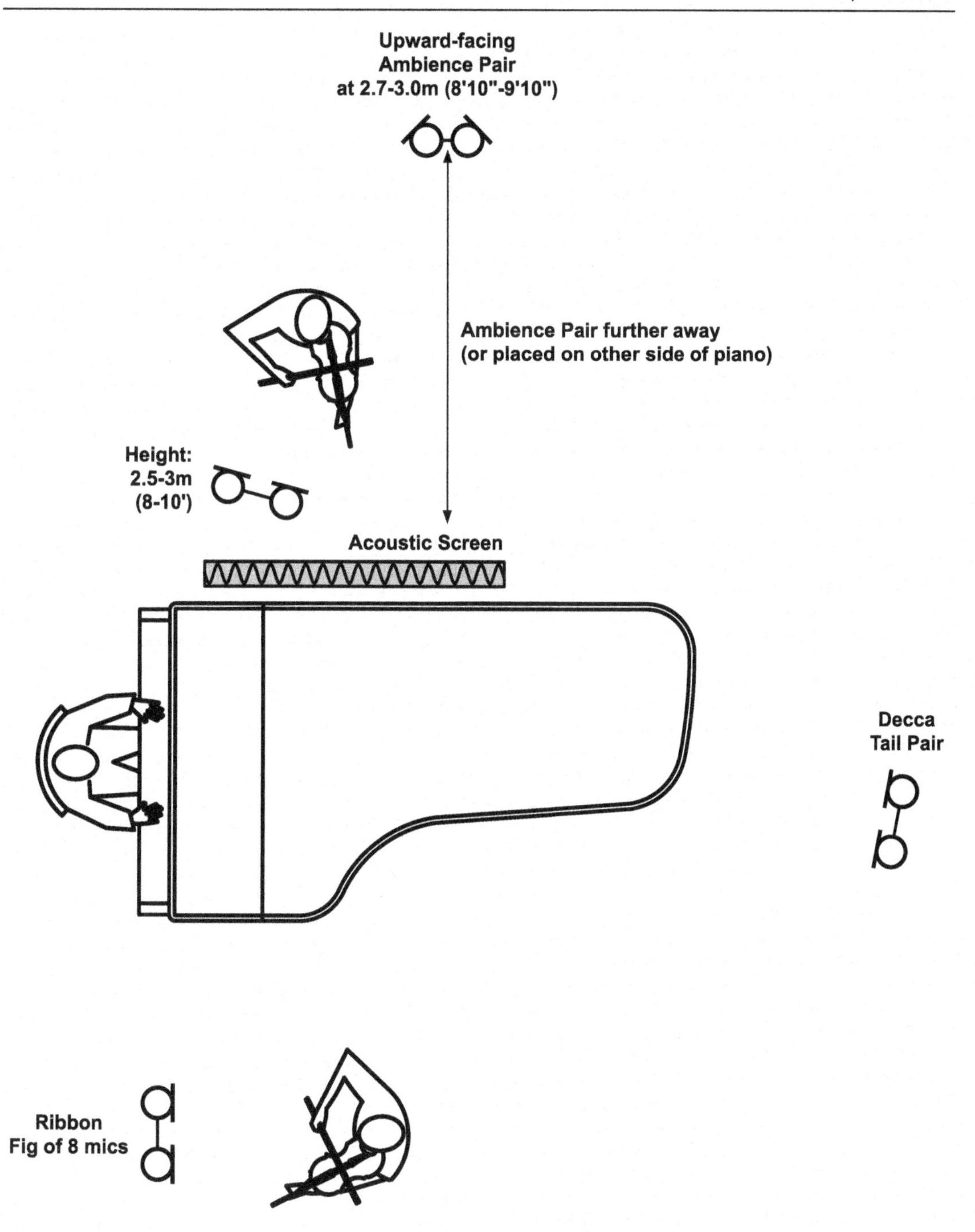

Figure 7.2c Studio layouts 3 and 4 for violin and piano – one using fig of 8 ribbon microphones to discriminate against the piano with the violinist in front, and the other with the violin on the 'wrong' side of the piano

loud piano on their side axes, and this is likely to be rather muddy in tone, so it is not a good component to add to the piano sound.

The position shown in Figure 7.2b is better for both mutual audibility and getting the violin microphones further away from the piano. The dotted line arc shown in the diagram indicates a range of positions for the violinist, depending on where they would prefer to stand; some might like to be far enough around the arc to have a view of the piano keyboard. The line of sight is good between the players in this position, and they should also be able to hear each other well. The distance between the piano and the violinist cannot be less than about 1–1.2 m (3′4″ to 4′) because below this, the benefit of distance on the ability of the microphones to discriminate against the piano starts to be lost. However, if the players are moved much further apart than about 1.2 m (4′), communication and therefore musicality and performance can suffer. If they are moved a great deal further apart, any spill that is present will start to sound too reverberant and have a detrimental effect on the overall sound.

Allow the artists to play together for a while and adjust themselves to a comfortable position along this arc. The violinist will usually take the music stand and place it off to one side to allow for movement and communication with the pianist. Once you and they are happy that they can hear each other properly, you should mark on the floor where the violinist should stand to ensure consistency between takes. A 0.5 m straight line of gaffer tape to stop them getting closer to the pianist should suffice. Once you have seen how the players like to angle themselves for performance, you can then adjust the violin microphones with the aim of getting a good blend of HF straight off the bridge and the wider radiation. In order to really find out how the microphones sound at a height of 2.5–3 m, you would need a stepladder, but even so, walking around in an arc around the front of the player and listening at your natural ear height will give you a good idea of where the tone sounds best. One advantage of placing the microphones at height has already been mentioned: it allows the players to be quite close horizontally while still getting the microphones far enough away from the instrument. The other advantage is that there is less variation in tone from the player's lateral movements if the microphones are looking down on the player than if they are at violin height. Violinists vary in the amount of physical moving around that they do while playing, but microphones placed high up should achieve a more consistent tone despite any movements.

Figure 7.2c shows two alternative positions in the same picture. One is the same as Figure 7.2b but with the player sideways on to the piano, and fig of 8 ribbon microphones are being used to discriminate against the piano in this case. The other shows the violinist placed on the 'wrong' side of the piano, behind the raised lid. From the player's perspective, some audibility will be lost; the violinist will lose contact with the top end of the piano because of the raised lid, but will still hear a lot of bottom end muddiness coming from underneath the piano. Visually, the players can see each other well, but the loss of audibility might make musical communication more difficult, so not so many duos would like to play in this position. From the point of view of microphones, this layout does help with discrimination against the piano on the violin microphones, so it has that in its favour, although much of the piano LF will be picked up regardless. The discrimination might be helped even further by placing an acoustic screen in between the violin and side of the piano, parallel to the hinge edge of the piano. However, this will only affect HF, much of which is stopped by the lid anyway, as LF will still diffract effectively around any screen. It should be possible to filter off some of this LF from the violin microphones, as long as nothing higher than about 190 Hz is affected; however, the screen will reduce the violinist's ability to hear the piano even further, and this is likely to be an uncomfortable playing environment for the players. Thought should be given to whether the reduction of spill is a price worth paying for a potentially worse performance.

Other things that might be useful additions to the violin and piano scenario are a small square of carpet under the pianist's feet to reduce pedalling noise from the pianist's foot and the floor. It will not have much effect if you have a squeaky pedal, and this needs fixing before the recording goes ahead. Some violinists can be noisy footed, stomping in a rhythmic fashion while playing, and another square of carpet underneath their feet might help with this. Avoid the carpet being too large, or the HF response of the room will be affected, as the carpet will absorb HF reflections and reduce the HF content of the natural reverb.

7.2.2 *Placing violin microphones*

When you are placing the violin microphones, you are trying to achieve three things: a really good violin sound, as much discrimination against the piano as possible, and good quality piano spill. The microphone placement has to work for both instruments.

The piano is going to be picked up by the violin microphones on around the 90° (side) axis. Therefore the microphones used need a good off-axis response with low colouration, and this will be a really important factor in your microphone choice. If the off-axis sound has an uneven frequency response, you will hear it in the piano sound. Listen to the piano microphones alone and then add in the violin microphones. If the piano starts to get muddy and coloured at this point, the problem lies with your violin microphones; you should also listen to these microphones alone to hear what they are doing to the piano sound. Given that the piano will be on the side axis of the microphones, this is another scenario where using ribbon fig of 8 microphones could be very effective in discriminating against the piano if placed with the side axis pointing along the keyboard, as shown in Figure 7.2b. This would work well for a violinist who is standing at the well end of the arc of possible positions.

7.3 Cello and piano in concert

Figure 7.3 shows a normal concert layout for cello and piano, with the cellist seated in the well of the piano and facing the audience.

The microphones for the piano and ambience can remain the same as for the violin and piano concert recording outlined earlier. Similarly, the important thing to avoid is picking up too much piano projecting off the piano lid on the cello microphones. With the cello, however, it is a lot easier to place these microphones lower down out of the line of the piano lid reflections, so the situation is an easier one to manage. In this case, 'low down' means about 75 cm (2′6″) off the ground, looking towards the bridge; microphones in this position will be visually acceptable from the audience point of view and will also produce an acceptable recorded sound. Floor reflections will usually enhance the sound if the cellist is on a stage rather than on a riser that might have a specific resonance. (See Chapter 4.)

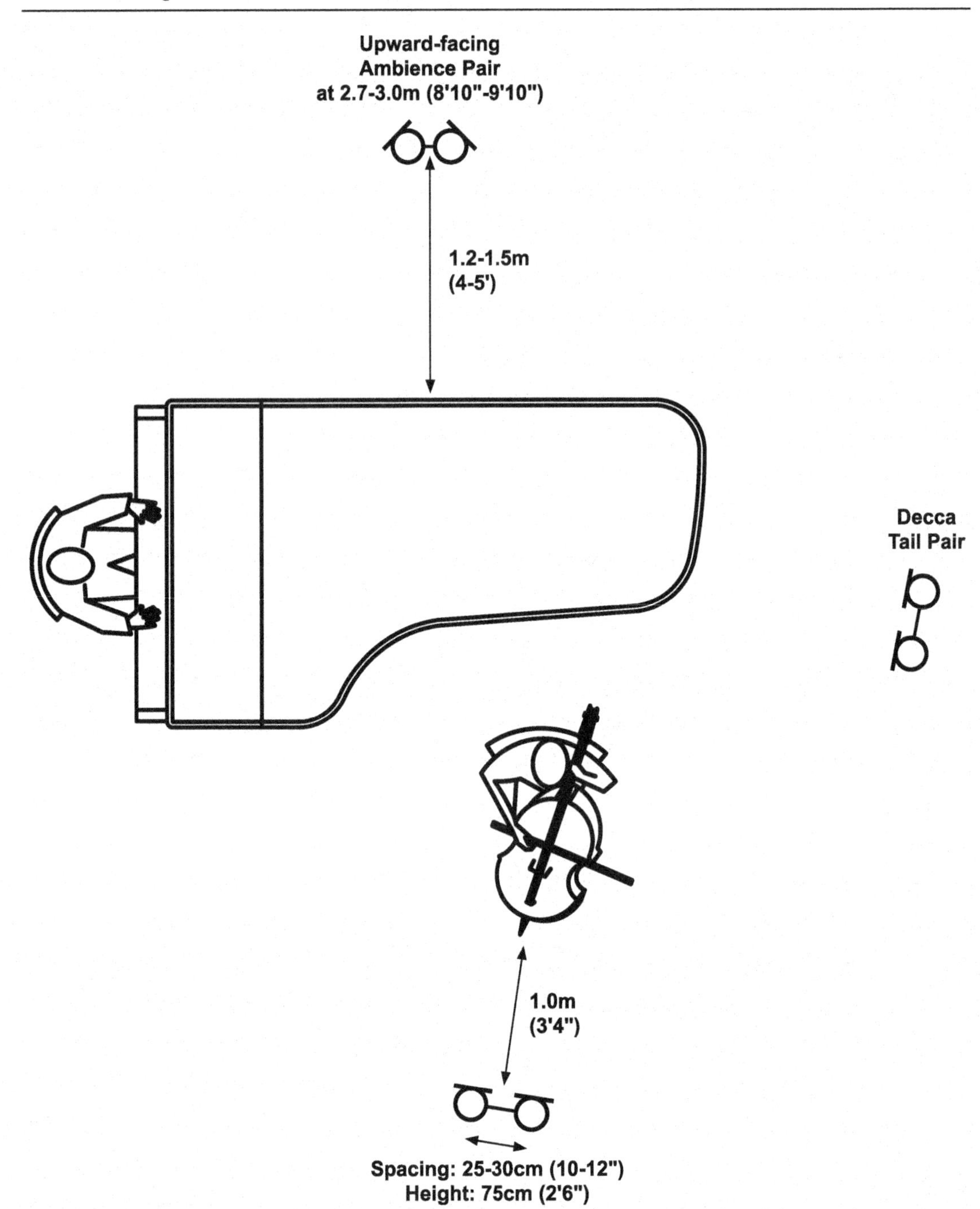

Figure 7.3 Concert layout for cello and piano

7.4 Cello and piano: studio layout

As with the violin and piano recording, moving the soloist out of the piano well where the piano is too loud is a good idea. Once again, a tail pair of omnis are shown on the piano along with an upwards-facing pair of ambience microphones. There is a similar range of cello positions that could be used, all facing towards the piano and along an arc about 0.9–1.2 m (3–4′) away, as shown in Figure 7.4a. For all positions discussed herein, the cellist will need something to stop the spike slipping; using a small piece of carpet under the player can help this, as long as it is not too big so that the floor reflections are very damped down around the instrument. The feedback from the floor can be a useful component to the cello sound.

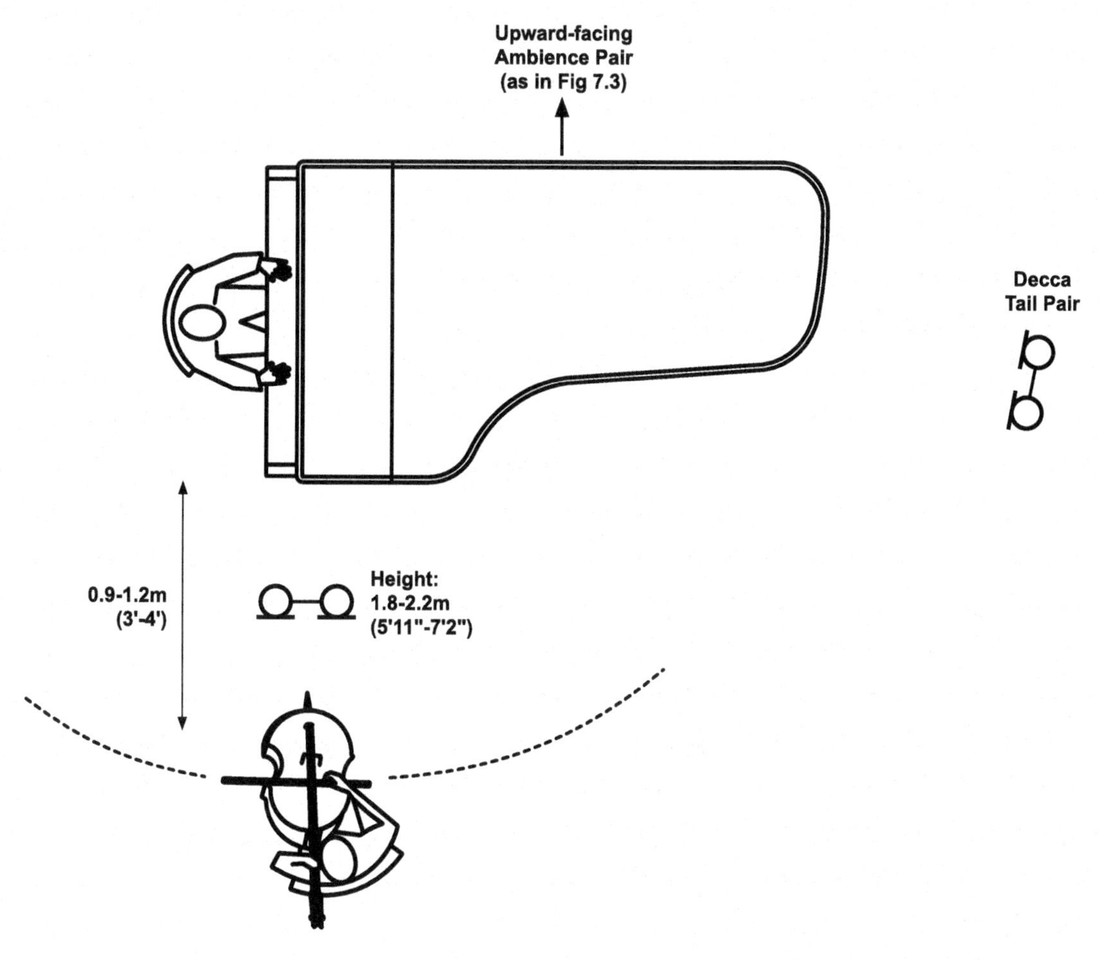

Figure 7.4a Studio layout 1 for cello and piano

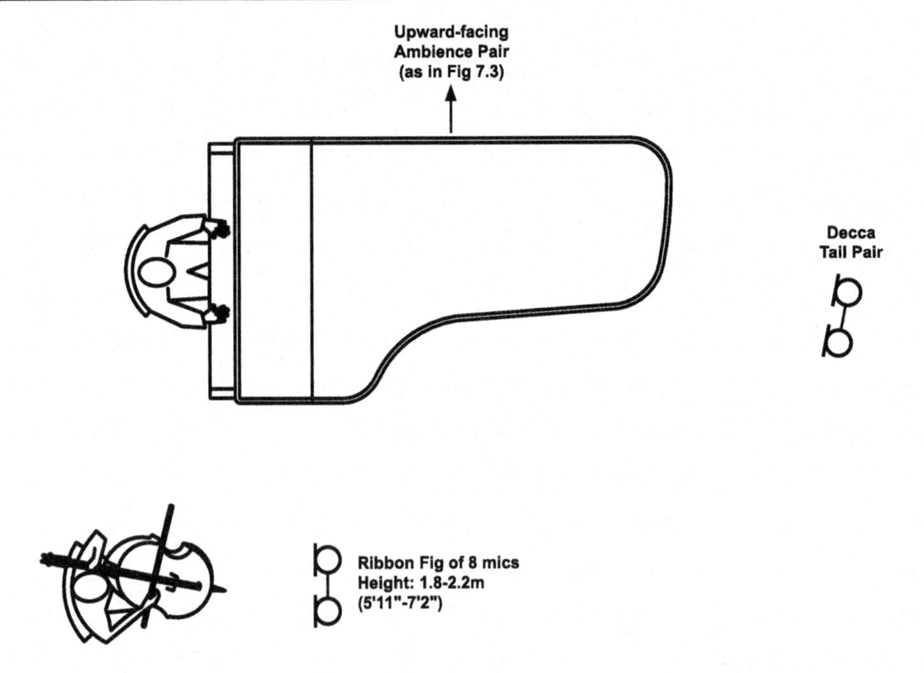

Figure 7.4b Studio layout 2 for cello and piano, using fig of 8 microphones to discriminate against the piano

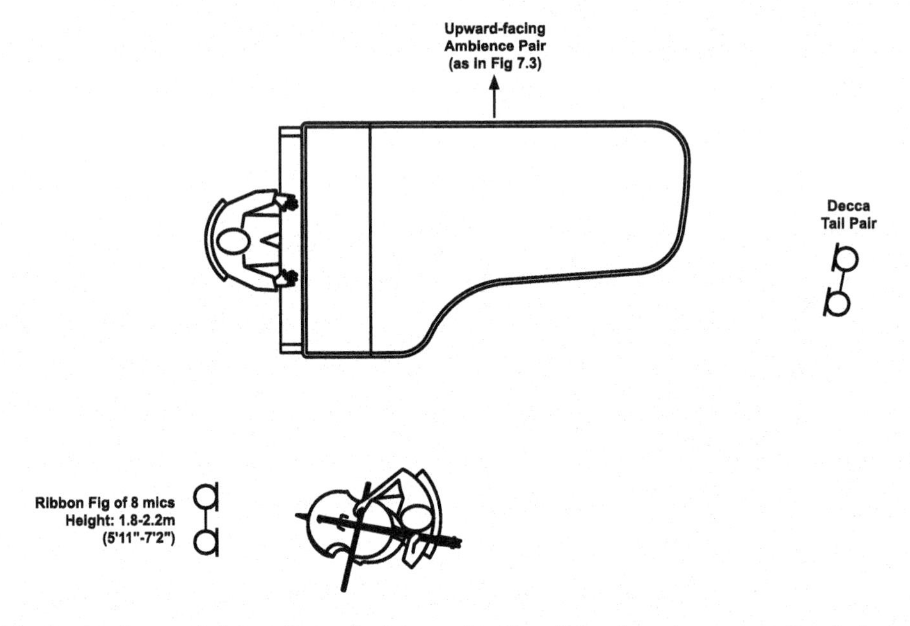

Figure 7.4c Studio layout 3 for cello and piano, using fig of 8 microphones to discriminate against the piano. This layout provides the players with better eye contact than that shown in Figure 7.4b.

Figure 7.5 Lynn Harrell on cello and Igor Kipnis on harpsichord, recording Bach and Handel sonatas in Henry Wood Hall, 1987[1]

Photo: Mike Evans; courtesy Decca Music Group Ltd.

The best position for the players will be the one in which they can maintain the closest contact in terms of being able both to see and hear each other well. It is important to prioritise player contact over acoustic separation because it is so central to getting a good performance. Figure 7.4b and 7.4c show more positions with the cello more sideways to the piano, one with the players seated facing in the same direction and one with them facing each other. In both positions, fig of 8 microphones can be used to discriminate against the piano.

Figure 7.5 shows Lynn Harrell in a cello and harpsichord session in a position that enables good communication with the harpsichord player. The harpsichord microphones are similar to a Decca piano tail pair, and the soloist's pair can clearly be seen.

7.4.1 Microphone position

For the position of the cello microphones, there is no reason to have to keep them low as for the concert performance, and so they can be placed at a greater height to give an appropriate sense of distance to the cello sound without having to be horizontally far removed. The cello is lower down when played, and so the microphones do not need to be at the height used for violin;

around 1.8–2.2 m (5′11″ to 7′2″) would be a good place to start if using cardioids for any of the positions shown in Figure 7.4a. The microphones are placed in the plane that lies perpendicular to the front face of the cello's body, as this is where the most HF will be radiated (see Figure 7.6 to clarify this). They should face the cello and be directed away from the piano, even though the rear axis will be pointing up to the ceiling and the piano will be more on the side axis. A cardioid that is warm sounding, with no HF lift or bright top end, will be best suited to the cello.

There is some argument for using omnis on the cello for their extended LF response, but there are a number of things to think about before committing to this. If omnis are used, the cellist will need to be physically a little further away from the piano to avoid picking up too much piano on the cello microphones. To reduce too much reverb pickup, it is necessary to move in closer to the instrument with omni microphones, and this brings with it several potential problems; the tone quality could become localised to part of the instrument, instrumental noises could be picked up disproportionately, and a slightly compressed quality could result. The risk using this method is that control over the cello will be lost when the pianist plays loudly and increases the piano spill onto the cello microphones. Any fader changes to the cello microphones that might be needed in mixing to keep the cello from being drowned in the louder sections will also affect the piano significantly, and so it becomes impossible to independently control the cello level. Using omnis on the soloist as well as the piano is most successful if the players are very good at balancing themselves acoustically and the piece does not have a large dynamic range. Given a more expansive piece, or players that might not always balance themselves perfectly, the margin for error is much narrower than if you use cardioids. Considering the overall set-up, the omnis that are used on the piano will pick up the cello warmth and richness, and so the reduction of these qualities from the solo cello microphones if cardioids are used will not have as dramatic an effect on the cello tone as might be first thought.

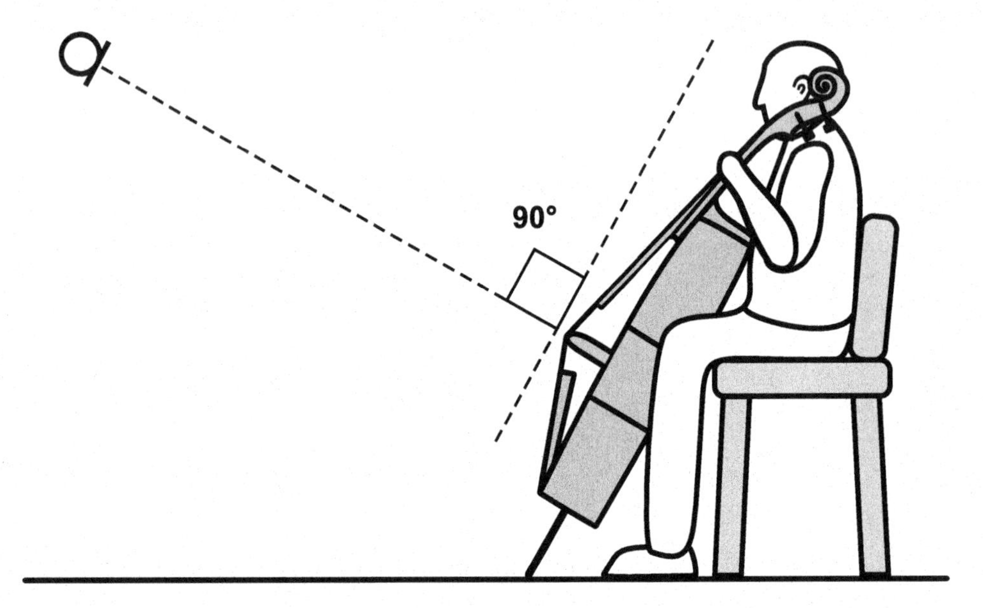

Figure 7.6 Showing what is meant by 'perpendicular to the cello's face'

If the cellist is in the position shown in Figure 7.4b and 7.4c, then using fig of 8 microphones with the side axis towards the piano will give some good separation as well as producing a warm and flattering sound.

When discussing the violin and piano earlier, we looked at the possibility of placing the violinist on the 'wrong' side of the piano. This position is not particularly useful when recording the cello; there is little in terms of separation to be gained, as the piano spill on the far side of the piano is mainly muddy lower end. This sort of coloured spill is a disadvantage when recording any instrument, but it is particularly unhelpful with cello as it cannot be EQ'd from the cello microphones in the way that it could be done for the violin or another higher pitched instrument.

7.5 Woodwind and piano

One of the important considerations when recording wind instruments is to avoid picking up too much key noise, and this means not getting too close. Using microphones with an HF lift will exacerbate the situation, and in a concert situation where microphone placement is often a compromise, ribbon microphones (Royer R-121 or Coles 4038) with their HF roll-off could be useful. Avoiding closeness is also important in obtaining a good overall tonal balance; too close and some part of the spectrum will be overemphasised. The end-blown woodwinds radiate their HF in an increasingly narrow cone shape centred on the axis of the instrument, and where there are open holes, much of the lower frequencies are able to radiate out of these. This means that, close up, the tone changes a lot with position. As discussed in Chapter 4, the side-blown flute radiates fairly equally to the front and back, so this opens up more potential microphone positions when combined with other instruments. Breath noise is a problem with flute in particular, and microphones in front will emphasise this. Any muddy LF piano spill can be removed from the higher winds with a high-pass filter as long as care is taken to ensure the filter's effects are not felt higher up the spectrum. As a guideline, the lowest frequency of the flute is around 250 Hz and that of the oboe 240 Hz. The clarinets have a more extended range, so the lowest note of the B-flat clarinet is around 135 Hz and the lowest note of the bassoon is 60 Hz.

In a live situation, there are a couple of options for woodwind microphones (see Figure 7.7). To avoid problems with audience sight lines, one method would be to use a pair in front of the instrument at about 1.2 m (4′) high, which is low enough to be out of the way but not so low as to be in line with the bell (where excessive HF and air noises will be picked up). If the microphone is still too high from the audience point of view, the microphone is better moved to one side rather than lowered in height to avoid being on-axis to the bell. The main danger from this low pair is that it can be too close to the instrument and thus emphasise key noise (and also breath noise in the case of the flute). Unfortunately, it will not usually be possible to rig a pair higher up on a stand because of the visual intrusion, so moving further back might help reduce key noise. The other alternative microphone set-up would be a more distant spaced pair of omnis or wide cardioids (depending on distance and how live the room is) suspended above and in front of the player at about 2.75 m (9′) high and 0.9–1.2 m (3′ to 4′) in front. This will behave almost like a main pair; it will give some space around the soloist whilst reducing the pickup of key noise. Bassoons can be particularly prone to key noise, and the more distant microphones are likely to be the better option in this instance.

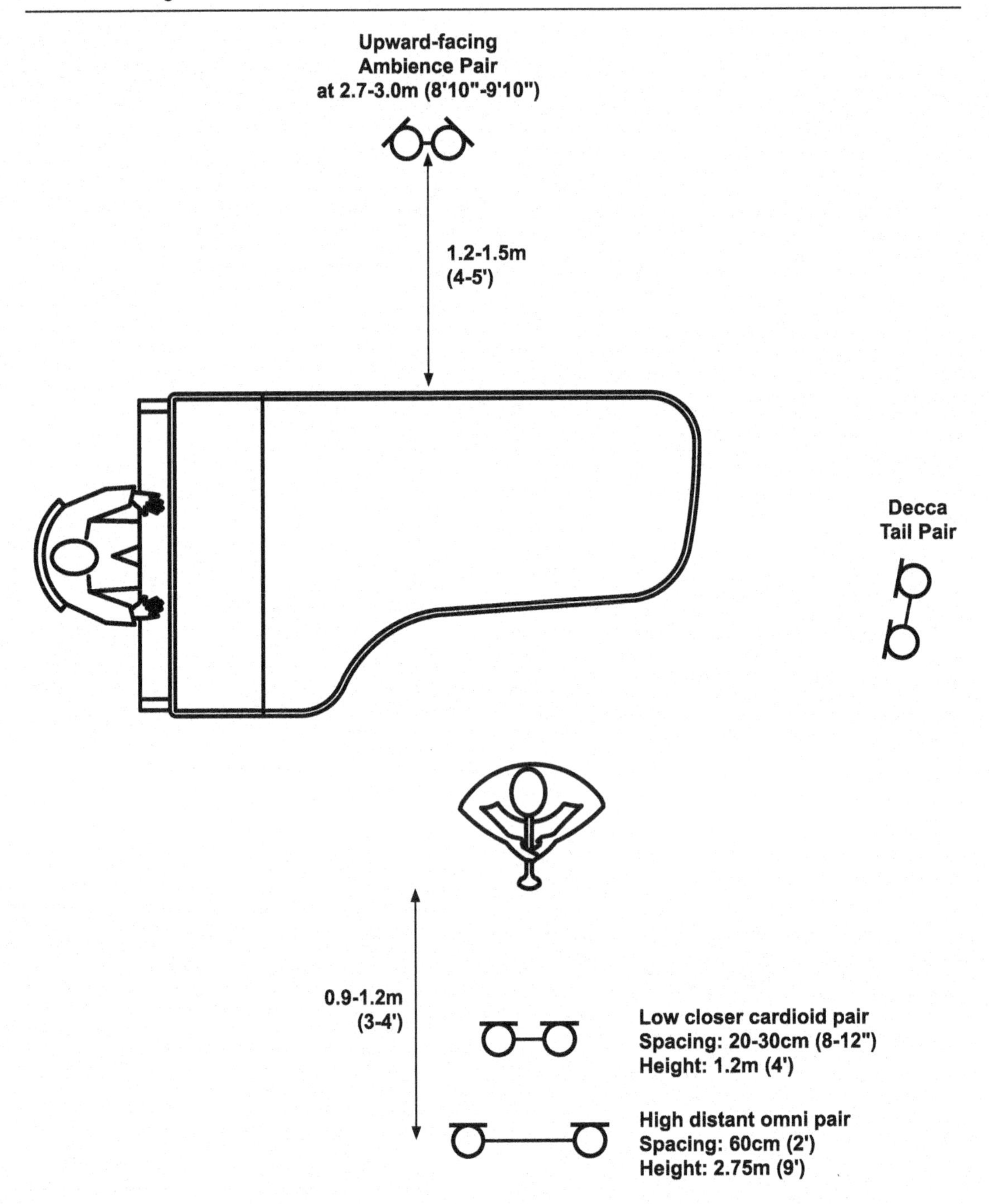

Figure 7.7 Woodwind and piano concert layout showing both closer and more distant pairs on the soloist

If there is an opportunity to use both of these pairs, then a useful balance could be struck by using the more distant pair to provide most of the sound and width to the source, and the closer pair (or single microphone) to provide some additional focus while avoiding excessive key noise. The 1998 EMI recordings from Saratoga with Martha Argerich and Michael Collins performing Bartók made use of both a close and more distant pair in this way.

Care should be taken to use shock mounts if you have them for the close, low mics placed on the stage, as this will physically decouple the microphones from the staging and avoid very low frequencies being transmitted mechanically to the microphones. If you don't have access to shock mounts, it should be possible to high-pass filter these microphones, as the piano microphones will provide some good LF content to the overall sound.

The approach to recording wind and piano in the studio can be exactly the same as that for the violin outlined in section 7.2. The soloist microphones are high but not too far back from the players, which yields a well-rounded sound and reduced instrumental noise.

7.6 Brass and piano

The most pressing problem recording brass and piano is one of level; brass instruments are capable of generating very high SPLs, and achieving a satisfactory balance in terms of level and perspective can be difficult. For a concert recording of trumpet or trombone and piano, the main difficulty will be getting reasonable piano level so that some detail can be added to the part to enable it to be heard. You will need to be a bit closer than usual with the piano microphones, remembering that if you need to get some cardioids in the well of the piano, using three of them (LCR) will give more even coverage. (See sections 6.6 and 5.7.1).

Solo brass and piano differ from more traditional chamber music ensembles such as string quartets and piano trios, which are designed to be heard in an intimate setting. Therefore, it might be more difficult to decide on the desired recording presentation; should it be presented as intimate chamber music or as something more suited to a larger space?

The techniques outlined earlier in the chapter can be adapted for solo brass and piano provided the potential for soloist spill onto the piano microphones is borne in mind. To help with this, getting a little more distance between the instruments is a good idea, provided the quality of the spill does not get too reverberant.

Figure 7.8 shows a position suitable for trumpet or trombone, both forward-facing brass.

In this position, at least 1.2–1.5 m (4′ to 5′) away from the piano, good contact can still be maintained with the pianist, and ribbon microphones can be used with their side axes aimed at the piano to aid a degree of separation. As when recording a brass ensemble (see Chapter 13), the microphones for the soloist should avoid being directly on-axis to the bell; the HF from a brass instrument's bell is increasingly narrowly beamed as the frequency increases, and any small player movements will result in large changes in tone. Place the microphones about 30 cm (1′) higher than the bell – about 1.2–1.5 m (4–5′) off the ground if the player is seated, and higher up if they are standing. This will also avoid any possibility of wind damage to ribbon microphones, especially from the trombone which produces moving air. In the case of the trumpet, whose lowest note is around 165 Hz, much of the LF spill from the piano onto the soloist's microphones can be filtered off. In the case of trombone, greater care would need to be taken as the lowest note is an octave lower, at around 80 Hz. In both cases, the instrument is facing away from the piano tail

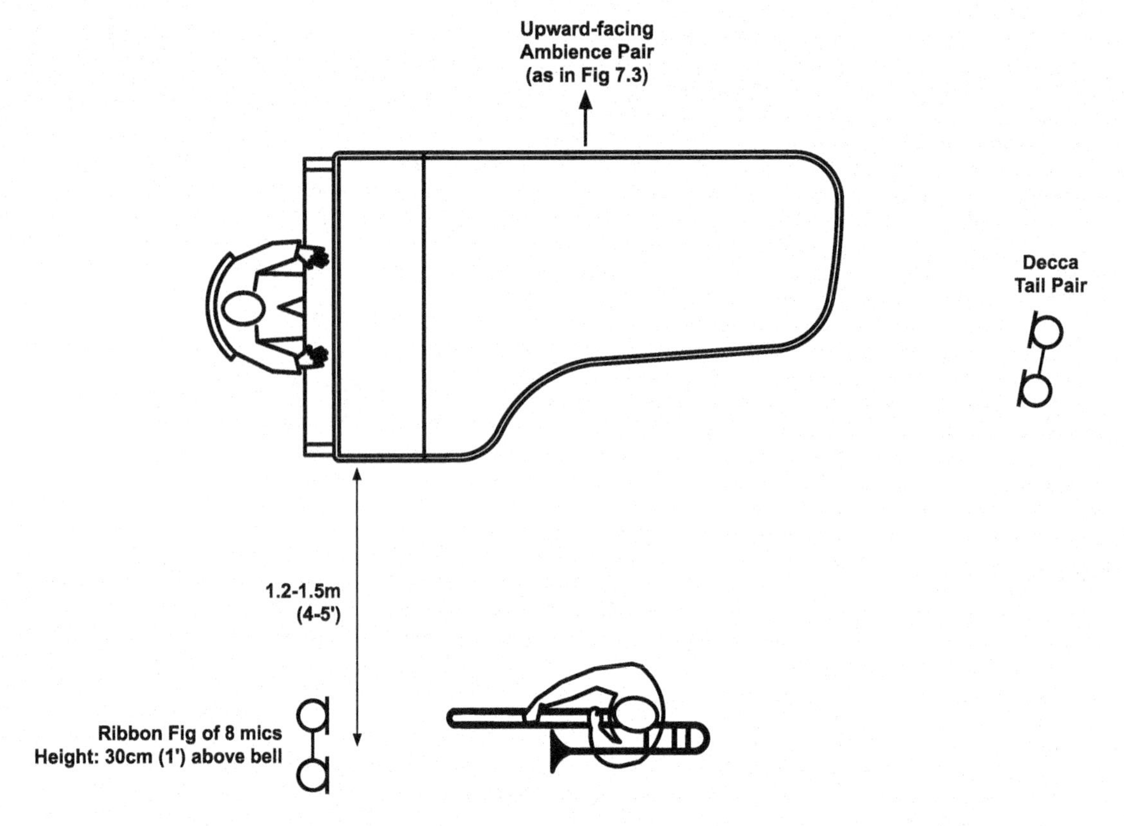

Figure 7.8 Trumpet or trombone and piano in the studio

pair, but there will still be distant sounding spill on the piano microphones. This can be embraced as providing a sense of space to the sound of the soloist.

The French horn can be approached a bit differently because it fires sideways and backwards from the player's right-hand side. Figure 7.9 shows the horn seated in such a way that the player has good eye contact with the pianist, and the instrument is firing away from the piano microphones.

The horn microphones are placed behind the player, at about 1.2–1.5 m (4′ to 5′) back from the horn and just high enough to be looking down to the bell, thus avoiding being on-axis to it. The tone quality from the horn can be changed by moving around in an arc behind it; the more the player's body is in the way, the less HF content is apparent in the sound. Although the horn microphones are pointing towards the piano in this position, their low height and the shielding effect of the player's body reduce the problem of piano spill. Microphone choices can include cardioids, wide cardioids, and ribbons.

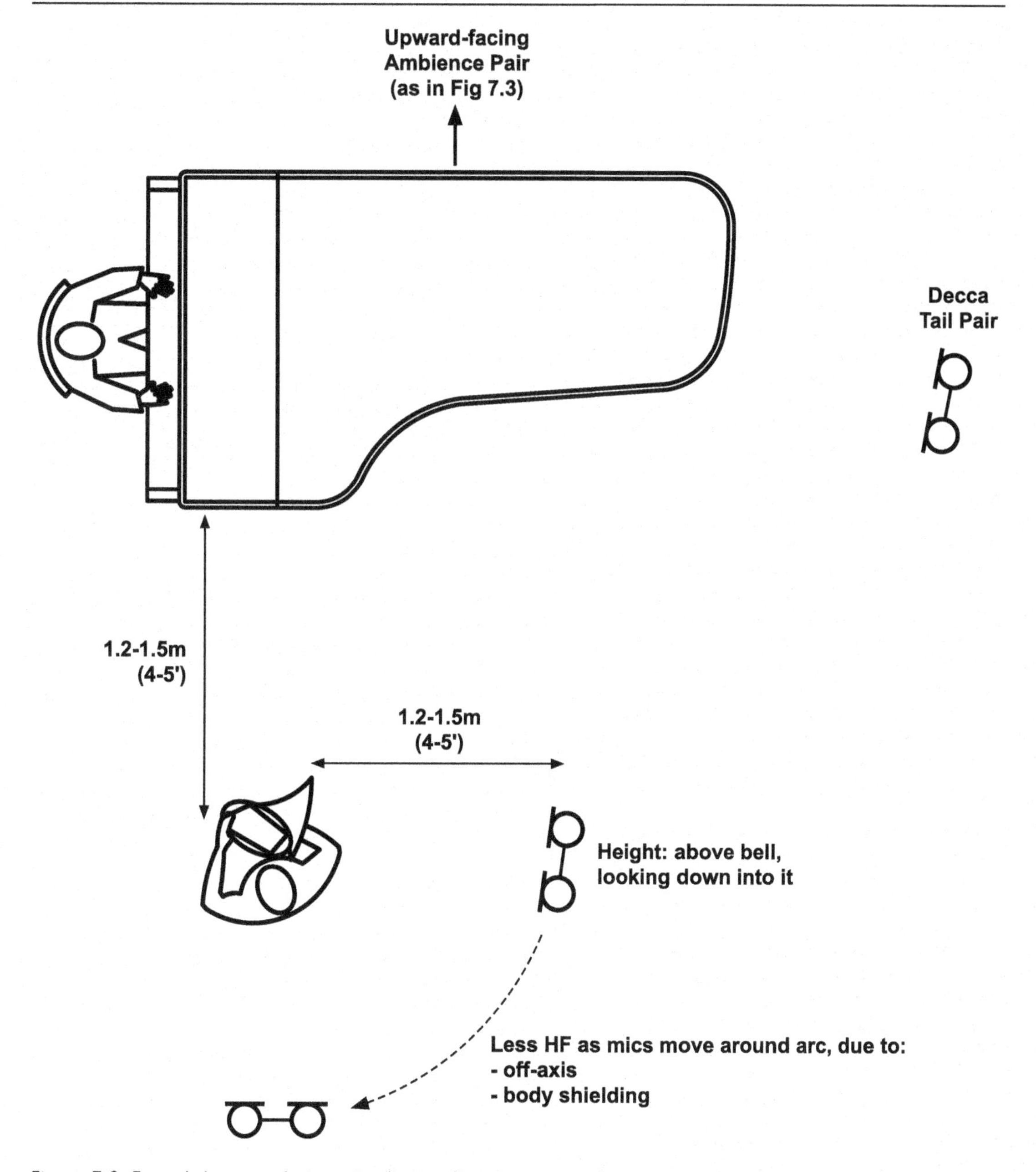

Figure 7.9 French horn and piano in the studio

Note

1 J.S.BACH/HANDEL/Sonatas for Viola Da Gamba/Lynn Harrell/Igor Kipnis/DECCA (1988) UPC 028941764614.

Chapter 8

The Decca Tree

There are many methods that can be used to capture the main body of an orchestra's sound (usually referred to as the 'main pickup'), and the stereo methods outlined in Chapter 3 can be used as an overall orchestral pickup. However, none of these approaches the real-world dominance of the Decca Tree for classical or film orchestral recording. Although there are plenty of voices that raise objections to the use of spaced microphones in favour of co-incident techniques, the warmth of the extended LF pickup of omnidirectional microphones and the lush and spacious string sound that is obtained with the Decca Tree mean that the technique is favoured in the commercial world. At Abbey Road, a core Decca Tree array of three Neumann M50 microphones is kept permanently rigged in both Studio 1 and Studio 2 and is used on almost every orchestral session. The website ScoringSessions.com contains a great pictorial record of film scoring sessions – whatever the film, wherever the studio, be it Air Lyndhurst in London or the Newman Scoring Stage in Los Angeles, there will almost certainly be the same collection of three microphones hanging over the conductor's head.

However, there is much confusion, misinformation, and misunderstanding about the Decca Tree, and it is unfortunately very easy to get it wrong. The purpose of this chapter is to explain what the tree is, why it is what it is, how to use it, and how to avoid the many potential pitfalls.

8.1 What is the Decca Tree?

Ask most engineers this question and the answer will be clear: an array of three omnidirectional microphones arranged in a triangle. But as we shall see, this is not necessarily the case. It needn't be three microphones, and they needn't be omnis – in fact it might even be better if they are not.

Given the possible variations, it is perhaps better to think of the Decca Tree as an array of between three and five microphones, which can be summarised in the table in Figure 8.1.

The key point to note here is that it is good idea to consider the complete set of microphones, including any outriggers (and double bass microphone, see section 8.1.1), as one complete 'tree' array. So you should make all adjustments, such as microphone position, gain, and pan whilst thinking of and listening to the whole set.

No. of mics	Physical arrangement	Notes
3	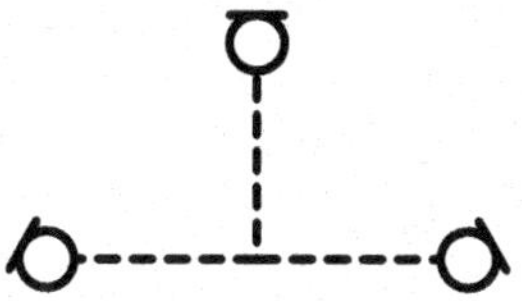	The basic tree, typically used for small orchestras
5	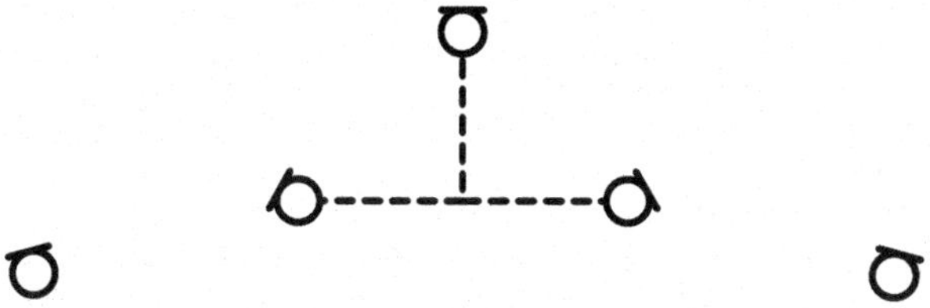	The basic tree with two outriggers, typically used for large orchestras
4		A variation on the tree with outriggers, typically used for large orchestras

Figure 8.1a–c Table showing Decca Tree variations

8.1.1 A note about double basses

There is something wonderfully visceral about the sound of the double bass, but as noted in Chapter 9, it can lose impact with increasing distance in a way that does not happen with the upper strings. For this reason, it can be necessary to put one or two ancillary microphones on the double bass section, and it is helpful to think of these as part of the tree system that forms the bulk of the string sound.

For the purposes of this chapter, we will include a single omnidirectional microphone at a height of about 2.1 m (7′) aimed at covering the whole section. Assuming that the orchestral layout is standard (with the double basses on the far right-hand side of the orchestra), the mic should be panned fully right (the same as the right outrigger), and the gain would normally be about 3 dB less than the gains of the tree microphones. For alternative orchestral layouts (such as with the basses in the centre), the panning should be altered accordingly. See section 9.6 for more detailed discussion of double bass section and principal microphones.

8.2 The three- and five-microphone trees

This section is quite detailed and prescriptive on the 'correct' positioning of tree microphones in the interests of getting up and running quickly. There is usually limited time available to get a satisfactory balance, and there are logistical difficulties and dangers of adjusting the tree mid-session – for example, you might need two people to move it safely. You are of course free to adjust and vary as much as you like when you have the time; the suggestions here are a summary of many years of experience with different ensembles and venues.

8.2.1 Dimensions and initial positioning on an orchestra

The horizontal plan dimensions for the basic three-microphone tree are shown in Figure 8.2. The typical Decca approach is to mount the array at a height of between 2.74 m (9′0″) and 3.20 m (10′6″). Some engineers will use the tree at greater heights of up to 4 m (13′), but at this distance the tree is acting more as an ambient pickup than a main pickup and usually requires the addition of a lot of close spot microphones on individual groups of instruments to get sufficient direct sound. In this arrangement, although the rig looks like a Decca Tree, it is not really being used as one.

The basic triangle is usually mounted so that the front/centre mic is positioned 15–30 cm (6″ to 12″) in front of the orchestra (towards the audience), measured from the outside chair legs of the lead violinist and cellist. For the use of a smaller, three-microphone tree on a string quartet or trio, please see Chapter 12.

The three-microphone tree is sufficient for a small chamber orchestra up to 8/6 (eight first violins, six second violins), but once the orchestra is wider and there are more than four desks of first violins, outriggers need to be added in order to obtain more detail from the outer edges of the ensemble.

Outrigger microphones should usually be set at the same height as the basic triangle. They are placed to the left and right of the triangle, and a good starting position is to put them in line with the backs of the chairs of the second desk of violins and cellos – this will normally work out as being between 320 cm and 335 cm (10′6″ and 11′) from the centre line. They should be placed

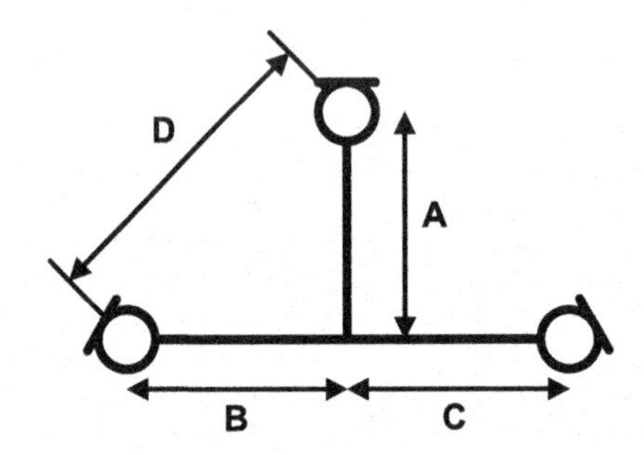

Figure 8.2 Dimensions of a basic Decca Tree

further back from the front of the orchestra – usually somewhere between 142 cm and 158 cm (4′8″ and 5′2″).

If the microphones being used were perfectly omnidirectional, it wouldn't matter at all in which direction they were pointing. But for the tree system to work well, giving the ideal combination of excellent imaging coupled with spacious sound, the microphones used should be somewhat directional and need to be pointed carefully. (See section 8.4 for discussion of suitable microphone types.) For a five-microphone tree array used on a large symphony orchestra as shown in Figure 8.3, the microphones should be pointed as shown in Figure 8.5. Figure 8.4 shows the five-microphone tree in use in the Sofiensaal in Vienna.

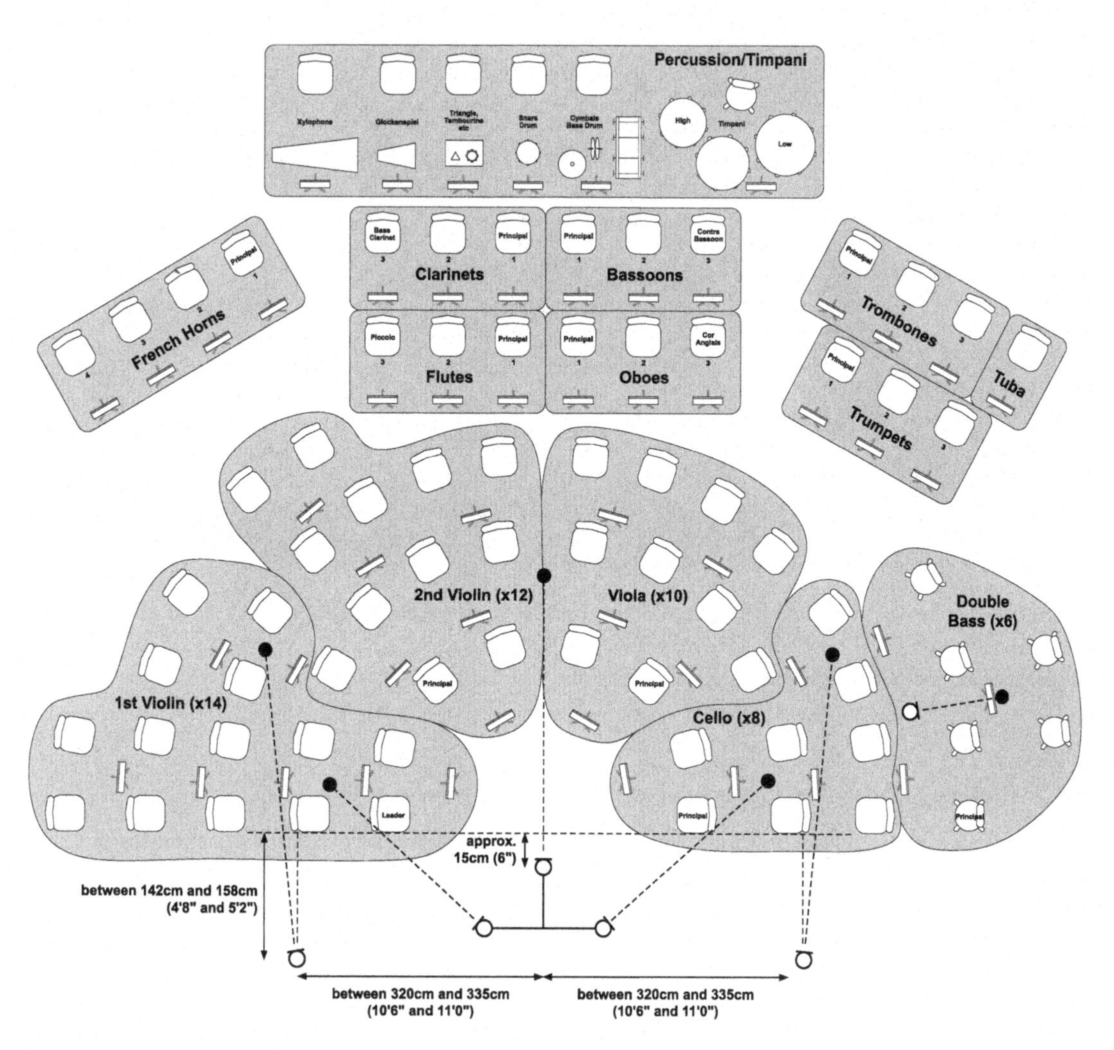

Figure 8.3 Using a five-microphone Decca Tree on a large symphony orchestra

Figure 8.4 Recording the Vienna Philharmonic Orchestra conducted by Sir Georg Solti at Sofiensaal, Vienna, in 1978 using a five-microphone tree

Photo: Elfriede Hanak; courtesy Decca Music Group Ltd.

8.2.2 Panning and gains for three- and five-microphone trees

For the Decca Tree technique to coalesce into something special, the gains and panning need to be set correctly. It is a common mistake to think that, in a five-mic Decca Tree array, both the outriggers and triangle left and right microphones should be fully panned left and right. In fact, it is much better to pan the triangle microphones in a little, and getting the value right makes a big difference.

Unfortunately, there is no common and consistently accurate way of describing pan responses other than to specify the gain response in dB for each feed to the left or right mix bus. However, many mixers and DAWs use other measurements to describe pan positions, such as a percentage

Microphone	*Aim the microphone at the eyes of a 170 cm (5′6″) person standing in the following positions:*
Left Outrigger	Between the chairs of the first violin desk that is three rows back and one in from the front of the orchestra
Left Triangle	Between the chairs of the second desk of first violins
Centre Triangle	Between the chairs of the second desks of second violins and violas
Right Triangle	Between the chairs of the second desk of cellos
Right Outrigger	Between the chairs of the cello desk that is three rows back and one in from the front of the orchestra

Figure 8.5 Where to point the microphones for a five-microphone Decca Tree array on a large orchestra

	Microphone					
	Left		*Center*		*Right*	
Desired signal gain to send to stereo mix bus	*Left*	*Right*	*Left*	*Right*	*Left*	*Right*
	0dB	−inf	−3.0dB	−3.0dB	−inf	0dB
	Fader	*Pan*	*Fader*	*Pan*	*Fader*	*Pan*
Pro Tools[1]	0.0	<100	0.0	> 0 <	0.0	100>
Pyramix[2]	0.0	Left	0.0	Centre	0.0	Right
Reaper[3]	0.00	−90.0	0.00	centre	0	+90.0
TotalMix FX[4]	0.0dB	L100	0.0dB	0	0.0dB	R100

Figure 8.6 Gains and pans for the three-microphone tree

[1] Pro Tools 2019.12 pan law – Session Setup: Pan Depth: −3.0 dB.

[2] Pyramix 11 pan law – Mixer Settings: Virtual Room/Stereo Pan Law: Sin/Cos law, Constant Power, −3 dB centre, Panner Type: Pan/Balance.

[3] Reaper 5 pan law – Project settings: Pan Law −3.0 dB, Gain compensation: Off, Pan Mode: Stereo balance/mono pan, Preferences: Pan fader unit display: −90 dB.. +90 dB.

[4] RME TotalMix FX 1.5 pan law – fixed at −3 dB centre.

scale from 1% to 100%, a scale from 1 to 64, or a scale based on the clock face ('pan it left to 9 o'clock'). We have measured the response of a number of common systems, and Figures 8.6 and 8.7 give practical panning and fader positions for Decca Tree applications. The desired gain levels in dB for the left and right mix buses are shown for each microphone, along with the equivalent fader and pan positions within Avid Pro Tools, Merging Technologies Pyramix, Cockos Reaper, and RME TotalMix FX.

It has been assumed that the microphone amplifier gains have been adjusted to be the same for all microphones in the array. Some of the systems shown have the option of multiple pan laws, so we have shown the optimum pan law settings for Decca Tree applications in the notes.

Note that in Figures 8.6 and 8.7, the gain of centre mic is –3 dB with respect to the outer left and right microphones. On most analogue mixers, this means that the three faders can stay at the same level as each other, since most analogue pan pots give a 3 dB attenuation for centre-panned signals. This centre mic level has been found to give the best results, but of course it is possible to raise or lower it if necessary.

If the centre mic level is too high, here are some of the things you might notice:

- The feeling of depth reduces – rather than getting a clear and realistic contrast between close sources and distant sources (e.g. front desks of strings and the woodwinds), everything will start to sound closer.
- The feeling of width reduces – the stereo spread moves closer to mono.
- There is a low frequency boost for the strings, and they might sound boomy.

If the centre mic level is too low, here are some of the things you might notice:

- The second violins and violas will not be loud enough in the mix.
- There will be a hole in the middle of the stereo image, with everything appearing to come more from the left and right extremes.

The centre microphone gives the engineer control over detail and level of the centre of the orchestra, but this can come at the expense of LF clarity. The four-microphone tree detailed in section 8.3 has the benefit of simplicity and produces clearer LF, but it is harder to place correctly to retain a clear orchestral centre. One technique aimed at obtaining the best of both techniques is to use a three- or five-microphone tree, but apply an HPF (around 110 Hz) to the centre microphone so that the LF is coming from the outer microphones only. (See also section 17.3.)

	Microphone									
	Left Outrigger		*Left Triangle*		*Center Triangle*		*Right Triangle*		*Right Outrigger*	
Desired signal gain to send to stereo mix bus	*Left*	*Right*	*Left*	*Right*	*Left*	*Right*	*Left*	*Right*	*Left*	*Right*
	0 dB	−inf	−0.4 dB	−13.5 dB	−3.0 dB	−3.0 dB	−13.5 dB	−0.4 dB	0 dB	−inf
	Fader	*Pan*	*Fader*	*Pan*	*Fader*	*Pan*	*Fader*	*Pan*	*Fader*	*Pan*
Pro Tools[1]	0.0	<100	−0.2	<72	0.0	> 0 <	−0.2	72>	0.0	100>
Pyramix[2]	0.0	Left	−0.2	72% L	0.0	Centre	−0.2	72% R	0.0	Right
Reaper[3]	0.00	−90.0	−0.10	−11.1	0.00	centre	−0.10	+11.1	0	+90.0
TotalMix FX[4]	0.0dB	L100	−0.1	L72	0.0dB	0	−0.1	R72	0.0dB	R100

Figure 8.7 Gains and pans for the five-microphone tree (see Figure 8.6 for footnotes)

8.2.3 Adjusting the Decca Tree placement: what to listen for

It is quite rare for a tree array to be used on its own for a full orchestra, despite the fact that it normally provides the overwhelming majority of the final sound. Additional microphones are used for further detail, focus, and control of perspective, and are discussed in Chapter 9. However, because the whole balance starts from the sound of the tree, it is essential to get it correctly positioned, and time should be spent on listening to the tree on its own at the start of the session. 'Tree' in this context means the complete array, including any outriggers and double bass microphone(s). Before adjusting the physical placement of the tree, be sure to set the panning and gains according to section 8.2.2 so you are listening to something appropriate.

Firstly, listen to the balance of the string sections; there should be an even balance of level and an equal stereo spread of all the string parts. Pay attention to the string sound alone and ignore the other elements of the orchestra. These might sound unduly recessed, but they will eventually be supported with ancillary microphones and can be ignored for now.

Photographic or lighting analogies can be useful when thinking about the sound produced by the tree array, or by any microphone being used to pick up a group of players. We can think of a microphone as 'lighting up' or being 'focussed on' a section of the orchestra, with its strongest light or centre of focus being on the players that it is pointing at. For example, the 'centre of focus' for the left triangle microphone will be the second desk of first violins. There should be a sense of depth of field such that the first and third desks are also in focus but the fourth and fifth desks are increasingly less clearly detailed, giving a pleasant sense of perspective to the sound. This 'depth of field' will be obtained at a particular microphone distance: closer in, only a single desk will be in focus; farther out, many desks will be equally picked up but none of them with sufficient detail. Where the orchestra is wider, outriggers are used to make sure that the desks of strings further out to each side do not become so lacking in detail that they feel too far away in the perspective. If the orchestra was plunged into darkness, and the tree array microphones magically transformed into spotlights, these lights should produce a good spread over the whole width of the string section, with the more distant strings towards the back of the orchestra receding away a little into the darkness. A string section that contains some natural sense of depth will sound larger and more realistic than one where all the players have close microphones and are presented at the same distance, resulting in a very 'flat' sound.

The height of the tree is a critical factor, as it affects the ratio of direct to indirect (reflected) sound. Different halls, with different reverberation characteristics, will need different heights, and different orchestras in the same hall might also need different heights; the tree system can be very revealing of flawed playing. As a general rule, the better the orchestra, the lower the tree can safely go without the string tone becoming unpleasant. The choice of repertoire is another important factor, as some pieces suit a more distant sound than others.

If the tree is too low down, here are some of things you might notice:

- You can hear separate desks of strings, rather than a pleasantly homogeneous whole.
- The string sound is too 'edgy'.
- The string sound is too dry, with not enough reverberation.
- There is too much high-frequency content (HF) in the strings; they sound scratchy.

If the tree is too high up, here are some of the things you might notice:

- There is too much indirect room sound, and not enough direct sound; everything feels a little too reverberant.
- You feel like you have moved a few too many rows back from the orchestra.

Outrigger heights are usually set the same as the main triangle, but there is some room for adjustment – perhaps up or down 7–10 cm (3″ to 4″) of the main height. Adding height gives more of a feeling of space, and lowering height is useful if the outriggers are too reverberant. (See 8.4.1 for discussion of the option of using different microphones for the main tree and/or outriggers.)

The next thing to consider is the balance between the strings and the rest of the orchestra. If this feels well balanced in the room, but listening to the tree feels as if the strings are underpowered, the tree needs to be moved. The natural inclination here would be to move the tree closer in; however, somewhat counter-intuitively, experience suggests the opposite. Moving the tree closer to the orchestra (not altering its height) produces an impression of a reduced body of strings; the sound becomes more confusing, and the first violins and cellos have effectively been moved more around the sides of the centre microphone. This makes it a less effective pickup because the microphones used are not truly omnidirectional but favour sources on their front axis over those on the sides. (See section 8.4 for discussion of the microphone models and characteristics needed for the Decca Tree to work at its best.) If more strings are needed, you are better off moving the tree a little out towards the audience.

One final word on positioning the tree – it is important to make sure the orchestra doesn't move during the course of the session. If you are working on a stage or floor where the front desks of strings are able to creep outwards (without falling off the stage!), then there is a danger that as a session progresses, this will indeed happen and will start to ruin your carefully adjusted string sound. If you experience a loss of string clarity during a session, it is worth checking that the gradual outward migration is not the cause. Prevention is a much better approach, and the studio floor can be marked with a line before the session starts to indicate the front edge of the orchestra which the outer legs of the players chairs should not cross. Mark it using something that is very noticeable but won't damage the floor when removed – black insulation tape is ideal – and if you approach the leaders of the first violins and cellos with some gentle words of explanation, this barrier, more psychological than physical, will seldom be crossed.

8.3 The four-microphone Decca Tree

Many of the most celebrated recordings made with a Decca Tree didn't use the three- or five-microphone techniques that we have concentrated on so far. For example, most of the multiple award-winning recordings made by John Dunkerley of the Orchestre Symphonique de Montréal (OSM) under Charles Dutoit were captured using a four-microphone tree, as shown in Figure 8.8. The technique was developed by Kenneth Wilkinson, who used it almost exclusively from about 1977 onwards. It is a much more difficult technique to get right, as the spacing of the microphones is critical, but it can give a wonderful clarity to the result, particularly at the lower end of the frequency range. (See section 8.2.3 for a note about improving LF clarity for the three- and five-microphone trees.)

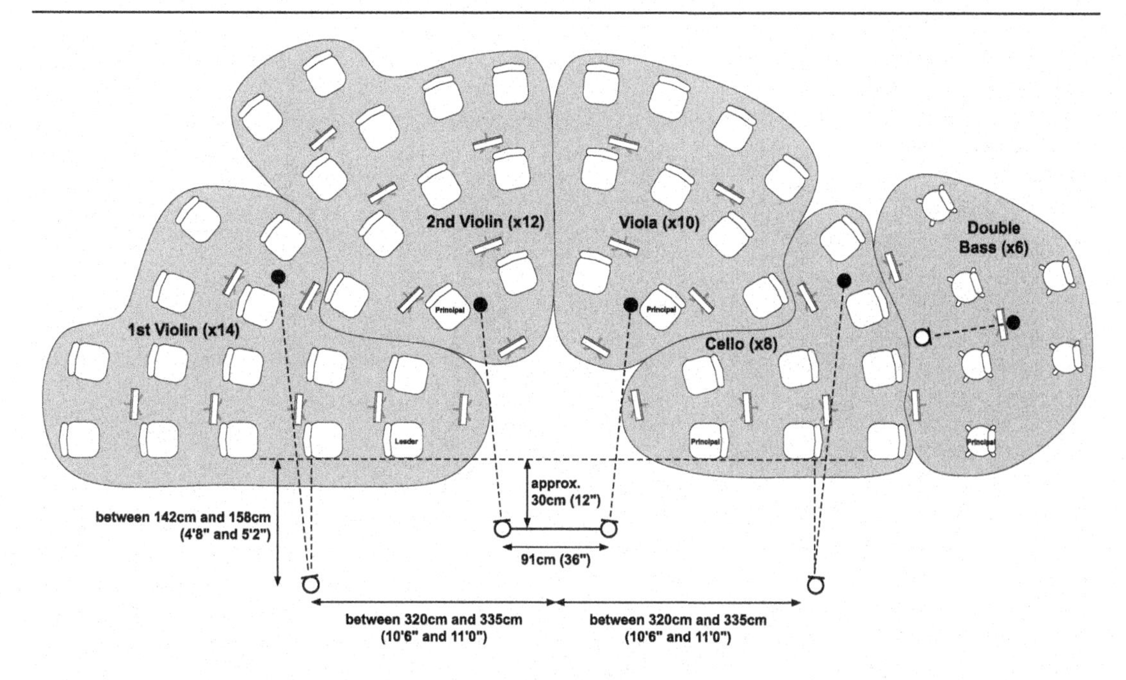

Figure 8.8 Four-microphone Decca Tree on a large symphony orchestra

The two main microphones are spaced 91 cm (36″) apart and positioned at normal tree height 30 cm (12″) back from the front desks. If the spacing is any wider, there is a danger of a pronounced hole in the middle effect; any narrower and it becomes too mono. The microphones are pointed slightly outward, aimed at points between the front desks of the second violins (left microphone) and violas (right microphone). The outriggers for the four-microphone tree are positioned and pointed just as for a five-microphone tree.

The optimum panning and gains are shown in Figure 8.10. Note that the panning of the left and right microphones is more central than on the five-microphone Decca Tree, with an approximate 7.5 dB level difference between microphones rather than the 13 dB difference on the five-microphone array. Figure 8.9 shows the Chicago Symphony Orchestra being recorded with a four microphone tree.

8.4 Microphones for the Decca Tree

The most favoured microphone for the Decca Tree by far is the Neumann M50, although Neumann stopped making them decades ago. While big orchestral studios have a set of them in their collections, their rarity means they command a high price on the open market, if they come up for sale at all. Figure 8.11 shows a CM50S, which is a modern rebuild using the same capsule.

The M50 is described as an omnidirectional microphone, but in reality, its directivity pattern is very far from a perfect omni above 1 kHz. As is characteristic of all pressure-operated microphones, it has an extended bass response that is virtually flat down to 20 Hz. It is omnidirectional

Figure 8.9 Recording the Chicago Symphony Orchestra with Sir Georg Solti at Medinah Temple using a four-microphone tree (1977–1979). From this angle, the right outrigger appears very close to the right tree microphone; the left outrigger gives a better impression of the spacing. Note that the main left and right microphones are not placed at the extreme ends of the tree arms.

Photo: Courtesy Decca Music Group Ltd.

at low frequencies, but as the frequency rises above 1 kHz, the M50 becomes more and more directional, moving gradually towards a super-cardioid type pattern by 10 kHz. On its front axis, the frequency response is flat to about 1.5 kHz and then starts to gradually increase to a peak of about +6 dB in the 8–10 kHz region (see Figure 8.12).

It is the combination of these characteristics that makes it so ideal for use in a Decca Tree:

- The smooth on-axis HF lift goes some way towards compensating for the natural attenuation of HF from distant sources; this gives greater reach into the orchestra and a very focussed and flattering sound to the string section.
- The extended bass response produces a wonderfully full and warm lower string tone.
- The increased directionality at mid to high frequencies offers advantages over a purely omnidirectional pattern in that there is some useful off-axis rejection and hence clearer stereo imaging. At the same time, it avoids the undesirable features of pressure-gradient cardioid microphones such as the reduced LF output and off-axis colouration.

	Microphone							
	Left Outrigger		*Left Main Pair*		*Right Main Pair*		*Right Outrigger*	
Desired signal gain to send to stereo mix bus	*Left*	*Right*	*Left*	*Right*	*Left*	*Right*	*Left*	*Right*
	0dB	−inf	−0.9dB	−8.5dB	−8.5dB	−0.9dB	0dB	−inf
	Fader	*Pan*	*Fader*	*Pan*	*Fader*	*Pan*	*Fader*	*Pan*
Pro Tools[1]	0.0	<100	−0.2	<50	−0.2	50>	0.0	100>
Pyramix[2]	0.0	Left	−0.2	50% L	−0.2	50% R	0.0	Right
Reaper[3]	0.00	−90.0	−0.19	−6.0	−0.19	+6.0	0	+90.0
TotalMix FX[4]	0.0dB	L100	−0.1	L49	−0.1	R49	0.0dB	R100

Figure 8.10 Gains and pans for four-microphone LR tree with outriggers (see Figure 8.6 for footnotes)

Figure 8.11 CM50S with the grille removed, showing the sphere on which the diaphragm is mounted

Photo: Peter Cobbin.

If you use normal, small-diaphragm omnidirectional pencil microphones on a Decca Tree, you will usually find that you get the expected rich lushness of the sound (due to the bass extension), but that the stereo imaging will be not be as precise as desired. Sound sources may seem to be vaguely 'over there somewhere' but with little accuracy and sense of perspective. This is a very common mistake and one of the main reasons why many engineers give up on the Decca

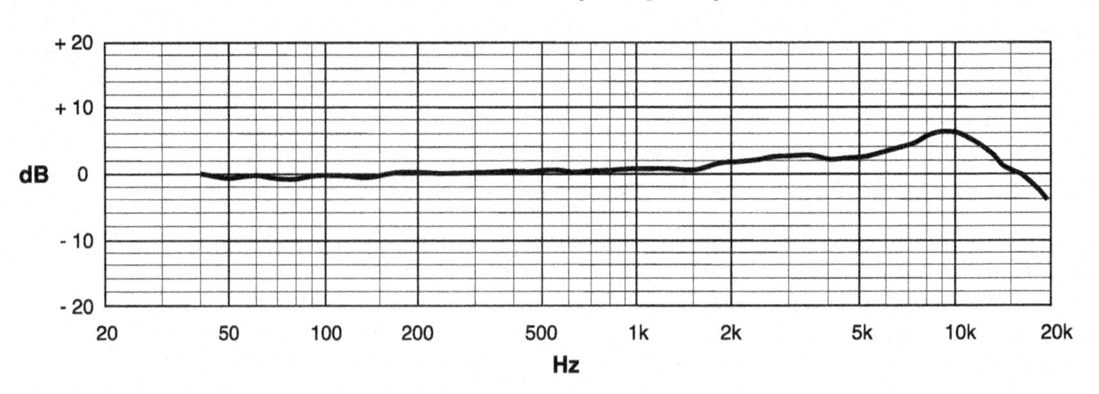

Figure 8.12 Frequency response of the Neumann M50 omnidirectional microphone

Source: Courtesy Georg Neumann GmbH.[1]

Tree technique after a few attempts. It is the directionality – the 'not-really-an-omni-ness' of the M50 – that makes the system work.

So what is it about the M50 and similar microphones that gives them this useful directional characteristic? Figure 8.11 shows a CM50S with the grille removed, and it can be seen that the diaphragm is mounted on the surface of a sphere that is larger than the diaphragm diameter. The sphere is 40 mm in diameter, and this has some particular effects on sounds approaching from different directions. At low frequencies, the wavelength of the sound is very long in comparison to the sphere dimensions (e.g. the wavelength of 200 Hz is approximately 1.70 m), and the sound can easily diffract around to the diaphragm; the sphere presents no effective obstacle at all, and the microphone will respond equally to sound coming from any direction. As frequency rises and wavelength decreases, the size of the sphere starts to become significant; the shorter wavelengths cannot diffract very well around it, and they will also begin to experience reflections from the surface. In this way, the microphone becomes increasingly directional at short wavelengths, with reduced output from the sides and rear. When compared with a typical 13 mm (0.5″) diaphragm pencil omni, which is designed so that its body does not exceed the diaphragm diameter, the directionality of the M50 begins lower down in the frequency range. The spherical shape also produces a particularly smooth, frequency-related rise in pressure build-up in front of the diaphragm for sounds approaching from the front (resulting in a gradual HF lift on-axis).[2]

The sphere-mounted diaphragm is still in use in the omnidirectional Neumann M150 and TLM50, but these are also expensive, and there are more affordable options. Given the effects of sphere-mounting on a pressure-operated diaphragm, it should in theory be possible to recreate some of this effect by adding a similar sphere to the pre-existing pencil omnidirectional microphone so that its diaphragm is flush with the surface. This forms the basis of the add-on spheres sold by DPA, Schoeps, and Neumann with names such as 'acoustic pressure equalisers' or APEs. These are designed to fit over true pressure-operated capsules to alter their response and directivity. (Large-diaphragm, variable directivity microphones such as the Neumann U87 or the AKG C414 are not suitable for this sort of modification as their omnidirectional output is created by the summation of the outputs from two pressure-gradient cardioids mounted back to back.)

Various diameters of APEs are commercially available, with 40 mm and 50 mm being most common. The sphere on the original M50 is 40 mm in diameter, but the enclosing metal grill casing has an additional effect on directionality at HF, and the net effect is approximately equivalent to using a 50 mm ball on a cylinder mic – which is why many engineers prefer to use 50 mm spheres rather than 40 mm. Microphones such as the DPA4006 have a tapering body, so add-on spheres have to be specially machined to fit these. For microphones with a simple cylindrical form, it is possible to make your own spheres; see Appendix 2 for a discussion of the modification of an inexpensive omnidirectional microphone.

The other important attribute when looking for alternative microphones to use on a Decca Tree is the HF boost. Some degree of on-axis HF lift is a feature of all omnidirectional microphones that are designed for use in a diffuse sound field (i.e. relatively distant microphone placement in a reverberant room) rather than those designed for free-field use (i.e. closer microphone placement in a dry space). Opinions vary, but several very high-profile engineers say that the best modern equivalent to the M50 is a Schoeps MK2H capsule fitted with a 50 mm sphere. See Figure 8.13a and 8.13b. Others prefer the Schoeps MK2S, which is brighter than the MK2H and has an HF boost starting at a lower frequency.

8.4.1 Alternative microphones for use in reverberant acoustics

In particularly reverberant acoustics, where you don't wish to capture so much of the indirect sound that it swamps the direct, it can be helpful to use more directional microphones for the outriggers and/or the main triangle. To keep the sound uncoloured and the blend and imaging consistent, it is important to use directional microphones that have a flat off-axis frequency

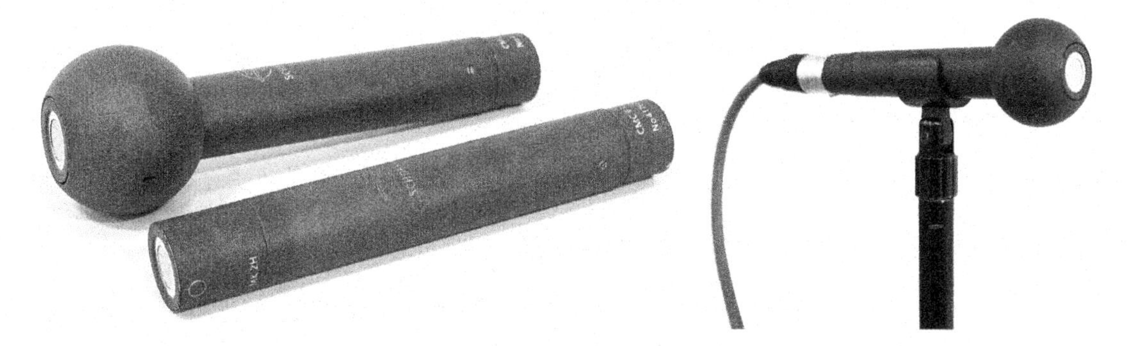

Figure 8.13a–b Schoeps MK2H with 40 mm spheres

Photos: Russell Mason.

response. For example, this means that the reduction in level of 6 dB at the sides that is exhibited by an idealised cardioid microphone should apply to all frequencies, and there should be a minimum of variation in the directivity pattern with frequency. This is actually very difficult to achieve, and the next best thing is to have off-axis frequency response that is smoothly varying rather than one that changes a great deal with a small change in angle of incidence. Cardioids with a good off-axis frequency response are rare, as they are difficult and more expensive to design; the usual trade-off will be a smoother response at the expense of reduced front-to-back rejection in decibels.

Perhaps the range of microphones with the least off-axis colouration are those produced by Schoeps, including the MK21 wide cardioid and the MK4 cardioid. This is a key reason why they have been so widely accepted in the classical music recording world, where one is almost always recording more than one instrument at a time, and the spill from off-axis instruments becomes a major part of the overall sound. The MK21 wide cardioids make excellent outrigger microphones in highly reverberant spaces, giving more 'reach' into the orchestra than classic M50s.

Another microphone with excellent off-axis response is the now-discontinued Neumann KM84 cardioid and its current replacement, the KM184; John Dunkerley has used KM84s as the main tree triangle in the reverberant chapel of King's College, Cambridge, with good results. The 1969 recording of Vivaldi's *Four Seasons* with Alan Loveday and Neville Marriner was recorded using a tree of M49 cardioids in St John's, Smith Square, London.

8.5 Mounting the tree

Rigging the Decca Tree requires good awareness of safe working practices. Because there is a heavy microphone array mounted almost directly over the conductor's head, and extremely close to the leaders of the string sections (who are probably clutching the most expensive instruments of the orchestra), it is absolutely vital that the mounting hardware, whether microphone stands or slinging ropes, is extremely sturdy.

Standard microphone boom stands are not robust or strong enough, and should never be used. Instead, it is well worth investing in a substantial stand from a reputable company, such as Manfrotto or ARRI. The conductor needs room to move about freely, so the vertical part of the stand cannot be too close to him or her. Therefore, the microphones have to be mounted on a horizontal boom arm with the vertical part of the stand kept at a safe distance behind the podium. With three potentially heavy microphones hanging off it, plus three cables, it is essential that this boom arm is counterweighted, so that there is no danger of it toppling forward. It is not sufficient to rely on the friction clutch on the microphone stand to hold the rig in place; it should balance horizontally even with the clutch undone.

When raising or lowering the tree, extreme caution is needed, as well as a sufficient number of helpers. The combined weight of the boom arm, three microphones, three cables, and the upper vertical sections of the main stand can be very heavy indeed, and could all too easily come crashing down onto anyone or anything nearby. All the major studios that have permanent Decca Trees use winch stands, which make it easy to raise and lower the stand in a controllable manner, and lock it at the desired height.

The original Decca Tree used sections of Dexion 225 slotted angle that were left over from building the racks in Decca's tape store! The regular spacing of holes in the metal made it easy to mount the mic holders at exactly the right dimensions. Figure 8.14 shows a tree of M50s mounted on a Decca Dexion Tree in the chapel of King's College, Cambridge.

Even though it is immensely practical, Dexion is rather ugly and industrial looking, and not something one would want to use for a live concert. For use where appearance matters, several companies make much more streamlined Decca Tree attachments, including Sabra-Som, Audio Engineering Associates (AEA), and DPA. Figure 8.15a–b shows a DPA commercial Decca Tree kit. These are not cheap, but they are good investments and a necessary addition

Figure 8.14 Decca Tree of M50s in King's College, Cambridge. Recording Verdi's *Four Sacred Pieces* in 1988[3]

Photo: Mark Rogers.

to your rig if you want to record live concerts in particular. A purpose-made Decca Tree rig will typically cost around the same as a good microphone, so when planning to purchase a set of microphones you will need to spend 25% of your budget on a rig and the remaining 75% on microphones.

If you do not have a tree frame of any kind available, it is possible to rig a tree using three stands. The centre microphone will need to go on a stand with a boom to reach forwards over the conductor, but the left and right triangle microphones can be placed on a single

Figure 8.15a–b DPA modular tree frame

Photo: Russell Mason.

stand each. This would be quite practical for a recording session although not acceptable for a live event.

8.6 Notes on the evolution of the Decca Tree[4]

It is often thought that the Decca Tree array was set in stone in the early years of Decca stereo recording, and no deviation from the 'standard' followed thereafter, but this is a very simplistic narrative, and something of a misrepresentation. The array outlined in Figure 8.3 was arrived at after many years of experimentation by the Decca engineers, beginning in the 1950s. It has existed in different forms from that time until the present day, and has always been open to experimentation and development by individual engineers according to the requirements of the time. Different microphones have been used through the decades, both cardioids and omnis, and the process of development was always rather organic, with the best ideas eventually winning out.

The move to stereo gradually happened during the 1950s in classical recording, and the earliest experiments at Decca were by Roy Wallace in 1954, who began to use three Neumann M49s (in cardioid mode) as the basis for an 'LCR' orchestral recording method using smaller spacing than the modern tree. When Neumann brought out the KM56 (cardioid) in 1956/1957, these replaced the M49s as they were smaller, lighter, and easier to rig. There was also a brief experimentation with the Neumann KM53 (omni), which has the same capsule as the M50 but without the additional directionality at higher frequencies.

In the meantime, Kenneth Wilkinson had continued making the mainstream mono recordings that were being produced in the early 1950s, using the M50s for their wonderful sound. When he began to get involved with stereo in 1957/1958, he thought about using the M50s in the new stereo techniques as well, and he introduced M50 outriggers in 1958/1959. For the main three tree microphones, he wanted a combination of the extended LF of the M50 and the directionality of the KM56. To this end, baffles were placed between the M50s to introduce some directionality at HF, and there were several high profile sessions where there was a 'shoot out' between the open KM56 tree and the baffled M50 tree to see which produced the best results. The photos in Figures 8.16 through 8.18 show some of the experimental variations that were explored. Figure 16.11 also shows an early baffled tree.

In these early days of stereo, loudspeaker quality was not as good as today, and these limitations meant that it was not so easy to perceive details in the sound, such as colouration due to baffles, or the fact that the M50 was actually already quite directional at HF. However, from late in 1963, Wilkinson began to try the M50s without baffles and discovered they produced good results: extended LF, good imaging, but without the baffle colouration. He increased the spacing between the microphones, and the 'classic' Decca Tree was born.

It is important to note that this does not mean it was adopted religiously and no other techniques used from that time. For example, Decca's Ring Cycle (1958–1965) used a KM56 tree with M50 outriggers throughout to keep the sound consistent. From 1967 to 1972, some Decca engineers used a six-microphone tree array (see Figure 8.19) and, as noted in section 8.3, a four-microphone tree was used in the 1980s and '90s for the OSM recordings. (See Appendix 1 for notes on tree variations for opera recordings.)

And a final word – why is it called a 'tree'? It is said that on a Mantovani session in Decca's Broadhurst Gardens Studio 1, Arthur Haddy said, 'It looks like a Christmas tree, boy!'

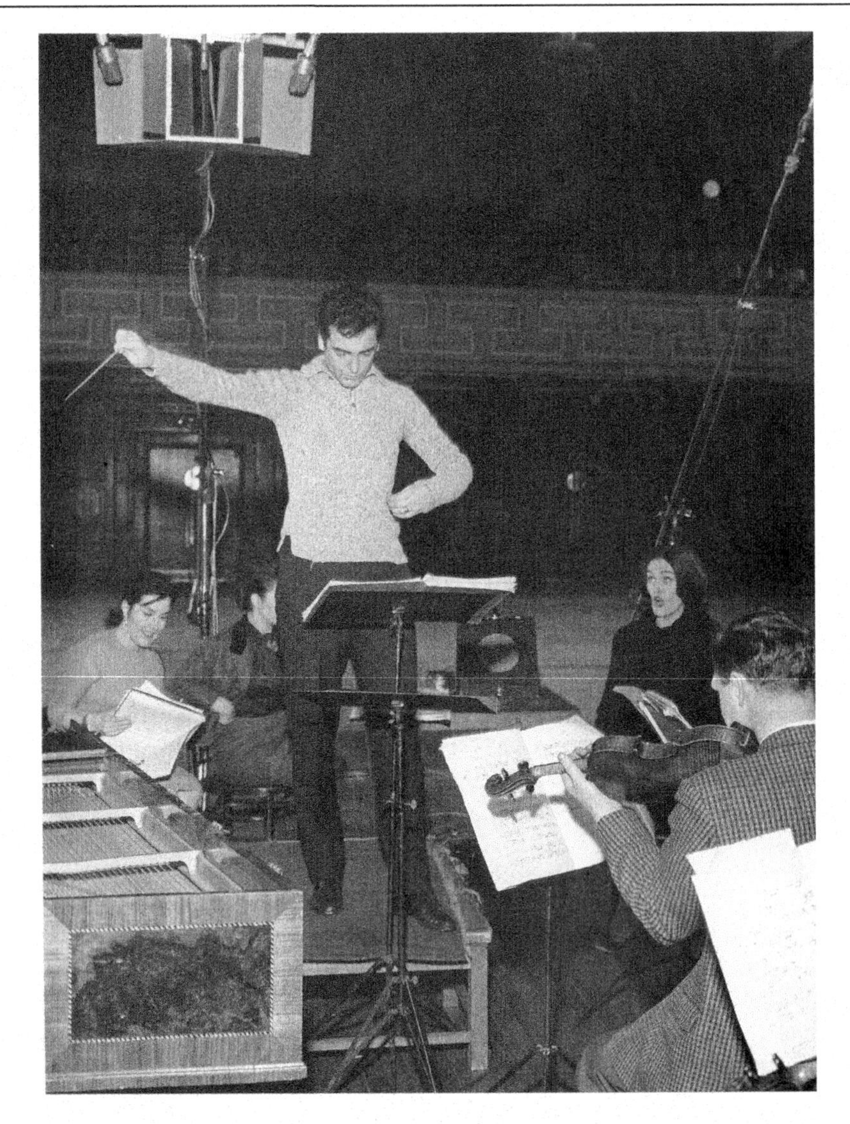

Figure 8.16 Baffled tree using a KM56 in the centre and baffled M50s on either side. Richard Bonynge conducting the LSO with Joan Sutherland recording Handel's Alcina in 1961 at Walthamstow Town Hall[5]

Photo: Courtesy Decca Music Group Ltd.

Figure 8.17 'Headed' tree with M50s enclosed in baffles on all sides. This shows Beatrice Lillie recording *Peter and the Wolf* with the LSO at Kingsway Hall in 1960.[6]

Photo: Courtesy Decca Music Group Ltd.

Figure 8.18 'Shoot out' between a KM56 tree and a baffled tree of M50s. The Accademia di Santa Cecilia, Rome, 1961.

Photo: Courtesy Decca Music Group Ltd.

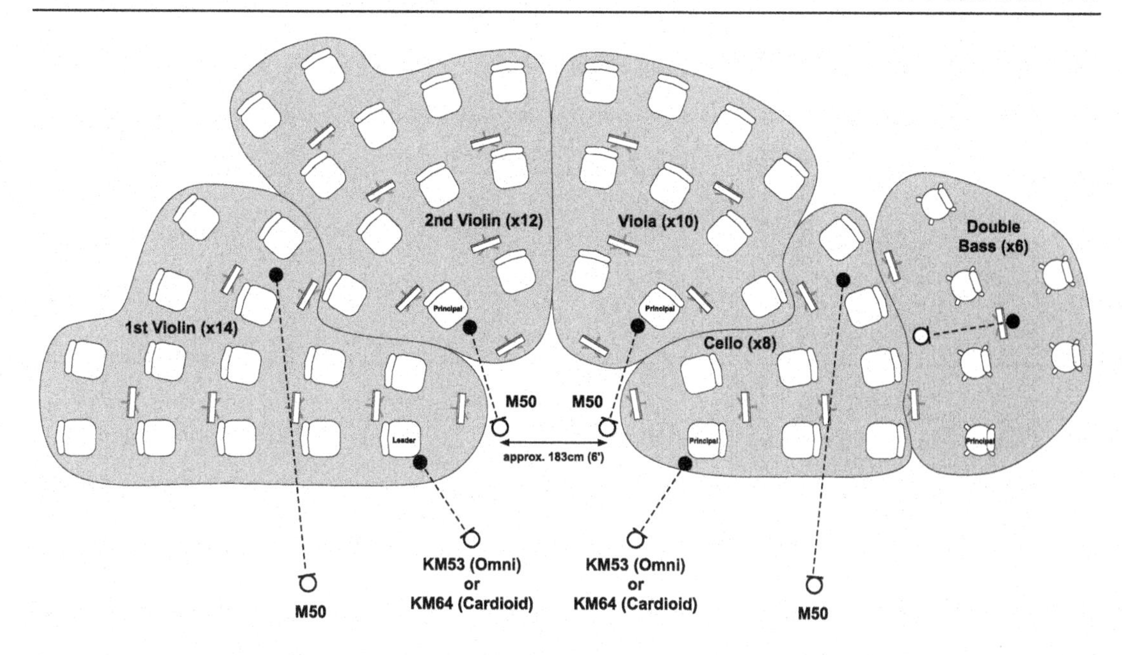

Figure 8.19 Six-microphone tree used by some engineers, 1967–1972

Notes

1 *A Pressure Microphone with Spherical Acoustic Surface*
Author: Dipl.-Ing. Stephan Peus (Georg Neumann GmbH)
Presented at the 16th Tonmeistertagung, Karlsruhe, 1990
https://de-de.neumann.com/product_files/7551/download

2 *The Diffraction Produced by Cylindrical and Cubical Obstacles and by Circular and Square Plates*
Authors: G. G. Muller, R. Black & T. E. Davis
JASA vol 10 July 1938

3 *VERDI/Four Sacred Pieces/Pater Noster/The Choir of King's College, Cambridge & Cambridge University Musical Society Chorus/Stephen Cleobury ARGO (1990) UPC 028942548022*

4 *Michael Gray has documented the development of the Decca Tree in his article here:*
The Birth of Decca Stereo
Author: Michael H. Gray
Association for Recorded Sound Collections (ARSC) Journal – Vol. 18, No. 1–2–3 (1986)
www.arsc-audio.org/journals/v18/v18n1-3p4-19.pdf

5 *HANDEL/ALCINA/Bonynge/LSO/Sutherland/Berganza/Freni/Flagello/Sinclair/Sciutti/Alva DECCA (1962) LP set 232–4*

6 *PROKOFIEV/SAINT-SAËNS/Peter and The Wolf/Carnival of the Animals/LSO/Henderson/Lillie/ DECCA(1960) CD 467 470–2*

Chapter 9

Ancillary microphones

9.1 What do we mean by 'ancillary'?

It is usually insufficient to rely solely on an overall orchestral microphone main pickup even though this will form the bedrock of the sound, giving the overall tone colour and sense of space and depth. To this end, additional 'ancillary' microphones are used, usually with the purpose of adding detail to sources that are furthest from the main pickup. These are commonly called spot microphones (UK), or accent microphones (US), but the use of the word 'spot' should not be taken to mean that the microphone is placed very close to the source with the intention of picking up a single instrument in isolation. Classical 'spot' or ancillary microphones are placed much further away than those used in pop/jazz/folk recordings. (See section 9.3.)

9.1.1 Which sections of the orchestra need ancillary microphones?

When considering which sections of the orchestra to bring into the light once the overall pickup is in place, there are a number of candidates, listed in priority order:

- Woodwinds
- Horns
- Brass
- Timpani
- Percussion
- Harp
- Celeste

9.2 Perception of orchestral depth and perspective

We think a lot about creating a good sense of orchestral width, but an impression of orchestral depth (front to back distance) is also extremely important to the art of creating a recording that feels believable and natural, even if it has been carefully manipulated to sound this way.

The perception of distance in a recording is built up from a combination of factors:

1. Reduction in HF and detailed close-up noises such as key noise, bowing noise, and transients – the less of this we can hear, the further away we think the instrument is.

2 Greater ratio of reverberant sound to direct sound – the more reverb, the further away we perceive the instrument's location to be.
3 Narrow image width, particularly of a single large instrument such as piano, or of a whole section such as the woodwinds. When a large instrument or group takes up only a narrow part of our listening stereo image, we interpret it as being further away.

Where these different factors reinforce one another to send us the same message, the sense of distance will feel quite real. Where they are in conflict (e.g. close instrumental noises, but lots of reverb and a narrow image) the sense of perspective becomes less believable and even confusing. To bring an instrument or orchestral section closer, we need to add some detail and closer sounds from the section to make the section less reverberant, and even to widen its image width a little.

The main question that arises then is, how deep should the orchestra appear to be? It needs to retain some depth to seem natural, but musical details should not be lost to an excessive sense of distance for the woodwinds, brass and percussion. When we watch a live performance, there are many visual cues in the concert hall to help us pick out details about who is playing and what is happening. If we shut our eyes, the listening experience changes a great deal, and likewise when we listen to a recording, none of the visual cues are present. This means that we need a little more aural detail from the rear sections of the orchestra, so the 'natural' perspective needs to be foreshortened a little. This does not mean presenting every instrument at an equal distance and flattening the perspective completely (such as happens with an infinite depth of field in a photograph), but neither do we want such a shallow depth of field that only the front strings are in focus. In order to foreshorten the perspective we will need to add some HF detail, increase the proportion of direct (non-reverberant) sound, and also possibly widen an orchestral section just a little.

It is ultimately the engineer's primary objective to make sure that the music is well served, and this means that musical details should not be lost in a wash of reverb or smudgy image. We have to make a fine judgement and arrive at a good balance between a sense of space around the orchestra and clarity of musical parts. If we can achieve both of these, then the music and musicians are well served and the listening experience can be a really moving one.

9.2.1 Effect of main pickup technique on perception of orchestral depth

The perception of orchestral depth is initially dependent on the actual distance between the main microphones and the various parts of the orchestra. If using a Decca Tree at around 3 m (9′10″) high, the microphones are fairly low down and close to the front of the orchestra, but a relatively large distance from the woodwinds. In addition, the spaced omnidirectional microphones do not produce a clear focussed image of the more distant instruments. There will be enough contact with the upper strings to feel engaging, but the rear strings, woodwinds, brass, and percussion will lack detail and focus, thus producing an exaggerated sense of distance between the front and the back of the orchestra. If the woodwinds, brass, and percussion are not mounted on risers, this can be exacerbated. However, if the main pickup uses directional microphones that naturally capture less reverberant sound (such as an ORTF pair), it will be mounted higher up and possibly

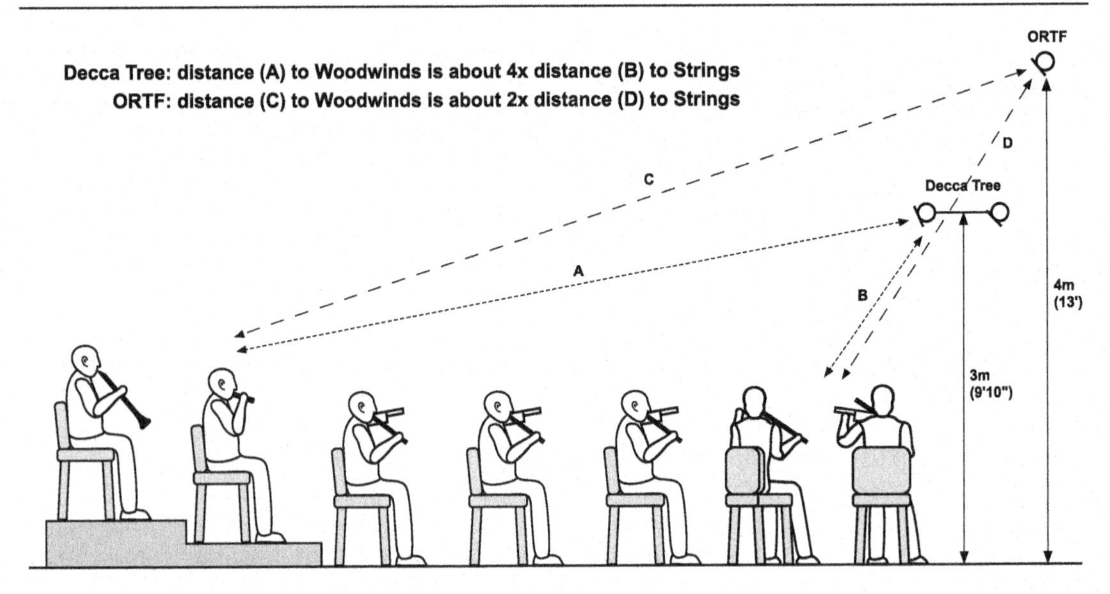
ORTF
Decca Tree: distance (A) to Woodwinds is about 4x distance (B) to Strings
ORTF: distance (C) to Woodwinds is about 2x distance (D) to Strings
D
C
Decca Tree
A
B
4m
(13')
3m
(9'10")

Figure 9.1 Illustration of relative distances of strings and woodwinds to the main pickup for a Decca Tree and an ORTF or co-incident pair

further back than a Decca Tree. The microphones can be placed at such a height and distance that the distance to the front strings and to the woodwinds are not too dissimilar (although maintaining some extra distance to the woodwinds is desirable in creating an appropriate sense of depth). Figure 9.1 illustrates this difference.

9.3 General notes on placement of ancillary microphones

To help to control the sense of perspective of the individual woodwind, percussion, and brass sources, additional 'ancillary' microphones are used. In an orchestral context, the engineer is often dealing with picking up a section rather than an individual, and the microphone placement has to reflect this. For engineers more used to working in pop music, ancillary orchestral microphones are more akin to drum overheads than to individual drum microphones.

The closer a spot microphone is placed to an orchestral section, the more the microphone will focus on the nearest individuals, resulting in uneven coverage. In the same way, the closer a microphone is placed on an individual instrument, the more localised and less natural-sounding the tone colour becomes that is picked up. Instruments do not radiate uniformly at all frequencies; their individual radiation patterns are an important consideration in placing a single microphone to obtain a good, natural-sounding tonal balance overall. This is best achieved with the ancillary mic placed further away; think in terms of 1.5–2 m (5′ to 6′7″) rather than 30–50 cm (12″ to 20″).

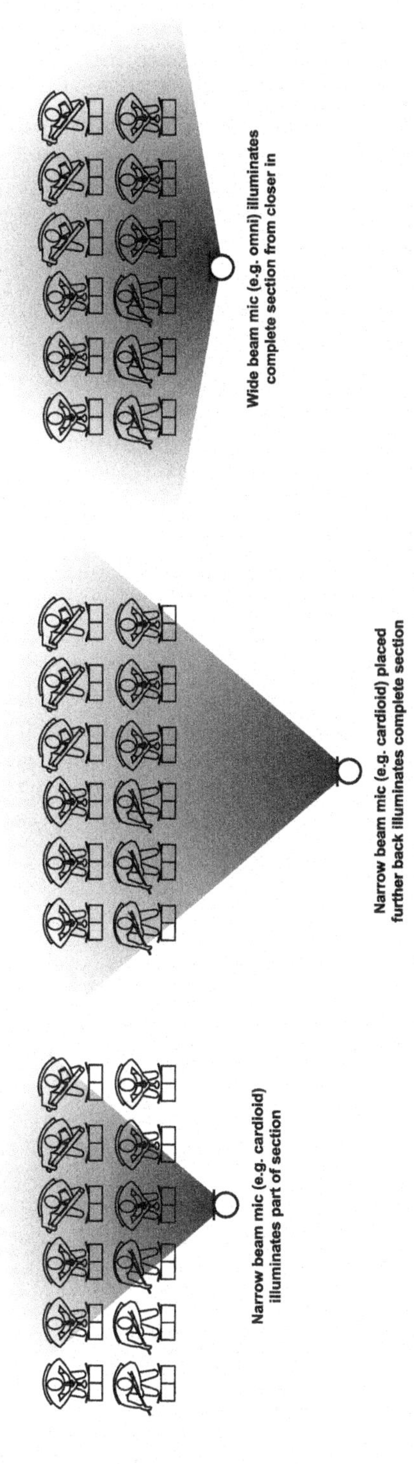

Figure 9.2 Illustration of the 'lighting analogy' for evenly covering a group of players

Closer placement often results in more intrusive key noise and other instrumental sounds, and these can make the ancillary microphone harder to blend with the overall pickup to create a realistic orchestral perspective. Very close instrumental noises present confusing cues to the listener; these sounds place them simultaneously with their ears very close to the woodwinds (for example) and at the same time, the main pickup places them further back in the room nearer the strings. If faded up too far, the close microphone signals could pull the woodwinds to a position apparently in front of the strings in the recorded image, which goes against our goal of creating an acceptable sense of perspective and feeling of a real space setting.

In a practical scenario, especially in a concert, the ideal placement of ancillary microphones may not be achievable and closer placement may have to be used. If this is the case, the task of balancing is made more difficult, and blending artificial reverb with the microphone source can be used to help with this. Choosing a microphone without a presence lift or even HF roll-off will also go some way to reducing the impact of excessive mechanical and key noise. If you have to place microphones close, then omnis or wide cardioids instead of cardioids will feel less zoomed-in.

A useful analogy when using ancillary microphones is to think of lighting areas of darkness by shining a floodlight or spotlight on the different areas of the orchestra. The lighting can be gradually increased in intensity (fade up the sound source), and it can be a narrow beam, medium, or very wide beam that takes in a larger number of surrounding players. A change of 'beam width' can be made either by choosing a different microphone pickup pattern (such as a hyper-cardioid for narrow, a cardioid for medium, or an omnidirectional for wide) or by moving the microphone closer or further away. Greater distance will incorporate more players into the 'floodlight', and a wide-beamed light (omni microphone) will need to be placed closer to the section than a medium-beamed one (cardioid) to concentrate on the same number of players. This helps us to think of the ancillary microphone as picking up and subtly highlighting a section of the orchestra rather than focussing exclusively on a single instrument. Figure 9.2 shows some examples.

9.4 Panning and levels of ancillary microphones

When deciding where to pan the ancillary microphones, the place to start is to match their panning to where their instruments appear in the image from the main pickup. This means that you can fade the instruments up and down without their images moving when you do so. However, it can be very effective to widen the panning slightly from this 'natural' position. This reinforces the sense of the section being a bit closer and also translates better into listening in a domestic setting where the loudspeakers may only be spaced 1.5 m (5′) or so apart. An orchestral section that is quite narrowly panned will feel very central under these listening conditions.

When judging an appropriate fader level, remember that the sections need to be brought closer, but not so far as to feel as close as or closer than the strings. It is usual practice to move faders while mixing according to what is needed in the musical balance and whether there is a solo line that needs subtly bringing out, and the aim is to draw the listener's ear to it without being crudely obvious in any way. In these instances, it works best if the engineer can anticipate the solos and begin the level change slowly before the important notes. There should never be any sudden increased presence of an instrument that practically shouts, 'Look at me, I've got a microphone!'

9.5 Woodwinds

This section will always be the first priority, for a number of reasons. In classical music, the woodwinds form the next most important musical dialogue in the orchestra after the strings, and this detail must not be lost. There are only a few of them, they are relatively quiet, they do not radiate as efficiently as the brass do, and they are seated behind the strings. Their individual tone colours are very distinctive, and they need the punctuation that is provided by the use of ancillary microphones to support them against the whole body of strings. The aim is to highlight the woodwind section, not to pick them out singly.

9.5.1 Woodwinds in two rows

Let's look at a normal double wind section, forming a grid of players 2 × 4. (See Figure 9.3.) The bassoons and clarinets on the back row might be on a riser of some sort. To act as 'floodlights', the microphones will need to be about 2.6–2.9 m (8′6″ to 9′6″) from the floor. These distances are based on the use of cardioids for their ability to discriminate against the percussion behind, or the violins and violas in front of the section. Omnis would have to be placed at between two-thirds and three-quarters of the height of the cardioids to produce a similar ratio of direct woodwind sound to reverb. However, microphones that are physically closer can result in more intrusive key noise.

It is important that the cardioids used have a smooth off-axis response; they will be picking up the sound of many other instruments around the woodwinds, and this spill needs to be of good quality. If the cardioids sound poor off-axis, they will colour the other instruments, and draw attention to themselves when brought into the mix. One of the most common problems when picking up woodwind sections is finding you have a nice woodwind tone sound but too much extraneous noise such as keys clicking, reeds leaking air, and keyholes leaking air and spit. When these 'close' sounds feel as if they are right in front of you, the illusion of natural perspective in the orchestra can be damaged. Avoid microphones that have any presence lift, as this will accentuate these sounds; a ribbon microphone (Royer R-121 or Coles 4038) might be a good alternative if you only have brighter sounding cardioids. When recording period instruments, you will find that they tend to be noisier than their modern counterparts. They have less sophisticated key systems and produce a lower acoustic output, which results in a higher ratio of noise to desired sound.

In order to 'light up' the section evenly using two microphones, and to avoid some players sounding closer than others, it would be best to cover the four players on the left with one microphone, and the four players on the right with another microphone (see Figure 9.3a).

Using a co-incident pair or ORTF pair to cover the section will lead to the closest players to the microphones being more in focus than the outer players, so the coverage will be uneven (see section 9.5.3). If you have enough microphones and desk channels, and you want additional control over the internal balance of the woodwind section, it is typical to use four microphones: two for the front row and two for the rear. See Figure 9.3b.

When using one microphone between four players, if the rear row is not on a riser, the microphones should be aimed at the back row; the bassoons are quieter than the oboes, and flutes radiate in a fairly omnidirectional way, so will be picked up anyway. The cardioid directivity pattern will give some level discrimination in favour of the back row, which should compensate for them being a little further from the microphones. If the rear players are raised up, then there is usually no need to discriminate in their favour in terms of the microphone placement, and the microphones can be aimed more between the players. See Figure 9.4a and 9.4b.

Panning of the wind microphones that have been outlined above would usually be around half L and half R, in line with the idea (discussed in section 9.5.3) of panning the section microphones slightly wider than the 'natural' width of the section as presented in the image from the main pickup. This enhances the clarity of individual lines, helps to reinforce the sense of the section being a little closer, and helps the recording translate well onto modestly spaced domestic loudspeaker set-ups.

For a discussion about using a coincident pair on a woodwind section, see section 9.5.3.

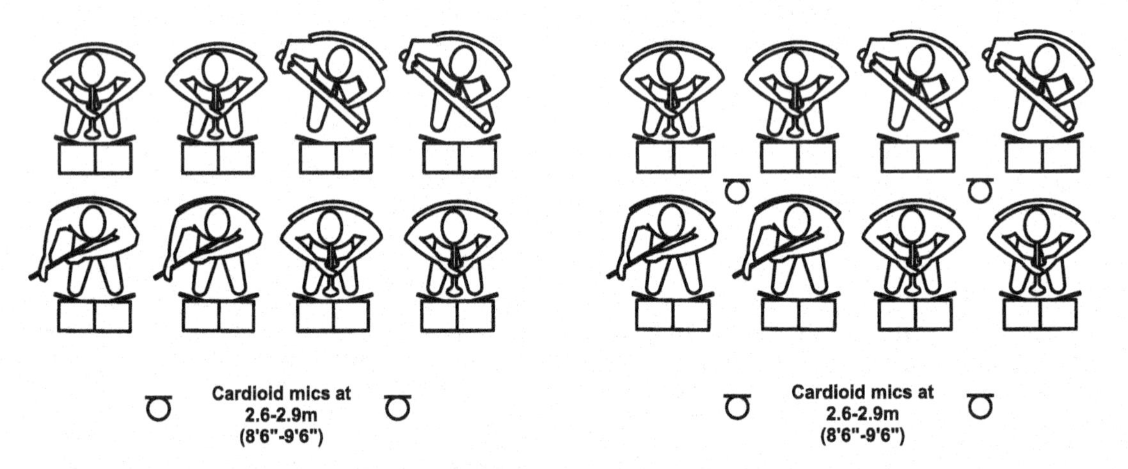

Figure 9.3a–b Woodwind sections in two rows, obtaining even coverage using two microphones and four microphones

Figure 9.3c Photo from Abbey Road showing woodwind section microphones for 2 × 4 grid

Photo: Carlos Lellis, Programme Director, Abbey Road Institute.

9.5.2 Wide woodwinds layout

The layout of the woodwind section is not always ideal as in the neat 2 × 4 grid shown in section 9.5.1. Where space is at a premium, they might be seated in a single row (1 × 8 players), or the section might be a large one and be laid out in a 2 × 6 grid. The key to picking up this wide layout is to go back to the lighting analogy and try to get even coverage of all the players. This is likely to be achievable with three microphones at a height of 2.6–2.9 m (8′6″ to 9′6″) (or more than three if the microphones have to be lower down; see section 9.5.3). The microphones would need to be panned approximately half L, Centre and half R depending on your desired final image position and how well this fits in with the image on the main pair. The left and right microphones might need to be panned a little more widely than this. See Figure 9.3d.

Figure 9.3d Photo from Abbey Road showing woodwind section microphones for wider section

Photo: Carlos Lellis, Programme Director, Abbey Road Institute.

9.5.3 Restrictions placed on woodwind microphone visibility or height

In a concert situation, you might find that there are restrictions on microphone placement for two main reasons:

- Restricted physical space that makes it difficult to fit stands in, such as in an orchestra pit, or in an adapted venue such as a church, where the orchestra layout has to be altered to fit the surroundings and can therefore be quite constrained.

- Objection to visibility of higher stands from the venue or other party, either as a matter of aesthetics or because the concert is being filmed and broadcast, such as the Proms series from the Royal Albert Hall.

To reduce the number of stands on a 2 × 4 woodwind section, an alternative to using single microphones on separate stands would be to use a stereo pair of microphones on one stand (e.g. ORTF or spaced omnis). However, if there is limited room to place the pair at an appropriate distance back from the woodwinds (because the rear strings are in the way), this can cause some problems in both image and perspective of the woodwind section. If a closely placed pair is panned fully left and right, the resulting woodwind section image will be too wide, so you will have to under-pan the pair or angle them inwards a little. The more intractable problem is that the central players will be physically closer to the pair than the outer players, and this will result in uneven coverage and hence uneven perspective in the final image. See Figure 9.5.

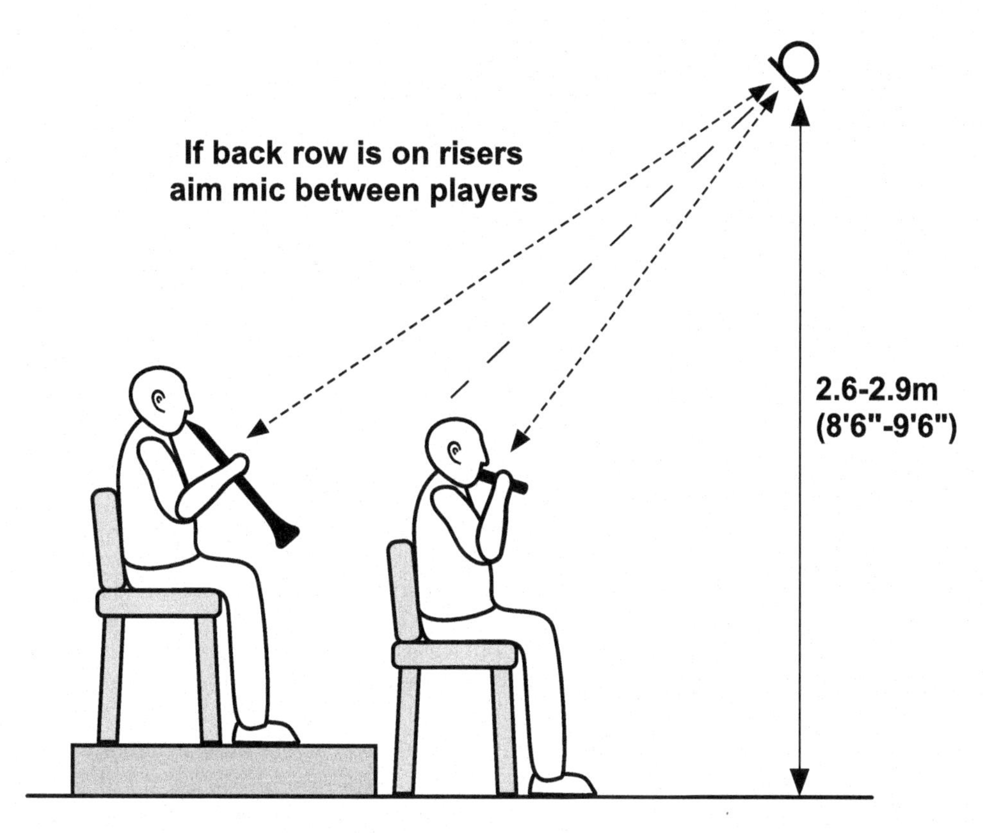

Figure 9.4a Sideways view of woodwind microphones with the rear row on a riser

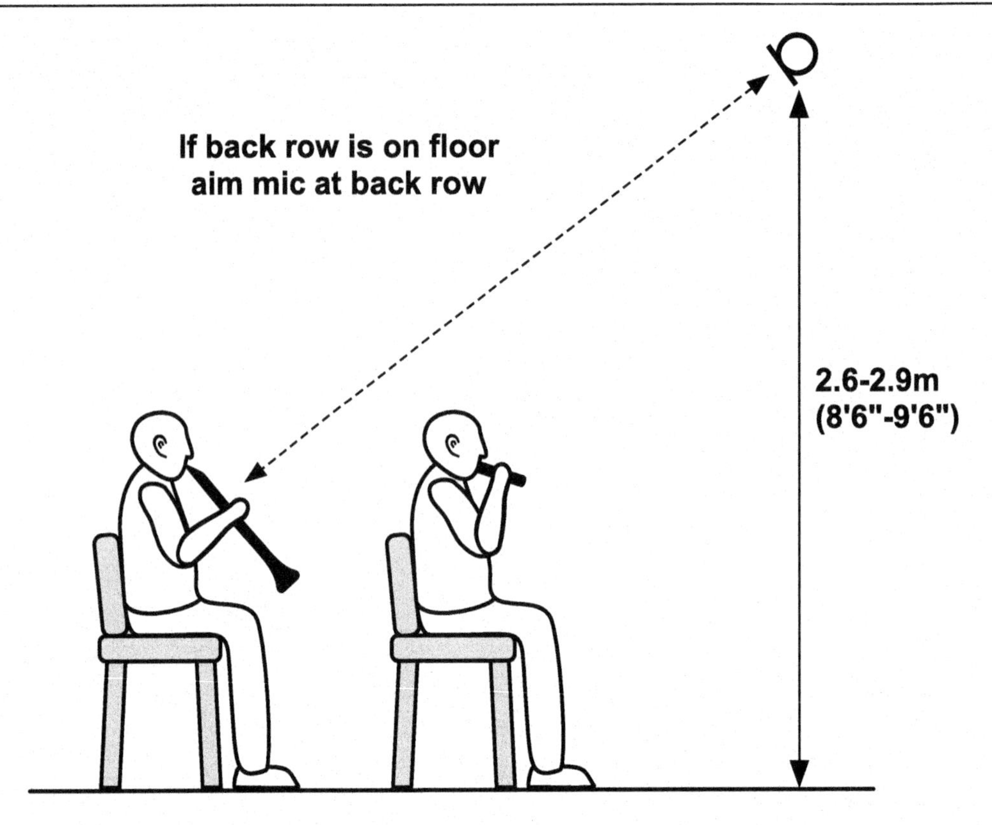

Figure 9.4b Sideways view of woodwind microphones without the rear row on a riser

To reduce the obtrusiveness of stands and microphones, you might be required to use them at a much lower height than the ideal. Remembering that the lower the microphones are placed, the fewer instruments they will cover evenly in terms of level and perspective. Therefore, maintaining consistency of coverage between players will require more microphones – commonly one microphone every two players. The more intrusive key noise that results can be ameliorated by choosing a microphone without any sort of HF lift or even with some HF roll-off such as a ribbon microphone. Care would need to be taken when blending these closer microphones with the overall stereo pickup so as not to bring the woodwinds too close to the listener.

When recording for a film score, a high degree of control over the orchestral balance is required in order that the mix can be tailored to fit in with the director's requirements at any stage of the proceedings. Therefore, instruments are treated more as individuals rather than sections.

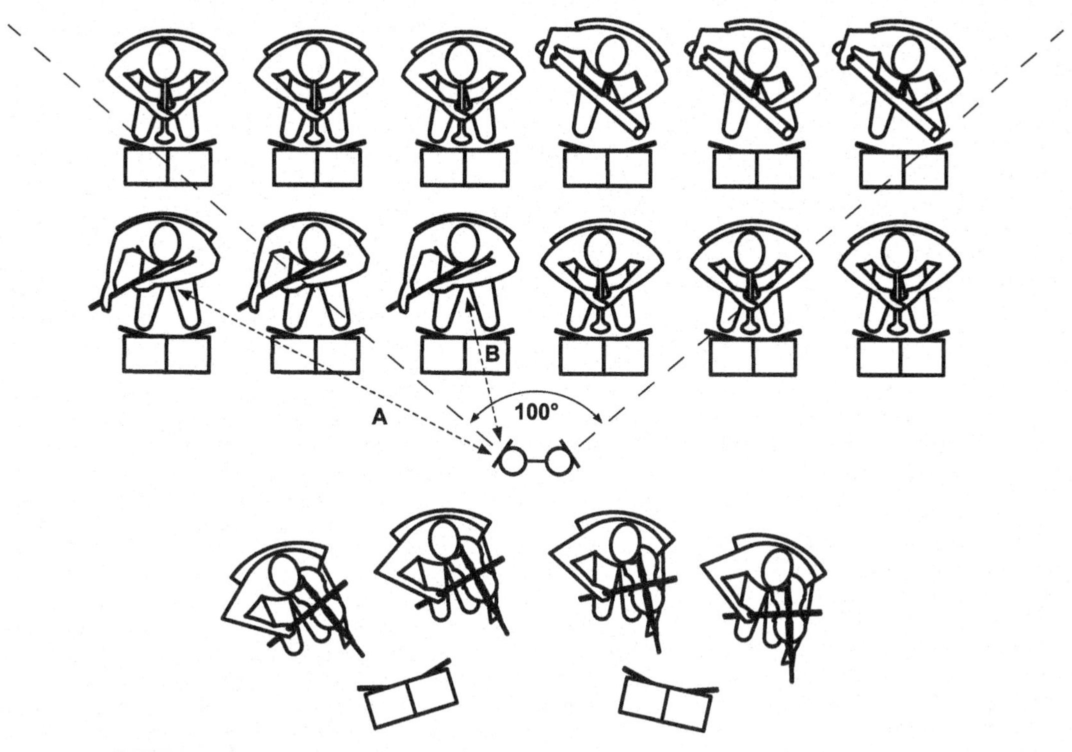
B
A
100°
ORTF Pair close to woodwind:
- very wide image (L to R stereo recording angle is 100°)
- large differences in distance between mic and players (e.g. A much larger than B)
- can't move pair further away because of the strings in front

Figure 9.5 An ORTF pair close to a woodwind section

9.6 Brass

Brass instruments radiate primarily from the bell, with increasingly narrow beam radiation at higher frequencies.[1] This means that the forward-facing brass are extremely efficient at making their presence felt, and they can carry right across the orchestra without feeling too distant on a recording. Add in the fact that they are probably placed on risers, and this is a section that you can often capture sufficiently well without the addition of ancillary microphones, especially if you are short of microphones, stands, or desk channels. The French horns are slightly different in that their bells are rear facing, and so they would be prioritised over the trumpets and trombones if you are short of resources. Musically, the horns are often scored along with the woodwinds, but they are capable of producing much greater SPLs.

Even if not always required, there can be some benefit in rigging some ancillary brass microphones, even if just to reduce the players' urge to regard the relatively distant main orchestral pickup as a target to be aimed at. The presence of brass section microphones nearer to the players can reduce any tendency to 'play out' in an effort to be heard. It is not uncommon to put out microphones for the brass section and end up not using them in the final mix. Use of the brass microphones will primarily be for the purposes of helping perspective rather than level. They are also useful for picking out solo lines, such as in the trombone sections in Rossini overtures.

9.6.1 Horns

The horns are usually situated to the left of the woodwinds, and in a concert hall they are very often in front of a reflective wall that will project the sound out into the audience. This is often unhelpful in a recording context, as the ancillary microphones will be close enough to pick up any colouration that comes from the first reflections off this boundary as well as the direct sound from the players. Using something absorptive behind the players, such as an absorbent screen at bell height, can reduce this effect for the purposes of recording.

Horns can be recorded from the bell side or from the front. Microphones in front of the player will produce a less edgy sound, as the radiation comes primarily from the bell and is heard a great deal by reflection. The higher frequencies that give the playing attack and raspiness are concentrated in a narrow beam directly from the centre of the bell. When trying to record horns from the front, you are more likely to run into space restrictions imposed by the string section, and a tightly packed orchestra will rule out this option.

When recording from behind, the tone colour will depend on the microphone placement and can be tailored to suit the repertoire. A microphone placed straight on to the bell will produce a lot of high frequency content, and a more edgy tone will result that might suit Shostakovich or Mahler, where the horns are more strident than mellow. For a warmer sound, it is usual to arrange the microphones to be off-axis from the centre of the bell to one degree or another, in order to reduce the HF content. Microphone choice would also be made with lower mid-range warmth in mind, and a warmer cardioid such as TLM170 or a ribbon microphone (Royer R-121 or Coles 4038 – provided it is not close enough to experience blasts of air) would be ideal. Using a cardioid or wide cardioid will reduce pickup of reflections from any hard surfaces behind the players, but a fig of 8 microphone would need to be orientated carefully to discriminate against reflections. Avoiding percussion spill onto the horn microphones is another factor in choosing directivity pattern. Where possible in a recording session situation, keep the horns away from the walls, or place an absorbent screen behind them.

With the players on a riser of 30–60 cm (1′ to 2′) high, a good place to start would be with the microphone(s) at around 2.1 m (7′) high, and looking down from a distance of about 1.5 m (5′) away. If it is not possible to get this distance back from the horns, some additional height can be added instead. Attenuation in the head amp of the microphones might be needed if you are a lot closer, such as around about 30 cm (1′) away.

Where the section consists of two to four horns, one microphone would be enough to just add some focus to the sound produced by the main array. Where there is a wider section of eight horns, two microphones would need to be used, each covering four players. Panning of a single microphone would usually be around two-thirds left, just to the left of the outer woodwind players. Where two microphones are used, panning will be about three-quarters L and half L.

9.6.2 Trumpets, trombones, and tuba

The remaining brass are usually placed over to the right-hand side, behind the violas, in a mirror image position to the horns which are on the left. For the purposes of recording and for the sake of the other musicians' hearing, it is preferable if they are not seated directly behind the wind players. If they are, then the woodwind microphones will pick up as much brass as woodwinds when placed in their ideal position, and their placement will have to be changed by moving them lower and closer to the winds to try and ameliorate this. This is perhaps a good moment to note that disturbing an orchestra's normal seating plans too much will not be helpful for their performance, so compromise in microphone placement might be needed. If there is space, having the trumpets (typically two or three players), trombones (typically two players), and the tuba sitting in a line allows the engineer to pick up the trumpets and trombones separately. A microphone needs to be used to capture the tuba transients really only if it has a solo part (common with composers such as Shostakovich and Prokofiev) (see Figure 9.7a and 9.7b). If there is not enough space for this single-line arrangement, the trumpets will usually be in front on the floor level, with the trombones and tuba up on a riser behind, with the tuba at the outer edge (right-hand side). See Figure 9.8a and 9.8b and Figure 9.9a and 9.9b. It is also the best case scenario if there can be plenty of space between the back of the strings and the brass in order to be able to set up the microphones at an appropriate distance away from the brass. If the players are on a riser, the microphones will be around 2.3 m (7′6″) high from the floor, and 1.2–1.5 m (4′ to 5′) back (which needs space behind the strings). These microphones will be looking downwards towards the instruments and off-axis from the bell. Where there are two rows of players, aim the microphones at the back row in the same way as you would when trying to cover a double-row woodwind section. Avoid being on-axis to a brass instrument's bell for several reasons:

- Avoidance of picking up the highest overtones, so producing a warmer tone.
- Avoidance of any blasts of air (important if you are using a ribbon microphone).
- Reducing the big changes in tone colour that result from the player's movement if the microphone is on-axis. The higher frequencies radiate in a narrow beam, and a small player movement can be very noticeable if the microphone is on-axis to the bell in the first place.

Again, directional microphones are preferred in order to reduce spill from strings and percussion (although the brass are so loud that any spill from surrounding strings is usually of little real

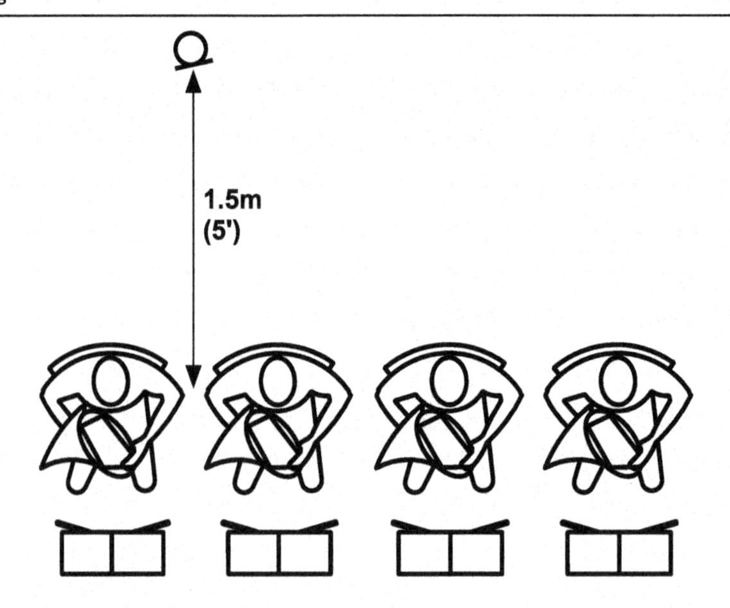

Figure 9.6a Microphone placement on four horns – plan view

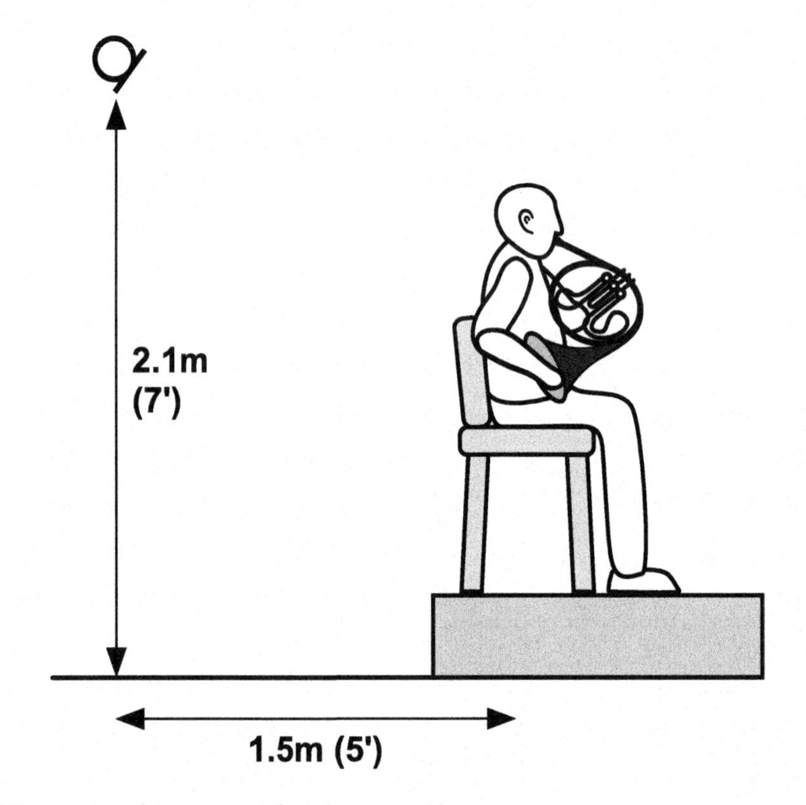

Figure 9.6b Microphone placement on four horns – side view

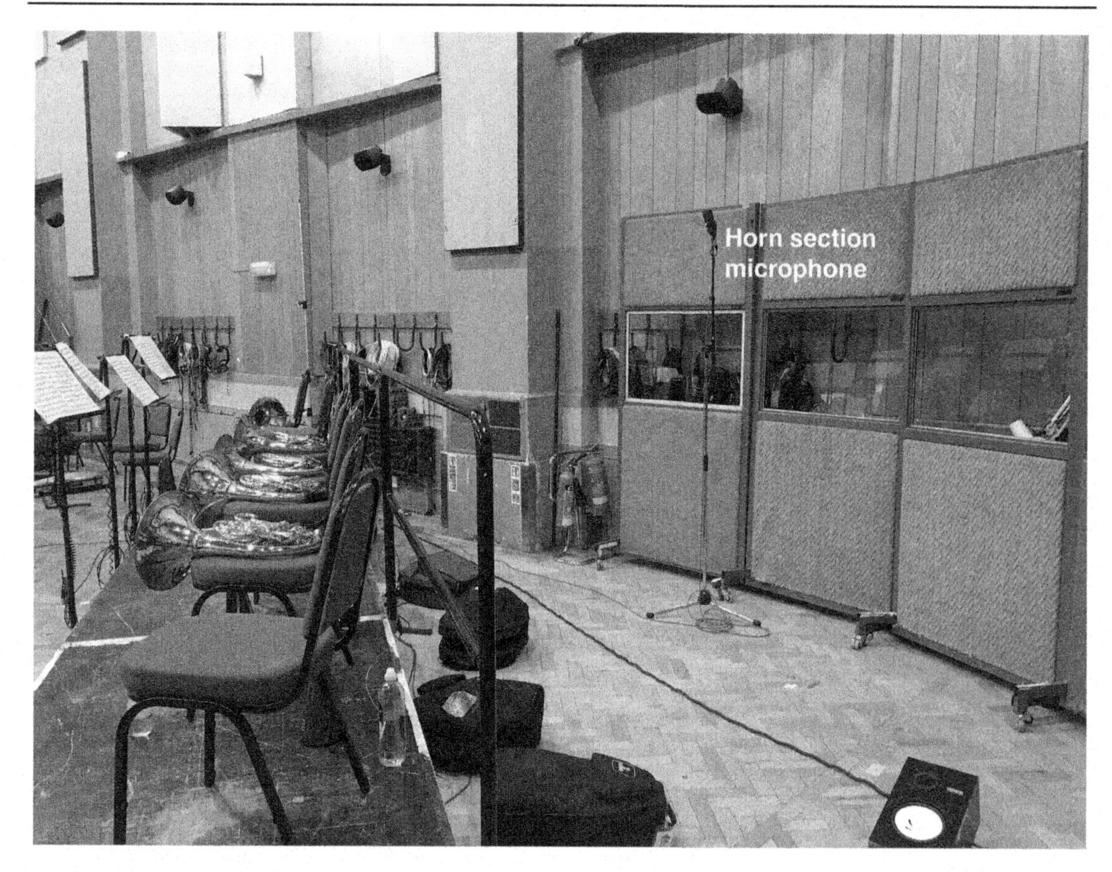

Figure 9.6c Photograph showing single microphone on four horns

Photo: Carlos Lellis, Programme Director, Abbey Road Institute.

significance). Ideal microphones for brass would have some natural warmth to their characteristics, such as Neumann M49 (if you can have a valve microphone), Neumann TLM170, or ribbon microphones (Royer R-121 or Coles 4038), which do not have a very extended HF response. On an important cautionary note, the fragility of ribbon microphones is often misunderstood. The ribbon microphone will withstand high SPLs but will be damaged by blasts of air, so should not be used on-axis to the bell, especially of trombone and tuba. If the tuba has a solo, it will need a spot microphone of its own, and this is best rigged from 60–90 cm (2′ to 3′) above the instrument and just off-axis from the bell.

Where you have two microphones covering a brass section, they would be panned so as to be in a mirror image to the horns in the image (assuming that this is where they were sitting for the recording). This means them being panned about 50% R and 75% R.

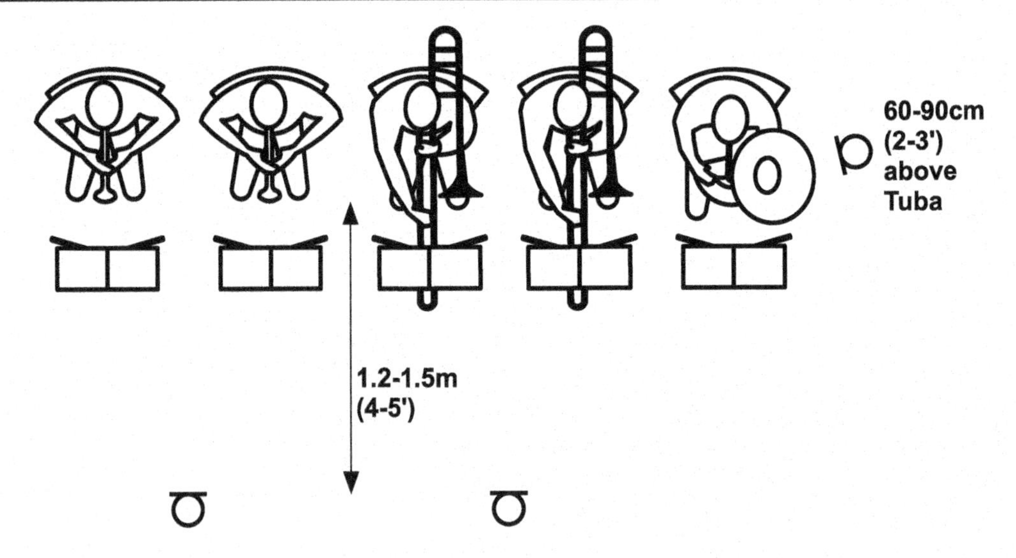

Figure 9.7a Brass in a single line on a riser – plan view

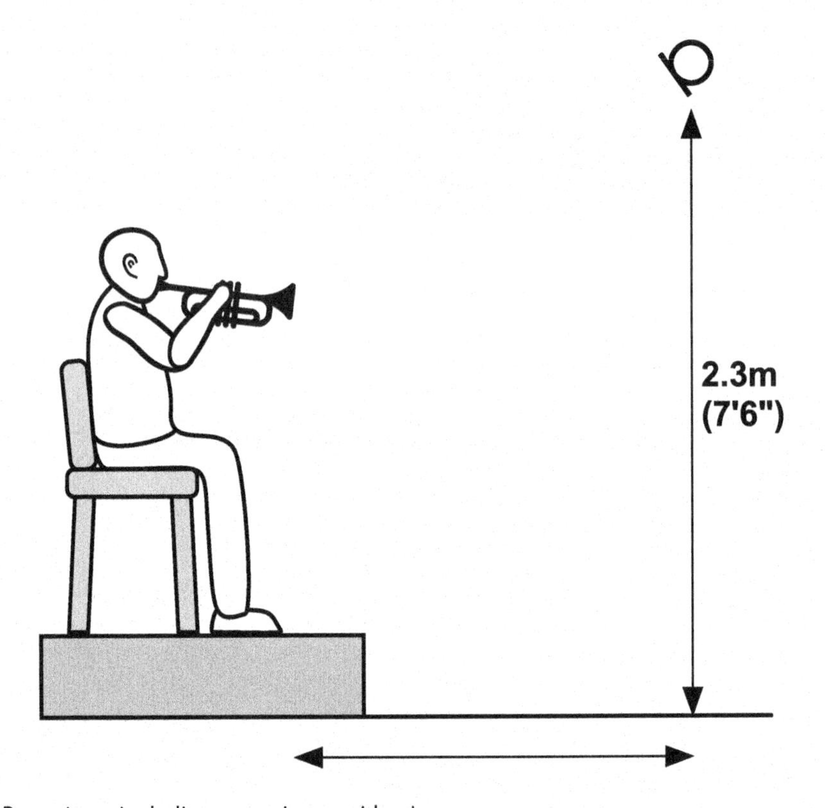

Figure 9.7b Brass in a single line on a riser – side view

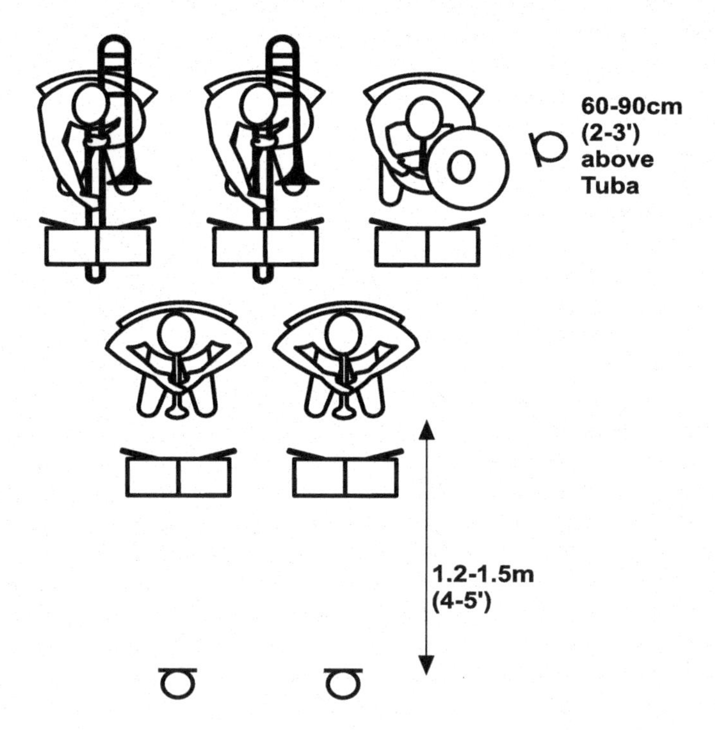

Figure 9.8a Brass laid out in two rows with rear row on a riser – plan view

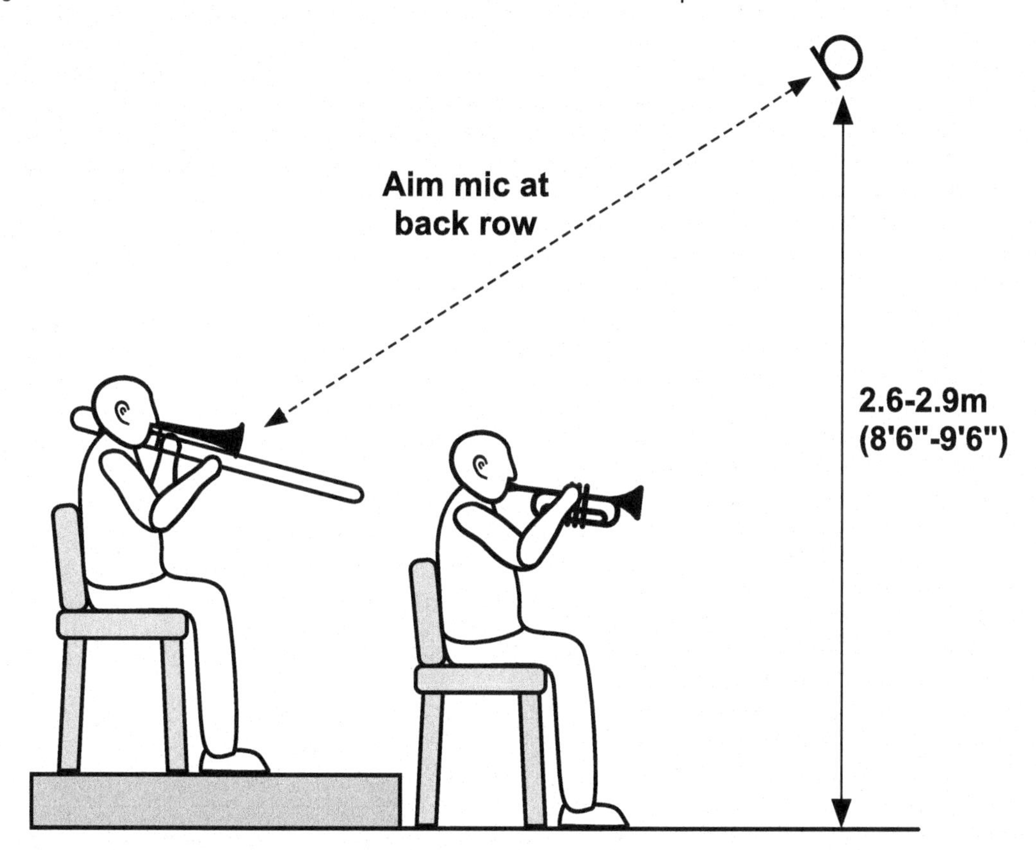

Figure 9.8b Brass laid out in two rows with rear row on a riser – side view

Figure 9.9a Brass section microphones

Photo: Carlos Lellis, Programme Director, Abbey Road Institute.

9.7 Percussion

This is a hugely varied section of the orchestra, and like the brass, there can be a tendency for the players to over-project because of the perceived distance from the main overall pickup. They are usually placed right at the back of the orchestra, and if it is a large section, they might take up a great deal of width. The sound of distant percussion on a main Decca Tree will have a rather unfocussed and distant image, and this needs to be addressed. Detail, clarity, and cleanliness of transients is important to percussion parts, and so ancillary microphones are used to add some subtle highlighting to the sound that is picked up by the main array, thus bringing the instruments closer in perspective.

Figure 9.9b Tuba microphone

Photo: Carlos Lellis, Programme Director, Abbey Road Institute.

Percussion players usually arrive early to set up, and it is a good idea to be friendly to them so that you can later ask them to adjust things to help you. At the setting up stage, see if you can get the bass drum in the centre where it will cause the fewest recording problems (see section 9.7.2). If a percussion part is too loud in the room and is all over every other microphone, it cannot be reduced in level easily without the means to screen off the percussion. The only solution is to talk to the players and ask them to adjust their playing to be quieter. If they are given spot microphones, they are more likely to be persuaded to play less loudly (and possibly therefore more musically) than if they feel that they have to try to play out in order to be heard by the main pickup.

We can usefully split this section into big drums and all the other percussion.

9.7.1 Marimbas, xylophones, and small percussion

While the other ancillary microphones that we have discussed so far have been aimed at covering a section of instruments and getting a good tone and even coverage of all of them, the approach to the percussion section needs to be a bit different. With woodwind ancillary microphones for example, we wanted some distance to avoid transients from key noise and to obtain a good overall tone, but with percussion it is a clean attack that we are trying to collect. It is usual to mic them fairly close (60–90 cm (2′ to 3′) away from the instrument) and only use a small amount of the signal to clean up the attacks of the notes. (See Figure 9.10.) In getting microphones in closer, spill from other percussion instruments can be greatly reduced. It might be thought that having all these microphones together on the percussion section would result in some serious comb filtering problems. However, because the spill is kept low compared with the SPL of the wanted signal, and the signals are only mixed in at a low level, few problems occur in practice.

Figure 9.10 Individual percussion microphones

Photo: Carlos Lellis, Programme Director, Abbey Road Institute.

The signal from these microphones is usually mixed in at 15–20 dB below the level of the main pickups, as the transient of the percussion can be perceived by the listener at fairly low levels. When mixing in these microphones, you should aim to keep them at very low level when they are not in use. This is particularly difficult if you are mixing live to stereo, and you will be best served by using a score (or a producer with a score to cue you in) to fade up percussion microphones as they are needed. If you leave them at a higher level all the time, you will be adding percussionists chatting and moving things in between moments of performance, as well as adding many sources of orchestral spill and muddying the orchestral image.

The ancillary percussion mics can usually be panned wider than they are actually situated in the image from the main pickup; in fact, they often work best panned quite a lot wider (but avoiding hard left and right panning.) This exaggerated panning will help clarity of single parts and reinforce the subtle foreshortening of the front to back distance, thus improving the translation onto small-scale domestic monitoring where the loudspeakers are only about 1–1.5 m (3′4″ to 5′) apart. As long as this panning is not overdone, there should be no conflict with the image from the main pair, as the ear will tend to fix the image of the whole instrument at the location of the transient. Wider percussion instruments such as marimba, xylophone, and glockenspiel can be picked up with a single microphone in an orchestral context; when they are heard from the audience, they are not perceived as very wide because they are placed at the back of the orchestra, and the aim is to only to capture the transients. For solo percussion work or a concerto, these instruments would need a stereo approach.

Remembering that *closer microphones usually means more microphones*, recording a piece with a lot of percussion means that you will need a lot of microphones to adequately cover the section. Mark the microphone stands so that you know which one is which if you find you have to move the microphones to follow the instruments. This is quite likely to happen as percussionists work together sharing parts, swapping instruments and moving them around in order to facilitate their playing of the piece, and you will need to keep an eye on the players in case things get rearranged during a session break.

9.7.2 Timpani and bass drums

Timpani are found even in baroque and early classical scores, and so a pair of them forms part of a very basic orchestral set-up. Where there are two timpani, a single microphone between them at about 1 m (3′4″) above the skin height should collect enough transient to give definition to the wallowing low frequencies that will be picked up on the main orchestral microphones. Where there are four timpani, one microphone between two drums will be needed (see Figure 9.11a). For recording session work, screens can be placed behind the timpani to prevent the generation of LF and MF reflections from nearby walls and corners (see photo in Figure 9.11b). These problems might occur if the orchestra is very large, pushing the outer edges (including the timpani and horns in particular) towards the walls of the room. Timpani used in authentic early music performances have more attack and a less refined sound than modern pedal timpani.

The large orchestral bass drum is a feature of the orchestra from the 19th-century repertoire to the present day. A big bass drum hit can take up a lot of headroom in a recording, and so if it is possible to get the bass drum placed centrally it will split this energy between the left and right channels of the recording. Placing it centrally will also avoid it having to fight with the brass

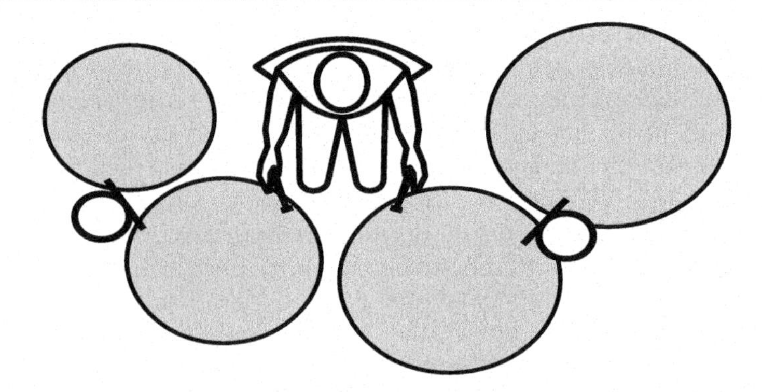

Figure 9.11a Microphone placement on timpani

Figure 9.11b Microphone placement and screens around the timpani from Abbey Road

Photo: Carlos Lellis, Programme Director, Abbey Road Institute.

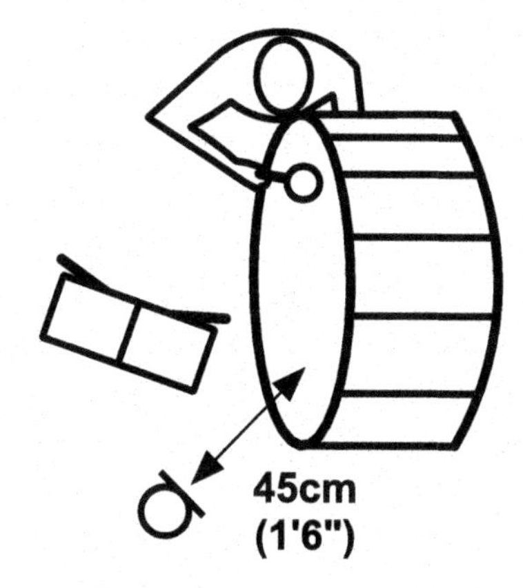

Figure 9.11c Microphone placement on bass drum

section. The microphone is best placed just above the instrument, looking diagonally downwards towards the skin from a distance of about 45 cm (1′6″) away, the aim being to pick up the beater transient without being too close to it. (See Figure 9.11c.) If the drum is a really heavy one, you might need to use the pad (attenuator) in the microphone at this sort of distance. If omnidirectional microphones are used for the main array, the LF component of the sound will be picked up by these, and the bass drum microphone only needs to be able to collect the transient. Therefore a cardioid would be adequate from the point of view of frequency range, although an omni could be used. This will have a greater disruptive effect on the overall image when it is faded up, and so care should be taken to keep it at a very low level until it is needed. Whatever you do, don't use a ribbon microphone for a bass drum; each drum hit causes a great deal of moving air, and this could severely distort or tear the ribbon.

If you are not able to add a microphone to the bass drum, you could consider turning the drum sideways so that it is projecting outwards towards the conductor. This can improve clarity, although there are potential downsides – if it is a big instrument, the pressure wave can be huge and overwhelm the sound, and the player will have to stand sideways. Better results will be obtained with the drum in the traditional sideways position where possible.

9.8 Double bass section

It is very common to add an ancillary microphone to the double bass section, even when the other string sections are not treated in this way. The bass line carries an important role in music of all kinds and underpins the harmonies and sense of key, and it is important that it has similar clarity and perspective as the violins. The smaller stringed instruments are quite directional in their radiation, especially of the higher frequencies that come from transients, overtones, and bowing noise. These frequencies are radiated in a narrow arc perpendicular to the instrument's front face, and when being played, the instruments are held so they are projecting these frequencies up and towards the main pickup. Double basses radiate much more omnidirectionally across their playing range, and although their bowing noise and higher overtones radiate more directionally from the front surface, these components of the sound are not directed towards the main array. Therefore, some of the immediacy of their playing and LF power can be lost, leaving the violins feeling closer and more immediate. In this way, the string section that underpins the orchestra has the potential to feel unbalanced on a recording.

A double bass section microphone or a microphone on the principal player is a good way of adding both some bowing presence and some additional lower octave (40–80 Hz) LF back into the overall mix. Of these two options, the section microphone would be the preferred choice to avoid focussing in on a single player, although both can be used for a large set-up. For a sectional microphone, a single microphone can cover four to six players if they are arranged in two rows: four in front and two behind. The microphone would be set in front of the players, at about of 60–90 cm (2′ to 3′) back, looking downwards from a height of about 2.1–2.3 m (7′ to 7′6″) and aimed at the middle of the section (see Figure 9.12). It will also work well if the microphones are looking into the section from the front of the orchestra (this might be the case if you are restricted in microphone placement choices, as was sometimes the case in St Eustache church in Montreal, the recording home of the Orchestre Symphonique de Montréal).

The microphone choice should be an omni (to provide the extended LF response needed), and some added HF lift will assist with the feeling of clarity and connection to the players as it will emphasise the bowing noise. Omnis with this sort of HF lift are those equalised for diffuse-field response; these include Schoeps MK2S, Neumann KM83, and Neumann M50.

A bass section microphone would usually be panned hard right in order to give the ear a cue for the basses being over on one side, but it should not be overused. It can be helpful to think of the bass section microphone as being part of the overall main pickup when you are balancing the core sound of the orchestral string section, and it might typically be used 2–3 dB below the level of the main pickup microphones. It is easy to overdo the amount of this microphone in the mix, as both the additional LF and the closer bowing noise are very viscerally appealing. Care must be taken not to overbalance when compared with the violins; the string section must still work as a whole.

A principal bass player's microphone would be placed lower than a whole section microphone, at about 1.2–1.5 m (4′ to 5′) high (see Figure 9.13). A principal microphone might be used in addition to a section microphone when there a solo in the piece or to replace the section microphone when it is too visibly intrusive in a concert, or when there is not enough space for the stand. When only one principal player is to have a microphone, care needs to be taken that the single player is not obviously pinpointed, and a pair of microphones could be used instead to give the player some width when panned apart (along the lines of the stereo spot microphone methods outlined in Chapter 4). Alternatively, adding an individual microphone to both principal players will provide wider coverage. Warm-sounding cardioids with good LF and some presence would be preferred if the microphone has to be placed closer in amongst the section in this way.

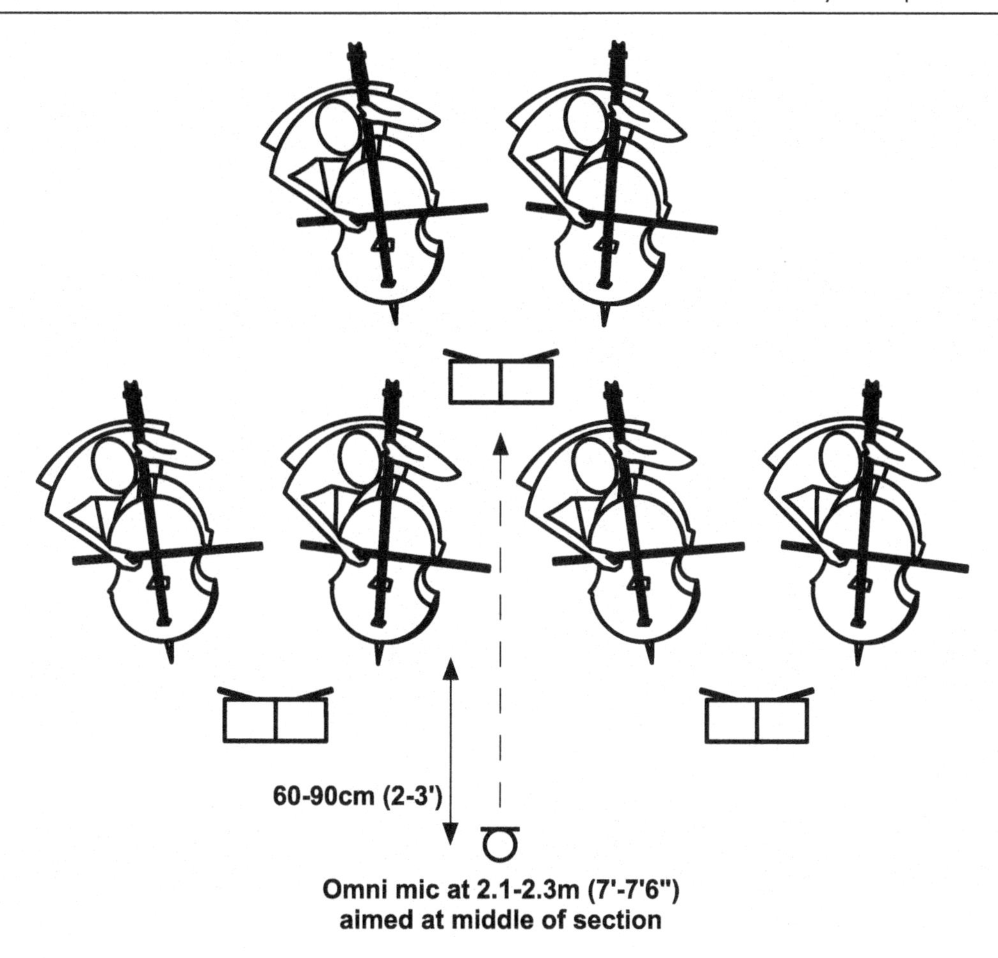

Figure 9.12 Placement of a bass section microphone

Neumann U47s or U67s would work well, but these are both valve (tube) microphones; transistor alternatives with slightly different characteristics would include the Neumann U87 or the AKG C414. The photo in Figure 9.14 shows some double bass section and principal microphone set-ups from Abbey Road.

9.9 Other string sectional microphones

Where a string section is large, additional string section microphones might be useful for capturing more detail from the back desks of the strings, particularly second violins and violas as they are not situated close to either the main array or the outriggers. They can also be used to help

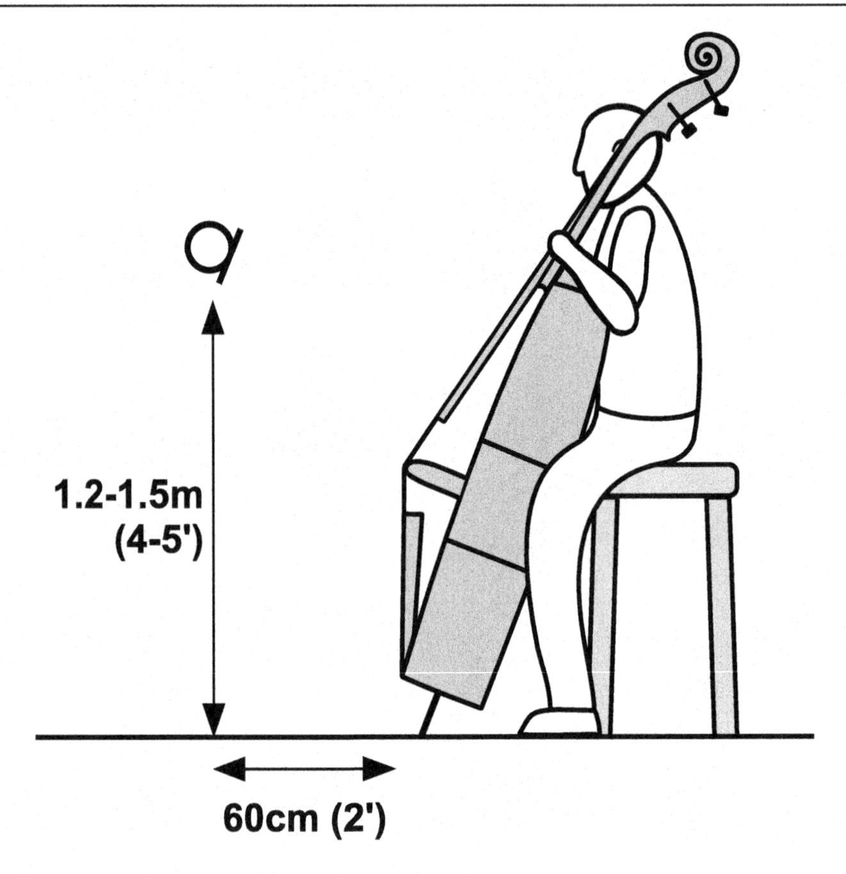

Figure 9.13 Placement of a principal bass player microphone

with the orchestral balance when the balance in the room is uneven between string parts. This can arise because of an inexperienced conductor or an imbalance in numbers and strength of parts in an amateur orchestra. It is advisable for the engineer to listen to the orchestra live and see how well balanced the string section is; a conductor such as Sir Georg Solti would always arrange an opportunity for the assistant to conduct while he stood at the back of the hall to assess the balance between parts. From where the conductor is standing, it can be hard to hear any imbalance when surrounded by the front string desks. Section principals in a professional orchestra might also be amenable to the engineer asking for a little more or less at certain points, although this approach should only be taken if you are very certain of the politics involved, and it goes without saying that you should never be high-handed. If you are unsure whether the orchestra is going to be balanced in the room, and you have the space, time, and permission to rig string section microphones, then they will give you more options if you need them later. There is no need to use them if they are not adding anything useful; you should never feel as if you have to use a microphone just because you have rigged it.

Figure 9.14 Placement of bass section and principal microphones from Abbey Road

Photo: Carlos Lellis, Programme Director, Abbey Road Institute.

Typically, the string section microphone will be a good cardioid, and would be placed at a maximum height of about 2.75 m (9′), where the tree is about 3.1 m (10′3″). (See Figure 9.15.) If the section microphones are higher than this, they will pick up too much brass; if they get much lower, more of them will be needed to cover the section evenly. If the string section microphones are overused, the natural sense of orchestral depth can be flattened out by pulling all parts of the strings to the front, so care should be taken to avoid this. Where section microphones are used on film recording or in a live orchestral theatre pit, they will be used at a lower height of around 150–180 cm (5′ to 6′).

For the necessary tight control over balance on a film scoring session, it is usual to use one microphone per desk of string players as well as section microphones. The film director can request changes in balance or even instrumentation at any time after the recording has been made, so close control is maintained over parts and spill during the recording to allow for significant remixes later on. The film approach to orchestral sound is different to classical recording

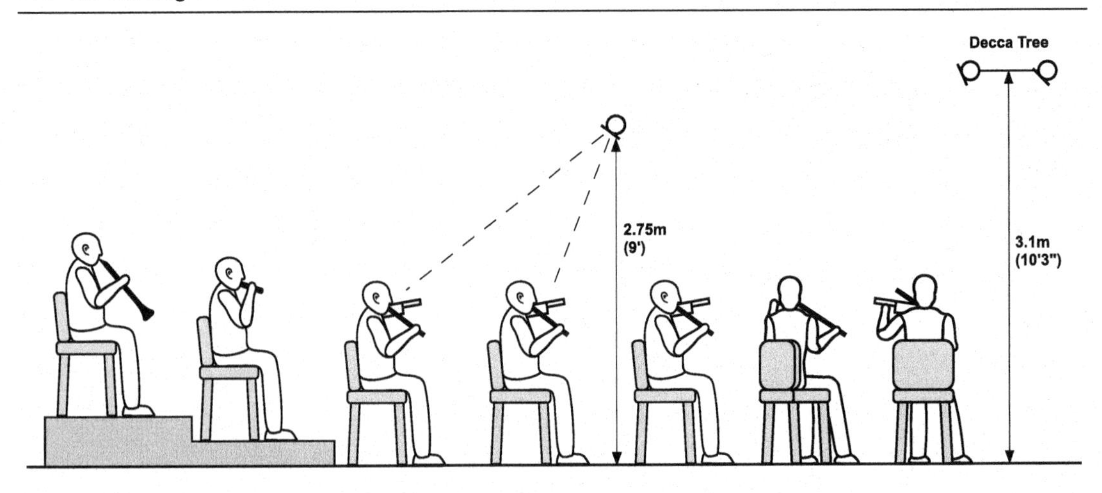

Figure 9.15 Placement of string section microphones

in that the orchestra is intended to be more immediate and enveloping, with less sense of being set in a real space with natural sounding perspective. This level of control with close microphone placement is not necessary on a well-balanced orchestra for a normal classical recording session or concert.

9.9.1 Leader: solo violin

Where the score calls for a solo line that is part of the orchestral writing, and not a concerto, the leader will play this. He or she sits on the outer edge of the front desk of the first violins very close to the tree microphones, and the simplest solution to the question of making the solo just a little more prominent is to ask the player to stand up for this part of the recording. This would bring the player just that little bit closer to the main array than the other violins, and give the player an appropriate amount of emphasis. However, it is now also common practice to put out an additional microphone if there is to be a principal solo. This would be placed in the same way as for a solo violin (see Chapter 4) but at a lower height of about 1.8–2.1 m (6′ to 7′) because the player is seated. (See Figure 9.16.) It would usually be panned about half L, exaggerating the actual placement a little in lateral terms, but this places the player definitely within the first violins in the recorded image, and will give it more impact in a domestic loudspeaker set-up. It is important not to leave this microphone open when not in use, as it will have an effect on the overall string section tone and image if left faded up.

9.10 Harp

The harp is another instrument that can get lost in an orchestral texture, and where it has significant solo lines to contribute it will ideally have a spot microphone assigned to it. This microphone needs to pick up the attacks of the notes in order to bring a little more presence to the

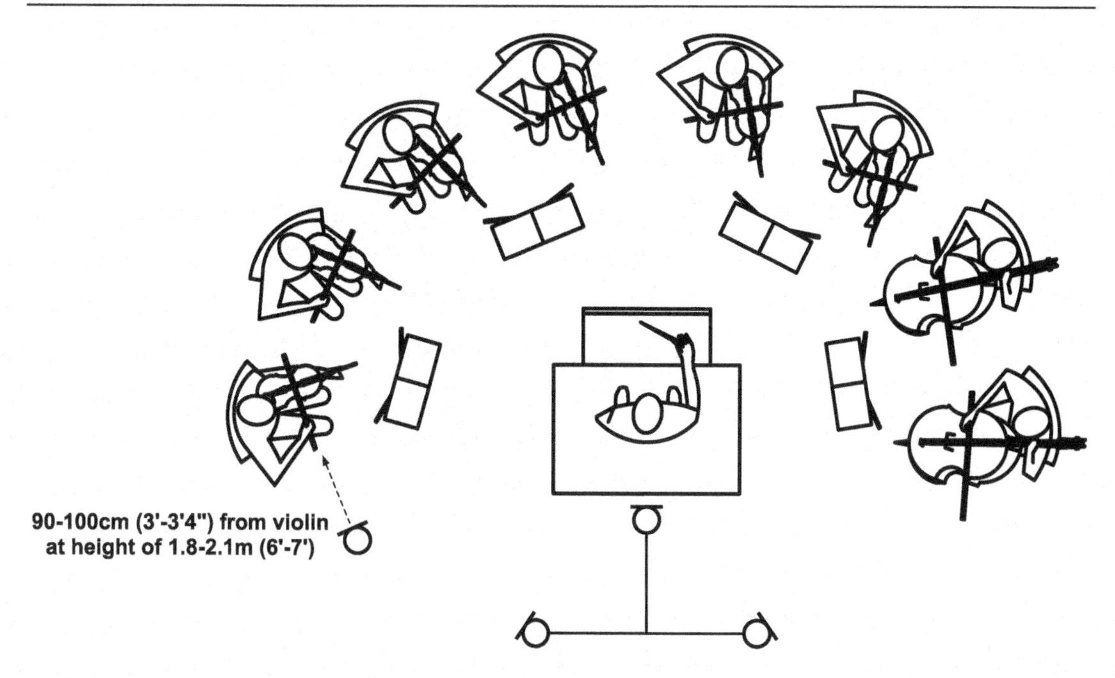

Figure 9.16 Placement of microphone for the leader of first violins for solos

sound of the instrument as it appears on the main pickup. The harp in concert position is usually situated somewhere at the back of the upper strings, towards the left-hand side, and between the horns and the woodwinds. The majority of the sound comes from the soundboard, which resonates and acoustically amplifies the much weaker sound of the strings. In addition, there are HF plucking sounds from both hands which are reflected off the body of the soundboard.

The player sits with the harp on their right-hand side to enable them to reach out for the left-hand lower notes. See Figure 9.17 for a view of the harp from above, showing two alternative microphone positions. The microphone position shown top left produces less pedal noise and HF reflections, and it can work very well to discriminate against the horns if the player is seated in a traditional orchestral position while recording. The microphone is placed on the other side of the instrument from the player, pointing forwards and down at the soundboard from a height of about 1.5–1.8 m (5′ to 6′). It is also very common while recording to move the harp from within the orchestra and place it on the other side of the main pickup, out in front of the violins between the main tree and outriggers. Its image appears partly left on the main pickup (where we would expect to hear it) and makes it easier to add an ancillary microphone without picking up horns and woodwinds as well.

When recording a live concert, microphone placement will often have to be more of a compromise as space can be more restricted and the orchestral layout altered. The microphone may have to be placed lower and further round towards the front of the instrument if it is to be visually discreet. One alternative microphone position is shown at the bottom left of Figure 9.17. A single microphone placed looking towards the player from the front end of the harp will pick

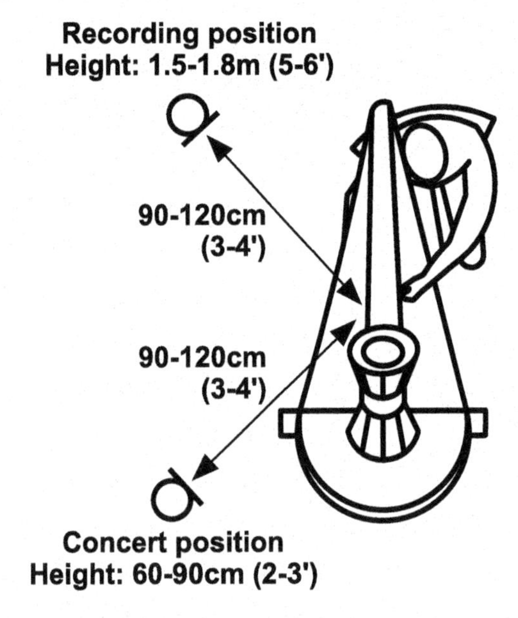

Figure 9.17 Alternative placements of a single harp spot microphone – plan view

up more string and pedal noise and HF reflections from the soundboard, but will work perfectly adequately to add a little definition to the part. A fig of 8 ribbon microphone (e.g. Royer R-121) can also be considered to discriminate against horn and woodwind spill depending on the particular orchestral layout being used.

9.11 Celeste

Finally, the celeste. This is a quiet instrument with a softer tone than the glockenspiel, and in a concert it will usually be placed towards the rear of the orchestra, either near the harp or with the percussion. It will mainly be used to double with harp, string, and woodwind parts as an additional tone colour unless it has a solo, in which case the orchestration should allow for its being heard (i.e. it cannot fight against brass and loud percussion). It will always need an ancillary microphone to capture the transients, and this can be placed to aim towards the open lid from a distance of around 90 cm (3′) and a height of about 1.2–1.5 m (4′ to 5′). The instrument can be approached as a small upright piano, but if the microphones are too close in, the mechanism will become too prominent in the sound. In a recording situation, it can be placed in front of the orchestra on the far side of the tree and outriggers, along with the harp and any orchestral piano part. Where the celeste has a part that needs to be played tightly together with xylophone or glockenspiel, it is usually better for the celeste to stay in the normal concert position near the harp and horns so the celeste and tuned percussion can hear each other.

Note

1 *Fundamentals of Musical Acoustics sections 12.2 and 20.8*
Author: Arthur H. Benade
Publisher: Dover
ISBN 0–486–26484-X

Chapter 10

Surround sound techniques

5.1 surround sound uses five discrete channels of audio plus a subwoofer which contains mainly LF effects. In this chapter, the channels will be referred to as LF, C, RF (left front, centre, right front) for the front channels and LS, RS (left surround, right surround) for the rear channels. LFE refers to the use of a subwoofer, which in cinema is used for 'low frequency effects', and it is a mono channel used for enhancement of very low frequencies. In orchestral recording, its use will usually be limited to some bass drum and bass spot signal.

10.1 Purpose of surround sound in classical music recording

The move from mono to stereo in the 1950s and 1960s meant that instruments could be spatially separated in a recording, making perception of detail in individual parts much easier. The move from stereo to surround provides a change in listening experience of a similar magnitude, although not simply because spatial separation of sources can be extended to include positions behind the listener. Although the rear channels are commonly used for sound effects located off-screen in film sound, classical recording in surround places the performers predominantly in front of the listener to reflect the real-life listening experience. It is certainly possible to make an orchestral recording that places the listener somewhere in the middle of the ensemble, and this has been tried as part of the exploration of the medium. However, given that most repertoire is designed to be heard from outside the ensemble to achieve the appropriate musical balance, surrounding the listener with players on all sides elevates the novel element of the technology over its use to transform a musical listening experience. In the spirit of always trying to make the recording enhance the music and the composer's intentions, experimental placement of performers that would not work for the repertoire in real life is outside the remit of this book. The use of surround sound has become most prevalent in opera recording (see Chapter 16), where the capture of live performance for cinema reproduction has naturally moved the audio into surround. The ability to place theatrical sound effects and offstage choruses into the rear channels has hugely enhanced the immersive nature of the listening experience.

The change from stereo to surround for classical recording involves something more than placing sound sources in the rear channels (although this is useful where the repertoire includes deliberate antiphonal or offstage effects). When we sit and listen to real musicians in a physical space, perception of detail in the playing is made easier by our having a secure sense of the location of each player that is unaffected by listener movement. There are many more location cues than can be reproduced in stereo, and hence the stereo listening experience is quite a fragile illusion that collapses if we move away from the central area between the loudspeakers. The surround listening experience can give the listener additional information about source location and the nature of the space, which makes aural perception closer to real life, requiring less concentration on the part of the listener to draw out detail from individual parts. The real value of surround sound for classical music lies in the subtle use of the technology to produce a realistic sensation of being in the room with the players. If stereo places the listener outside and looking into the room through the window, surround sound brings the listener right into the space by skilful use of reverb in the rear channels in particular. An additional benefit is that louder orchestral and choral tuttis (which can feel somewhat saturated in stereo) can feel more open in surround sound, as if there is always room for the sound to get louder.

If a recording is mixed in surround, it will not necessarily fold down well to stereo; it is likely to lose perceptible detail if this is done by simply summing the existing surround channels into left and right only. The two reproduction methods are different, and a separate stereo mix will usually be needed if one is required as part of the project deliverables.

10.2 Panning a Decca Tree in 5.1 surround

The core microphone techniques described for stereo orchestral recording can be rigged in the same way when working in surround, so we can look at the case of how to pan a Decca Tree and outriggers between five channels instead of two.

Figure 10.1 shows the layout of a set of five surround loudspeakers suggested by the ITU (International Telecommunication Union ITU-R BS.775–3). The subwoofer location is not included.

In order to pan between five loudspeakers (LF, C, RF, LS, RS), the desk or DAW used needs to have surround panning capability. Apart from a simple joystick arrangement for panning and sending different amounts of a signal to all five loudspeakers in varying quantities, there is a *divergence* control that is not found in a stereo system. This allows the engineer to avoid a source appearing to be very tethered to a single loudspeaker. For example, if a signal is sent to the LS loudspeaker and divergence is increased, the same signal will also be sent in increasing amounts to the RS and LF speakers.

Using divergence is a particularly useful technique to avoid the centre channel becoming over-localised when panning a Decca Tree. When working in stereo, a centrally panned signal (such as the centre microphone of the Decca Tree) will be sent to LF and RF equally and will appear as a phantom centre source. Where there is a discrete centre channel, panning the microphone centrally will send it to the centre loudspeaker only, and this can make the centre of the orchestral image collapse into the loudspeaker location. Many classical engineers will avoid using the centre channel for this reason, while others will use it with a degree of divergence applied. Divergence will send increasing amounts of the centre channel to the LF and RF speakers instead of the centre speaker alone, reducing the level to centre as the levels to LF and RF are increased.

When it comes to presenting an orchestral recording in surround, it can be very effective and enveloping to bring the sides of the orchestra partly around the sides of the listener almost in a horseshoe shape, placing the listener into something like the conductor's position. The instruments that are located towards the back of the orchestra and produce the sense of depth in a stereo orchestral perspective (woodwinds, brass, percussion, and so on) are left in the front

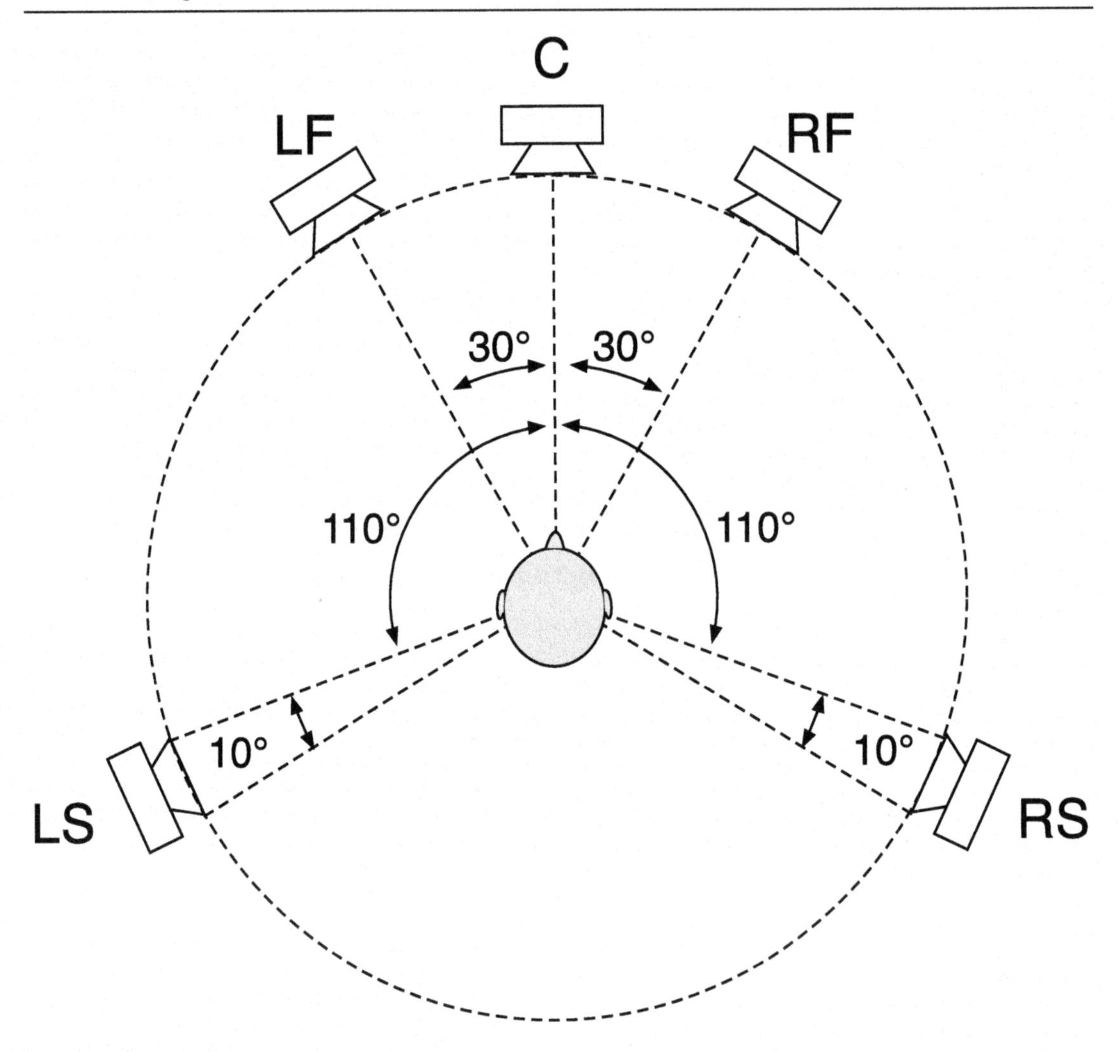

Figure 10.1 5.1 surround monitoring system layout

speakers, with only the string section being 'wrapped around'. Reverberation is used in both the front and the rear loudspeakers, taken partly from microphones rigged for the purpose (see section 10.3) and artificial reverb (see section 10.4).

In order to move the apparent location of sources beyond the LF and RF loudspeakers and around to the side of the listener, some of the signal needs to be sent to the LS and RS loudspeakers. Care needs to be taken here; because the listener is orientated with their ears across the left to right axis of the loudspeaker set-up, creation of phantom images works best either in front of or behind them. Sources panned down the sides (between LF and LS, or between RF and RS) are only located very imprecisely, and if too much signal is sent to the surrounds or the listener moves a little out of position, the image will tend to collapse into the rear loudspeakers, thus moving the orchestra behind the listener and breaking the illusion.

Bearing all this in mind, the five-microphone Decca Tree panned in surround would be approximately as follows:

- Centre microphone panned to the centre channel (or a phantom centre if preferred, or somewhere between the two using the divergence control*).
- Left and right outer tree microphones panned fully into the LF and RF speakers.
- Left and right outriggers brought around the sides to a position about 30% of the way towards the rear speakers (panning left microphone between LF and LS and right microphone between RF and RS.).

*Use of centre or phantom centre: for music-only mixing, it is usual to send something to the centre channel with some divergence. For orchestral film score recording, many engineers avoid using the centre channel altogether, as it is used for the film's dialogue. A centre orchestral channel can get in the way during the film dubbing ('re-recording') mix and end up being removed or reduced in level, thus affecting the orchestral balance.

Ancillary microphones such as woodwind section microphones can also be treated slightly differently in surround by sending a very little of their signal to the rear loudspeakers with the majority sent to the front. This will bring the players closer to the listener, but it should be minimal as not much is required to achieve the desired effect. Again, if it is overdone, the front image will fall into the surrounds. As noted earlier, the LFE channel can be used to enhance the bass drum and double basses by sending some of their microphone signal to this channel. This does not affect the stereo imaging, but just adds some very low frequency bass energy to the sound in the room.

10.3 Natural reverberation: additional microphones for 5.1 surround

Additional microphones are needed to collect reverb cues that can be used to create a realistic sense of space, and there is a body of academic research into the best ways of collecting signals for this purpose. There is also a variety of industry practices driven primarily by practical experimentation and trial and error.

The signals used for reverb in the surrounds (LS, RS) need to be different from those sent to the front loudspeakers. If the same signal is sent to the front and rear speakers (i.e. one signal sent to LF and LS and the other to RF and RS), the reverb localisation will tend to be at the side of the listener. Therefore, all methods used for collecting reverb for surround sound use four individual microphones that will collect four different but sufficiently correlated signals.

There are two ways to produce four useful reverb signals from a group of microphones: one is to use omnidirectional microphones that are set at different positions in the room; the other is to use directional microphones and point them in different directions, with or without a degree of additional physical separation between them. It is important to avoid picking up anything other than very low levels of direct sound on the rear ambience microphones; these signals will be sent to the rear loudspeakers, and if they contain a significant amount of direct sound, the main image can be pulled behind the listener. This is a particular risk if the listener moves or is not ideally placed in the centre of the loudspeakers. Therefore, if omnidirectional microphones are being used to collect room reverb, they will need to be placed well into the reverberant field at a good distance from the players.

When using spaced microphones to collect reverb, the principles of working with spaced microphones apply (see Chapter 3). If they are placed a long way apart, the resulting signals are too de-correlated to produce an enveloping stereo image and will produce reverb signals that tend to remain located in each loudspeaker. For better correlation that will produce a more realistic sense of space when panned between LF and RF, or between LS and RS, they need to be spaced at no more than about 1 m (3′4″). It should be noted, however, that a very widely spaced pair of omnis (at about the separation of orchestral outriggers but further back in the room) can be useful for filling in additional reverb at the *sides* of the surround loudspeaker array. One microphone will be panned between LF and LS and the other between RF and RS.

The following are two suggested methods using directional microphones. The first is the Hamasaki square,[1] which is an array of sideways-facing fig of 8 microphones, as shown in Figure 10.2.

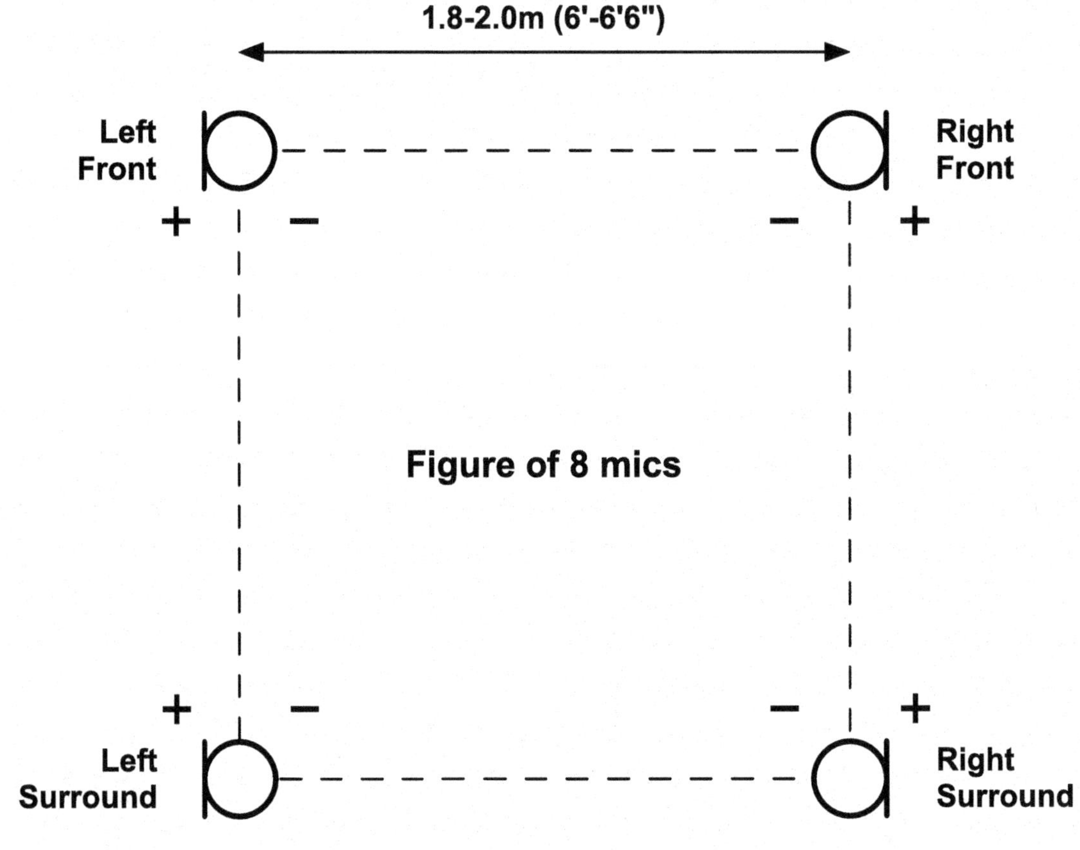

Figure 10.2 Hamasaki square of fig of 8 microphones

These are placed at some distance back in the hall from the orchestra, in the reverberant sound field, and they produce a signal rather like that from a surround reverb unit. The front microphones are used for front reverb (sent to LF and RF channels), and the rear ones are used for rear reverb (sent to LS and RS). This arrangement is used at the Royal Opera House, Covent Garden, for cinema and DVD recordings in surround sound (see Chapter 16). Taken as a whole, the rig is quite coherent in itself, and so it produces a good spread of reverberant sound in the image across the front and the rear of the listener.

Because the microphones are fig of 8s, they will discriminate against the direct sound from the orchestra, and any low level of direct sound that is present will arrive tens of milliseconds after the direct sound arrives at the main orchestral pickup. This means that the direct sound in the front loudspeakers dominates any direct sound in the rears both in level and in earlier timing (the precedence effect). The listener interprets the direct sound as coming solely from the front speakers, thus placing the orchestra in front of the listener.

Some engineers will choose to remove the timing differences between orchestral and distant ambience microphones by using delays. Adding delays to the orchestral microphones in this way can make the sound clearer by avoiding any slight smearing of transients, but it also has a tendency to make the recording feel 'flatter' in perspective. Leaving the difference in timing between distant ambience and orchestral microphones untouched tends to maintain a greater sense of

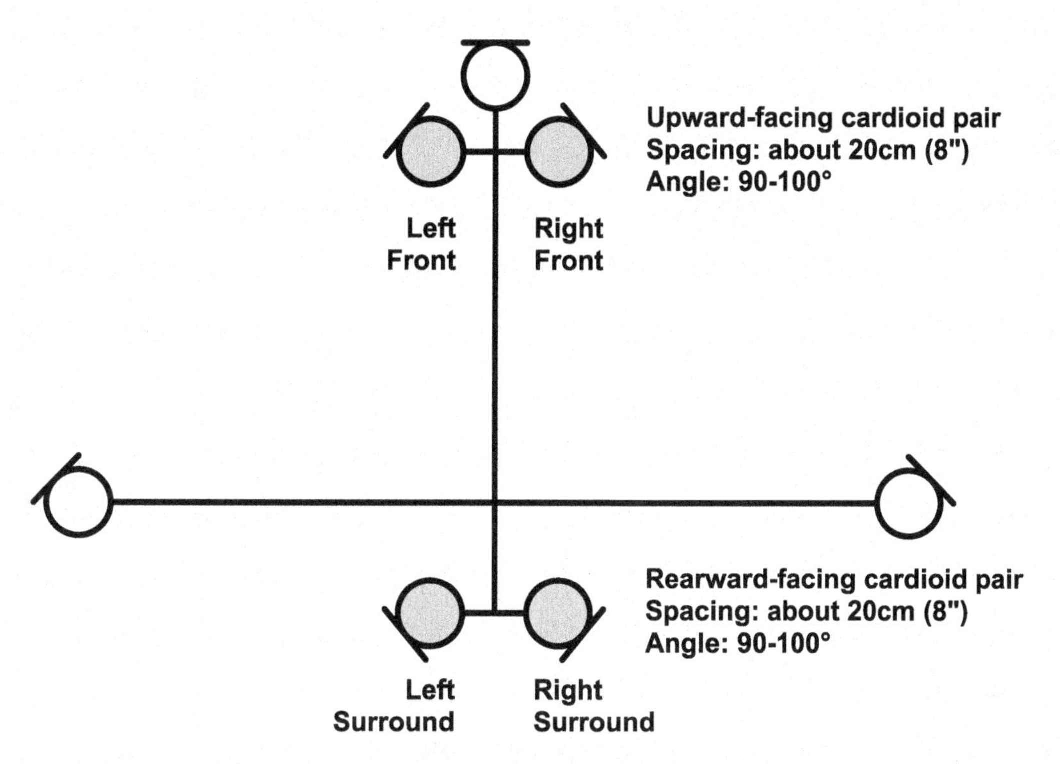

Figure 10.3 Decca Tree with additional KM84 pairs for LF-RF and LS-RS ambience

spaciousness. As seen in Chapter 16, the use of delays is essential when applied to very close radio microphones in live opera recording, but it is otherwise optional.

A second method of collecting ambient signals is illustrated in Figures 10.3 and 10.4, and it is a development of the upwards-facing ambient pairs used in stereo recording that we have seen elsewhere in the book (see Chapter 12 for an example). An upwards-facing pair of KM84s spaced at about 20 cm (8′) and angled at about 90°–100° is mounted at the front of the tree, and a similar pair of KM84s is mounted immediately behind the tree, facing away from the orchestra. The upwards-facing pair is sent to LF and RF, and the rear-facing pair is sent to LS and RS.

These microphones are close to the main pickup, so they provide time-coherent signals and produce a good ambient pickup from the strings in particular. Although there is a theoretical concern that picking up any significant amount of direct sound on microphones aimed at producing

Figure 10.4 Photograph of the array from Figure 10.3, although only the front ambience pair is clearly visible

Photo: Carlos Lellis, Programme Director, Abbey Road Institute.

reverb for the rear speakers will cause imaging problems, in practice this technique does not usually cause any difficulties.

The addition of a little brightness boost of about 2 dB shelf at 10 kHz to any ambience microphones, whether front or rear, can make the recording feel more exciting and will compensate a little for the acoustic loss of HF when sound travels larger distances.

10.4 Artificial reverberation in 5.1 surround

The room reverb collected by whatever surround microphone technique is used will almost certainly need to be subtly augmented by artificial reverb. This is similar to blending the natural room sound with some additional artificial reverb in stereo, an essential skill that is discussed in in Chapter 17. The signals from the surround microphones are very useful for reproducing the characteristic of the early reflections in the actual hall and can be blended in with a smaller amount of artificial reverb for the remaining tail of the reverb. A professional reverb unit such as a Bricasti M7 has a control for the balance between early reflections and reverb tail, so it can be skewed in favour of the latter if it helps with the sense of realism. It is usually preferable to use as much real reverb from the room as possible if the hall is good, as the effect will be more convincing.

There are very few reverb units that have surround capability, that is those that will generate five channels of reverberation appropriate to the panned position of a source within a surround phantom image. Common practice is to use a front stereo reverb unit and a rear stereo reverb unit, possibly of a different type. If there is a difference in quality between the units available, then the better one will be used at the front. It can actually be beneficial to have slightly different reverb in the front and rear speakers as it reduces the obvious signature of the artificial reverb, which can be noticeable even with the very best reverb programmes. Both reverb programmes need to be reasonably well matched in terms of RT60, decay shape, and colouration characteristics for this to work well. One way of checking that the front and rear reverbs are sufficiently well matched is to sit sideways to the monitoring, thus facing either the LF and LS speakers or the RF and RS speakers. From this position, we can utilise our excellent L-R lateral positioning discrimination to see if reverb decay retains a stable stereo image or whether it pulls to one side or another at any point during the decay. If it moves, it indicates that the front and rear reverbs are not sufficiently well matched and will need to be adjusted.

In deciding which microphones need to be sent to the front and rear reverb units, there are no hard and fast rules, but there are some general guidelines. The microphones designed for collecting hall reverb will not be sent to either reverb unit. For the initial set-up, as for stereo (see Chapter 17), all the orchestral microphones should be sent to the front reverb unit at the same post-fade aux send level. This can then be adjusted to suit if necessary; the microphones that might be best left dry are those that are being used to capture transients from an instrument for clarity and are only present at low level in the mix. These would include percussion microphones and maybe the double bass microphone.

When it comes to deciding what to send to the rear reverb units, this can be quite dependent on repertoire and situation. Bear in mind that the effect of adding more reverb to a signal (even if it is reproduced behind the listener) is to either make it feel bigger if the instrumental spot microphones are high enough in level (good for a concerto soloist) or to give it a sense of distance (good for woodwinds in some repertoire).

Taking a typical example from the Royal Opera House, the main orchestral microphones would be sent to the rear reverb, but the string section spot microphones would not. The woodwind microphones might be sent to the rear reverb for repertoire where they are playing a traditional orchestral role (suited to the 18th and 19th centuries), but where they have more individualistic solo lines and need to be more present (for 20th-century repertoire such as Prokofiev and Shostakovich), they will not be sent to the rear reverb. If the celeste and harp are used for ethereal effects, they will be sent to the rear reverb, and horns will depend on the effect required, so they might not be sent if it makes them too distant. Concerto soloists such as piano will be sent to front and rear. This will make the piano feel larger rather than more distant when taken in conjunction with an appropriate level of piano microphones in the mix.

Setting the level of the reverb returns is important, and it can be quite a delicate balancing act between front and rear reverb return levels. They will usually be in very roughly equal amounts (i.e. they are likely to be within 5 dB of one another), but care needs to be taken not to bring back too much rear reverb as it can quickly become very swimmy.

10.5 Offstage effects in surround: location of sources behind the listener

Sources that are panned between LS-RS behind the listener are perceived as a phantom image in a similar way to sources panned between LF-RF in front. The rear loudspeakers are spaced at a wider angle of 120°–160° rather than the 60° angle between the front speakers, and the practical experience of many engineers suggests that the phantom source location is not as focussed as it is when sources are placed in front. It does work sufficiently well, however, to indicate that a group of musicians is definitely located behind the listener.

In the context of an opera recording, offstage sources might include chorus, sound effects, or solo brass (such as fanfare trumpets or hunting horns). There are choices to be made, partly dependent on how clearly the source needs to be located. Sounds placed exclusively in the rear loudspeakers will be very clearly located behind the listener, but they can be brought around one side or the other by panning between the front and rear sets of speakers. The effect of this is a source that is less clearly localised, but this might be desirable for something like hunting horns. For the surround mix of the 2004/2005 EMI recording of *Tristan und Isolde*, the offstage horns were panned down the side of the listener. To bring a source that is located only in the rear speakers a short way around to the sides, a small amount of the signal has to be added into the front speakers. However, the HF from the rear speakers is partly blocked by the ears' pinnae, and the image can be quite quickly pulled around to the front with only a modest signal level. To prevent this from happening, an HF shelf of around 10 kHz can be added to the rear signal to compensate for the HF loss at the ears, and this will help to keep the image anchored mainly behind the listener.

10.6 Object-based audio: Dolby Atmos

Dolby Atmos[2] is the next generation of surround sound reproduction. It incorporates height reproduction by means of having loudspeakers in different horizontal as well as vertical planes, so forming a truly enveloping soundscape. It is designed so that encoded audio (in the form of a Dolby Atmos Bitstream) can be reproduced in a range of environments, with different numbers and locations of loudspeakers.

In traditional 5.1 cinema sound reproduction, there are multiple surround loudspeakers to cover the large auditorium, but they are all reproducing either the LS or RS signal. To avoid any member of the audience hearing sound from the nearest surround loudspeaker ahead of sound from the front loudspeakers, all the surround speakers have a delay inserted into their signal path according to the size of the individual cinema. The delay cannot be perfect for every seat in the house, so its calculation is a compromise between compensating for the worst seat (i.e. that which experiences the largest natural delay between the front and the nearest surround speaker) whilst not degrading the sound for the best seats.

To increase the realism and effectiveness of sound source location from around an auditorium, additional discrete channels of audio could be added, played from loudspeakers placed around the sides and above the listeners, thus effectively encasing the audience in a hemispherical array of loudspeakers. Provided the mixing engineer was able to mix the audio in exactly the same listening environment, with the same number of loudspeakers placed at the same distances, the results would be very effective. However, given that no two auditoria are alike in terms of dimensions and speaker location, taking surround sound to a more immersive and detailed level needed a different approach.

'Object-based audio' is such an approach, and involves treating each individual sound 'object' separately rather than mixing the audio together to form new audio signals. For classical recording, an 'object' could be a microphone source or single channel of reverb return. Once the mixing engineer has created a mix in a standardised Dolby Atmos mixing room, the panning location and level of each 'object' is encoded as data for the duration of the mix. When the time comes for this to be reproduced in a different listening environment, the desired location of each object is decoded and panned to its original position using the loudspeakers that are available. The reproduction environment (which might be a cinema, or a Dolby Atmos system in a domestic living room) is likely to have fewer loudspeakers which are placed in less ideal locations than those in the standardised mixing room, but the original panning locations are recreated using the available loudspeaker set-up, using phantom images where necessary.

The final mix therefore, is not stored as discrete channels of audio as it would be for 5.1 surround, but as a collection of individual mix sources (or 'objects') stored as audio files, each with an associated data stream of encoded panning positions. Any level changes during mixing are captured as part of the audio file, and the audio can only be played back with the aid of a Dolby Atmos renderer.

When working in Dolby Atmos, the engineer can pan things to different vertical locations as well, and this can be particularly useful for subtly enhancing the sense of a real enveloping space in classical recording. In 5.1 surround sound and stereo, reflections from the ceiling level of the hall are reproduced in the same horizontal plane as the direct sound and other reverb, but Dolby Atmos enables the engineer to locate these reflections in a higher plane that is above the orchestral image and also sits above behind the listener. The effect of this is that space is freed up for crescendos and large tuttis; they can keep getting louder without feeling overwhelming or congested, as they can in simple stereo reproduction. When all the reverb and acoustic energy is no longer concentrated in the horizontal plane but is spread all around the listener, the acoustic levels involved in the hall can be more realistically reproduced.

As with the move from stereo to surround, the temptation to locate sources all around for sake of demonstrating the technology is a potential red herring for the classical engineer. Aside from the wonderful possibilities for offstage effects, and for recording compositions that have a spatial component to the performance, the medium can be used much more subtly but to great effect for the enhancement of the sense of space essential to classical music performance.

At the time of writing, there is a lot of interest in remastering old recordings for release in Dolby Atmos, particularly opera material. This can be done with skilful use of reverb even from an original two-track recording, but where there are four-track and eight-track edited masters, the engineer has more to play with. Four-track masters were very common for 1960s and 1970s opera, with two tracks for the orchestra and another two tracks for the voices. This allows the voices to be pulled forwards during mixing for surround if necessary.

Notes

1 *Multichannel Recording Techniques for Reproducing Adequate Spatial Impression*
Author: Hamasaki, Kimio
Affiliation: NHK Science & Technical Research Laboratories, Tokyo, Japan
AES Conference: 24th International Conference: Multichannel Audio, The New Reality (June 2003)
Paper Number: 27
Publication Date: June 1, 2003
Permalink: www.aes.org/e-lib/browse.cfm?elib=12288

2 Dolby Atmos *www.dolby.com/us/en/technologies/dolby-atmos/dolby-atmos-next-generation-audio-for-cinema-white-paper.pdf*
www.dolby.com/us/en/brands/dolby-atmos.html

Chapter 11

Solo instruments and orchestra

Following on from our chapters on orchestral recording techniques, the next thing to consider is how to approach the addition of a soloist to an orchestral layout. A concerto might be written for any instrument and orchestra, and in the case of the concerto grosso, a small group of solo players is involved. From the recording point of view, we would like the appearance of orchestral perspective and depth to be maintained as before but with the addition of a soloist that is performing a central musical role. The soloist therefore needs a certain degree of prominence in the recording while retaining a believable perspective in relation to the rest of the orchestra. The soloist should feel as if they are located in the same space and should be placed just in front of the orchestra with a sense of appropriate scale between the two. For example, a piano should not fill the width of the image of the orchestra. This can be a challenge to get right, and it must also be remembered that the balance is never a static thing but is likely to need to be altered during the course of the piece depending on musical content.

The most common situation that the engineer is likely to come across will be a piano or violin concerto. The size of the grand piano means that its addition to the stage in a live situation results in the greatest disruption to the orchestral layout and to our ability to implement the usual orchestral microphone techniques.

The sections on non-piano instruments in this chapter begin with the violin, and this includes the most detailed discussion on spot microphones, imaging, and layouts. Even if you are looking for information on another instrument, it is recommended that you read the section on violin concertos first to get a more detailed overview of the principles involved. The sections on wind, cello, guitar, brass, and percussion build on the ideas discussed in the violin section, and you should begin to get a good feel for how these can be adapted to suit your circumstances.

Where a pair of microphones is discussed on the soloists in this chapter, it is assumed that they are spaced 25–30 cm (10″ to 12″) apart and forward facing unless otherwise specified. For more detailed discussion about placing microphones on individual instruments, see Chapter 4. This information should help you to make judgements about how the microphone placement might be adapted in a live concerto scenario while still getting a good tone from the solo instrument.

11.1 Piano concerto: studio layouts

The first thing to consider is where the piano is to be located. For a recording session where there is enough space and the engineer has some control over the layout, it is usual to place the piano approximately 0.9–1.2 m (3′ to 4′) in front of the orchestra, and behind the tree (see Figure 11.1). This is further away from the orchestra than the piano would normally be placed for a concert, but it does allow for reduced piano spill onto the outer tree microphones. If the piano is too close to the tree, there will be a high level of piano on the overall pickup. When the tree microphones are fully panned, an over-wide and diffuse piano image will result, and it is best if this can be kept from dominating the orchestra in level. The additional piano spot microphones will be used to centre the piano image and give more localisation and focus. While a greater distance between piano and orchestra is of benefit to the recording engineer, it impairs communication between the conductor and the pianist, and the final piano position is a compromise between these two conflicting requirements. You will need to do some negotiating with the musicians involved; the piano could also be placed a little off-centre over to the right so that the conductor is closer to the keyboard and pianist. (See Figure 11.1.)

To fit the main tree microphones in between the piano and the orchestra is awkward, and there are a couple of possible solutions. The most usual solution is to rig an upright stand with its base in the gap between piano and orchestra, with the tree rigged on a short horizontal boom on top, and with ample counterweights at the other end of the boom. Much less common is to use a really heavy duty stand with its base on the open side of the piano and the boom arm leaning right over the piano to get the tree into the right position. It is becoming harder to find such large stands in good working order, and it goes without saying that this is only practical for a recording session

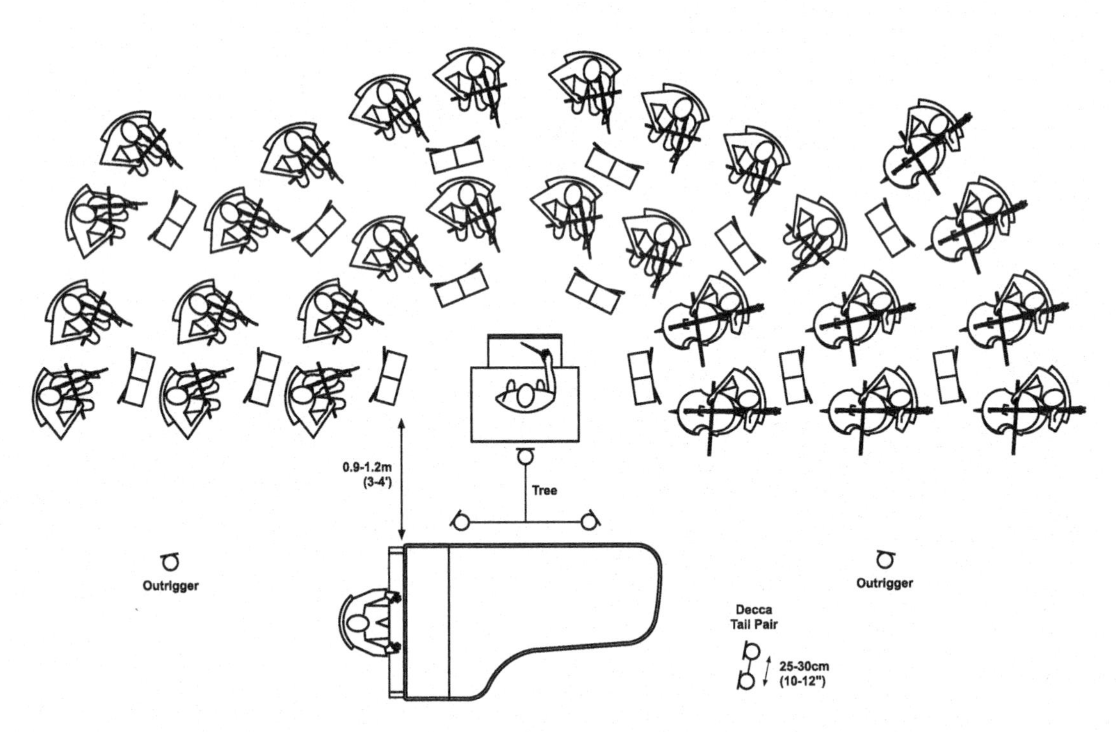

Figure 11.1 Session layout for piano concerto using a single piano pair

as it is extremely obtrusive. See Figure 11.2 for a photo of this arrangement taken in a Martha Argerich session in Montreal.

In addition to providing more stability to the piano's image, the additional spot microphones shown in Figure 11.1 are needed to provide appropriate detail and to make sure that musically important lines are not lost in the tutti sections. Their level in the final mix will be dynamic rather than static, depending on the changing musical content of the piece. The spot microphones can be placed in a similar position to the microphones we looked at for solo piano recording (see Chapter 5), and with a spacing of around 25–30 cm (10″ to 12″) so that they do not produce too wide an image when panned fully left and right.

In a situation where the piano is placed at least 0.9–1.2 m (3′ to 4′) back from the orchestra, omnidirectional microphones would be the first choice for adding depth and warmth to the lower end of the piano. Because of the microphones' positioning near the lower orchestral strings, the omnis have the potential to pick up a lot of cellos if they are too close. The presence of the cello spill will limit how much of these microphones can be used before their contribution to

Figure 11.2 A side view of the two main microphones of a four-microphone tree being boomed over the piano in St Eustache, Montreal. Martha Argerich and the OSM conducted by Charles Dutoit, recording Bartók's piano concerto in 1997.[1]

Photo: Richard Hale.

the orchestral image starts to pull the cello placement towards the centre and conflict with the cello image from the tree. The distance of the piano from the orchestra makes a big difference to whether the omnis will work; once this distance is reduced by only a little, to around 0.6–0.9 (1′ to 2′), the cello spill will be too high in level and the omnis will need to be replaced with wide cardioids (or cardioids as a final resort).

Where the piano has been placed over to the right to bring the keyboard closer to the conductor, this side orchestral spill problem is greater as the piano microphones will now pick up basses as well as cellos and drag these towards the centre if overused. If image confusion becomes a problem and it is not possible to increase the distance between the piano and the orchestra, it will be better to use directional microphones for the spots. In theory, even more directional microphones such as hyper-cardioids could be used, but only if their off-axis response is smooth, as they will still be picking up orchestral spill on their side axis. A flat off-axis response is something that will only be found in very high end hyper-cardioids as it is something that is difficult to engineer. A cheap pair is not likely to perform well enough in this particular situation, and the orchestral spill has the potential to be very coloured and unpleasant sounding. Ribbon fig of 8 microphones (Royer R-121 or Coles 4038) might also be helpful in this situation as, although they lack HF, their off-axis response is very smooth and their side rejection is excellent.

Microphones placed in the well of the piano are discussed in section 11.2 in the context of recording a piano concerto in concert. The sound will be rather close and dry, and so on a session where you have more control of the physical piano placement, using the tail pair would be a first choice.

11.1.1 Using two pairs on the piano

This is a more advanced and nuanced technique than using one pair, and it can offer greater control when used correctly. Rather than using a single tail pair, two spaced pairs are used, one of omnis and one of cardioids (or wide cardioids). See Figure 11.3 for their placement on the piano; the orchestral microphones remain the same as for Figure 11.1.

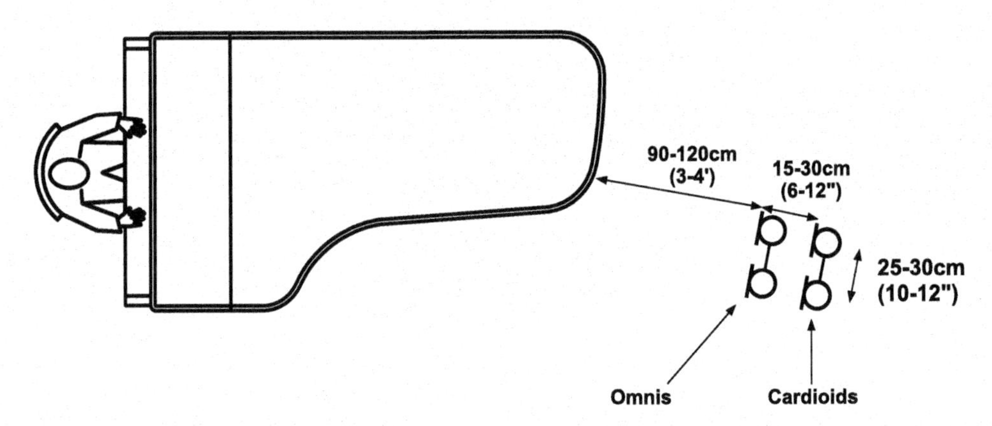

Figure 11.3 Session layout for piano concerto using a two pairs on the piano

The omnis are placed closer to the piano because they pick up more indirect sound and side orchestral spill, so placing them a little closer helps maintain the piano's relative level. Fig of 8 ribbon microphones could also be used as an alternative to the cardioid pair; they have two advantages: being able to point the null axes at the cellos and having a very smooth off-axis frequency response so that any spill has a good quality of sound.

The relative levels of the omni and cardioid pairs will control how much focus there is on the piano, and this will vary between pieces and within a piece as the musical texture changes. However, it important to base the sound on either one pair or the other and not have them contributing equal levels to the mix. This is because the spatial separation of the two piano pairs results in phase differences between the signals from each, and this can produce a comb filtering effect that is dependent on the relative levels of the two signals. The effect will be most pronounced where the two pairs are contributing equal levels to the mix, and cancellations of specific frequencies will occur where the distance between the microphone pairs is an odd multiple of a half wavelength. Making one pair or the other more dominant and restricting the other to a level at least 6–8 dB lower will reduce any colouration from interference between the two signals to insignificant amounts. Figure 11.4 shows the effect of summing two signals that are half a wavelength different in phase: when they are equal level (top) and when one is allowed to dominate (bottom).

If the cardioids are used as a basis for the piano sound, adding a little of the omni will bring in some warmth to the lower notes. If the omnis are used as a base, the cardioids will be brought in

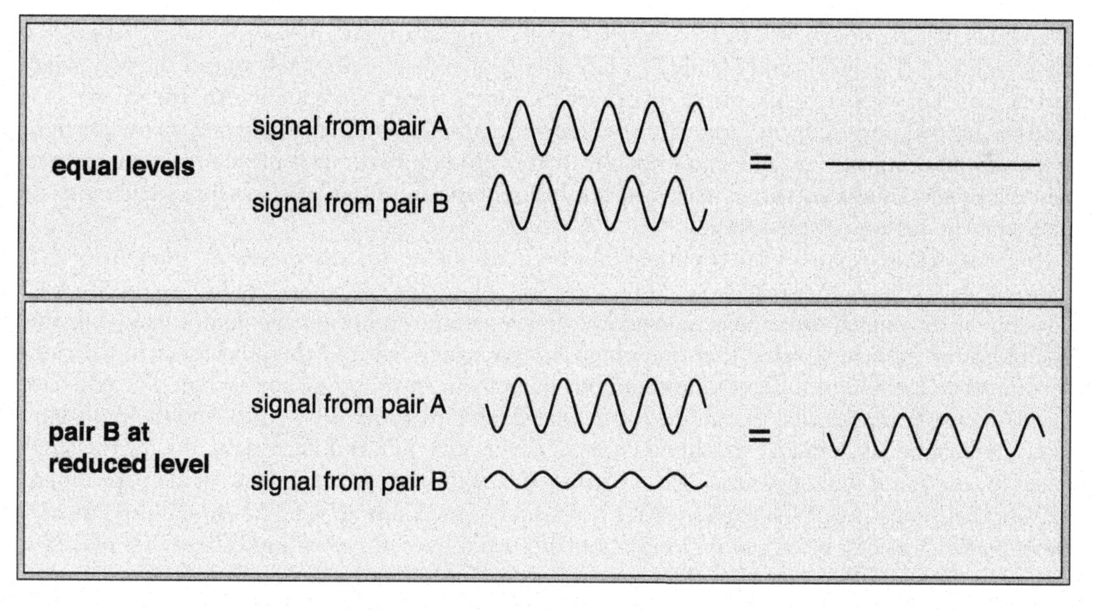

Figure 11.4 A single frequency that is delayed by half a wavelength between microphone pairs. The top indicates that selective cancellations can occur for some frequencies if each pair contributes equally to the mix. The bottom shows that if one pair dominates the mix, any frequency cancellation or boosting due to delays between pairs is kept to a minimum.

for clarity in loud tutti passages where piano detail might be lost. This would be normal practice for a large concerto such as by Liszt or Rachmaninov, where the orchestration is very full, and if the omnis were used to try to lift the piano detail further out of the mix, too much level would be required and the main orchestral image will be affected too much (as discussed in section 11.1). Therefore, lifting the cardioid level will be more effective at bringing out the piano detail without disturbing the orchestra.

On a mixer operational note, the four piano microphones would usefully be assigned to their own group so that their relative levels can be preserved while the orchestral versus piano balance can be adjusted during the mix. Always beware of using too much of the piano spot microphones and aim to keep the perspective between piano and orchestra believably natural in feeling (even if it is better than nature, and detail more consistent!). The piano should feel only marginally in front of the orchestra, and during tutti passages you should add just enough additional spot microphones to retain the detail without pushing the piano into the face of the listener.

11.1.2 Double piano concerto recording layout

For the best communication between players, and placing both pianos centrally, the sort of nested layout in Figure 11.5 is preferred. However, because this would mean that the lids are facing in different directions, shielding one piano from the tree more than the other, it is usual practice to remove the lids for the purposes of recording this layout.

A word of caution: when removing a piano lid, you will need more people than you imagine as they are very heavy. Dropping the piano lid into the mechanism while trying to remove it is something that no one wants to have to explain to the pianist, conductor, or the representative from Steinway. The lid should ideally be placed flat somewhere completely out of the way where no one will fall over it, walk on it, or otherwise inadvertently damage it. An alternative is to bend the hinged portion to 90° and store the lid vertically on its side, supporting its own weight, and again, making sure no one can knock it over, including orchestral players rushing towards their tea break. Unless you are a master at not losing things, put the locking screws back into the hinges on the piano for safekeeping.

On each piano, a pair of microphones can be used above the instrument at a height of 1.2–1.5 m (4′ to 5′) above the strings (i.e. in the position where the raised lid would have been). Each pair should be panned differently to separate the two pianos a little in the final image, but they should not be panned so hard that one piano is over to the left and the other over to the right. The aim is to be able to differentiate them but have them overlapping somewhat. The left-hand piano (microphones 1 and 2) would be panned hard left and about 30% right, and the right-hand piano (microphones 3 and 4) would be panned about 30% left and hard right. As for the single concerto, the fairly wide panning of the piano pairs will not result in an overwide piano image because their constituent microphones are not widely spaced, at around 25–30 cm (10″ to 12″). These piano pairs can be rigged on long booms to reach over into position. The tree is placed so that its outer microphones are on the orchestral side of the piano and not above the instrument. As before, this positioning will be a trade-off between the artists' preference for close communication and the engineer's preference for space between the piano and the orchestra. The tree stand in this instance will be best managed by use of a vertical stand, as described for the single piano concerto; using a very heavy duty stand to reach over two pianos is likely to be impractical.

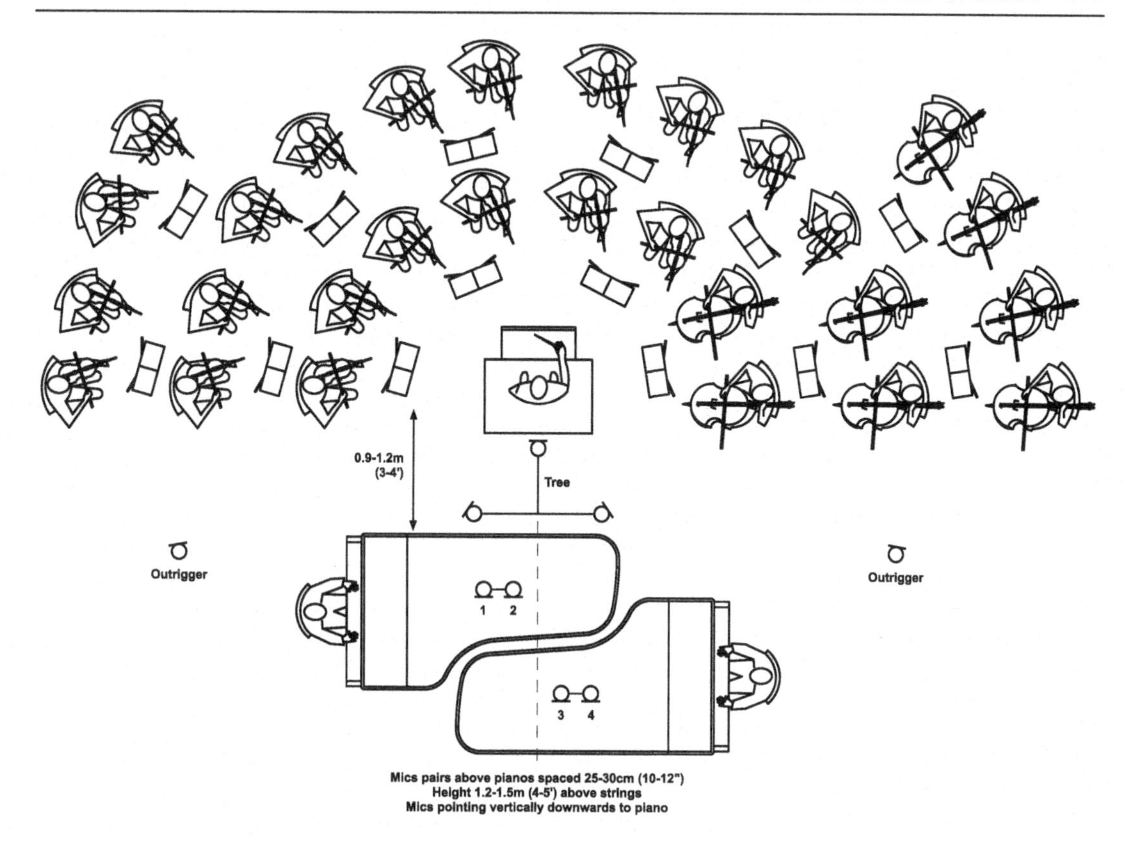

Figure 11.5 Layout for double piano concerto

11.2 Single piano concerto: concert layout

The layout in a live concert is often less than ideal for recording, with the piano partially embedded in the string section and a shallower stage than would allow for any further physical separation of the piano and orchestra. The most important difference from a studio recording is that the main tree pickup will be on the audience side of the piano, and this has implications for the overall orchestral pickup used. Figure 11.6 shows a typical concert layout for a piano concerto.

It can be seen from this that the awkward placement of the tree creates two problems: firstly, the proximity of the piano to the main pickup, and secondly, the main pickup having to be placed further back from the orchestra than usual. In this situation, replacing the three main tree microphones with cardioids (while retaining omnis for the outriggers) will reduce the level of piano on these microphones and compensate a little for the increased distance from the orchestra. The microphones should be arranged to point towards the orchestra, placing the piano on their side axes. It cannot be emphasised enough that microphones used in so many classical recording situations need a smooth off-axis response because they will often be picking up direct sound from all sides. In a concert situation, it would be very common to have an orchestral piece in one half

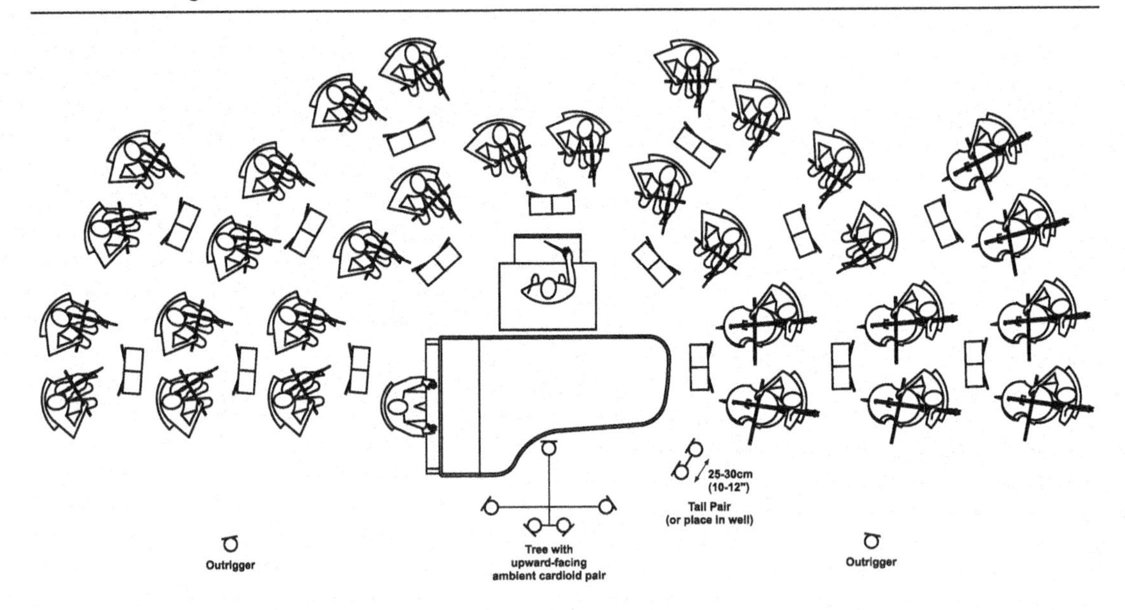

Figure 11.6 Piano concerto concert layout

and the concerto in the other. A practical solution to producing a satisfactory recording of both pieces would be to rig the tree frame with a set of omnis for the orchestral piece and cardioids for the piano concerto where the orchestra will have shunted backwards on the stage to accommodate the piano. The only way to really manage the tree in this situation will be to sling it, as any stand large enough to reach from the floor to 3 m (9′10″) above the stage will be far too visually intrusive.

A single piano spot pair can be placed as for a tail pair (using cardioids, wide cardioids, or fig of 8 ribbon microphones to reduce pickup of lower strings; the problems inherent in side orchestral string spill onto piano microphones were discussed in the studio layouts section 11.1), or they could also be placed in the well of the piano. The tail pair is more likely to cause imaging problems because of the proximity of the lower strings, but the pair in the well of the piano will sound somewhat dry and close, and will need some reverb to help it to blend in. Depending on the exact situation you find yourself in, there is trade-off to be made here when deciding which to use. Another way of helping the closer sound from piano spots placed in the well would be to add an upwards-facing ambient pair of good cardioids to the tree frame. (See Chapter 6 for discussion of ambient pairs.) These could be used as a piano ambient pair to reduce the relative dryness of the piano and would also avoid picking up much of the audience. They would cause no imaging problems as their image would be based around the same centre line as that of the tree.

11.3 Piano concerto conducted from the keyboard

This is a more common arrangement within groups that specialise in the performance of early music, where the tradition of conducting from the keyboard was part of the performance practice of the time.

The layout in Figure 11.7 shows the piano end on to the orchestra, with the lid removed. This positioning will necessitate some adaptation in the positioning of the overall pickup if you are using a tree because its usual position will place it directly above the piano where it will pick up a lot of hammer noise. A three-microphone tree in particular will have to be moved further back so that the centre microphone is also away from the piano hammers. A two-microphone tree could be used, such as in the set-up for Ashkenazy's recordings of the Mozart concertos with the Philharmonia[2] in the mid-1970s, although this can leave a 'hole' between the strings on either side. In this particular case, the woodwind microphones were used to fill this in.

With the tree set back in this way, string detail will be lost, and it is usually necessary to add in some microphones for the inner strings to help with perspective. Using cardioids or wide cardioids will help discriminate against piano and woodwinds, and these 'in-riggers' would be at a height of around 2.6–2.75 m (8′6″ to 9′). The only word of caution is that if you use too much of the string in-riggers, you can flatten the orchestral perspective and lose too much depth. Their use is commonplace in film music recording where the natural sense of orchestral depth is less important than a very full and enveloping string sound.

Having avoided too much hammer piano pickup on the tree, additional piano microphones can be added to obtain a good piano sound. There are two options to try to begin with: one above the piano strings (but not above the hammers) and one on a floor stand at tail pair height. (Both

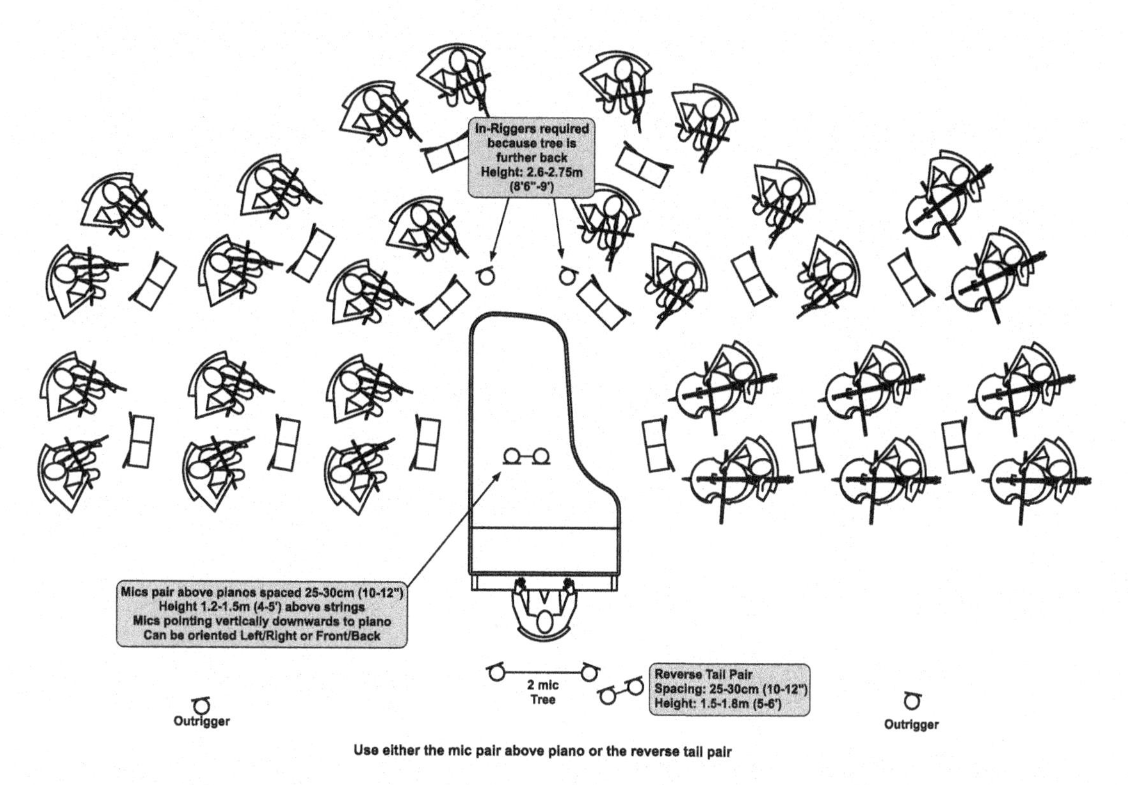

Figure 11.7 Layout for concerto conducted from the keyboard. Only one or the other of the piano pairs is intended for use.

are shown in Figure 11.6, but it is not intended that both should be used.) The narrowly spaced pair of wide cardioids (or cardioids) above the strings should be slung if this is a concert, but they could be stand mounted if on a session. They would be spaced around 25–30 cm (10″ to 12″) and positioned about 1.2–1.5 m (4′ to 5′) directly above the strings as shown. They can be orientated L-R as viewed from the audience, and can then simply be panned L-R as the image will overlay comfortably onto that produced by the tree. They can also be orientated one directly in front of the other (as one might place a pair on the piano if the orchestral pickup was of no consideration). In this case, the L-R panning could be either way; try both and see which works better.

The second option using a pair of omnis can be thought of as a 'reverse tail pair'. The spacing and height of the microphones is that of the tail pair but placed instead at the keyboard end of the piano, looking over the player's right shoulder to the upper range of the instrument. This technique was something that was used by renowned Decca engineer Kenneth Wilkinson; it does not seem very promising at first glance, but it actually works very well, and it was frequently used on Ashkenazy's Mozart concerto recordings mentioned earlier. Figure 11.8 shows Vladimir Ashkenazy conducting from the keyboard.

Figure 11.8 Ashkenazy playing the piano and conducting the Cleveland Orchestra in the Masonic Auditorium

Photo: Courtesy Decca Music Group Ltd.

11.4 A note on the size of the grand piano

It is not unusual to have to deal with a piano that is smaller than a full-sized concert grand piano, and apart from the instrument having a less good bass tone and smaller overall acoustic power, the approach to recording it will be essentially the same. Because the smaller grand piano is somewhat quieter, you will find that you either have to fade up the piano microphones a little more to get enough piano or move the microphones closer. The latter is likely to produce a better result given that there will be plenty of orchestral spill on the piano microphones, and fading them up too high could be detrimental to the orchestral image. However, when dealing with the traditional concert or studio layout with the piano sideways on to the orchestra, a shorter piano means that the tail pair is not so far from the centre of the orchestra, and so the disruption of the lower strings image should be reduced.

When using the 'end on' piano recording layout, a smaller instrument will result in less disruption of the central orchestral layout and so might be considered an advantage. The inner strings should not need so much support from string section microphones, and the woodwinds will be closer to the tree and therefore require less use of their own ancillary section microphones.

11.5 Violin concerto: studio layouts

Other solo instruments are easier to accommodate physically on the stage than the grand piano, and so the orchestral layout can remain undisturbed. One of the advantages of a recording over a concert layout is that the soloist and conductor can have much better eye contact during a recording. There are a couple of ways that this can be achieved (see Figures 11.9 and 11.10).

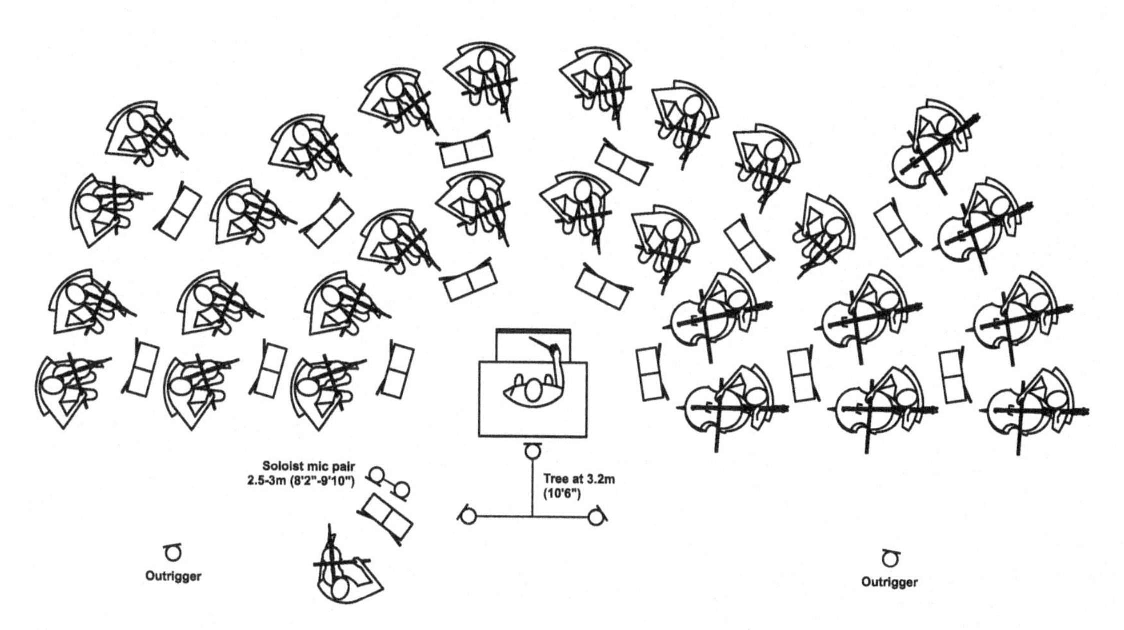

Figure 11.9 Violin concerto – reverse concert layout for recording

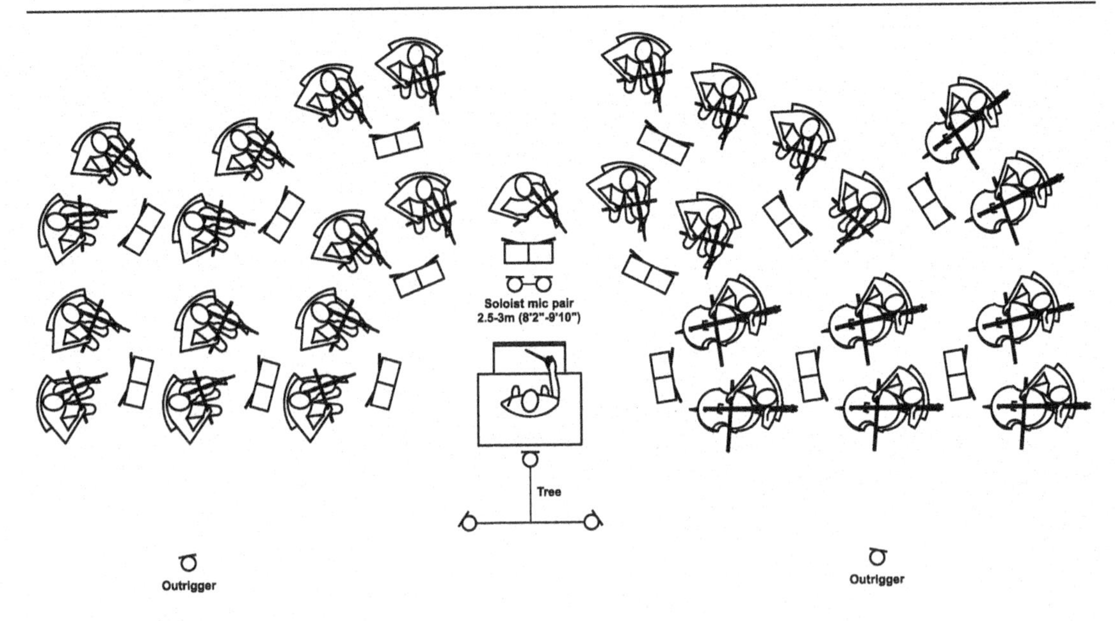

Figure 11.10 Violin concerto – soloist in orchestra recording layout

In Figure 11.9, the violinist stands a little behind the tree, and faces towards the orchestra and conductor in a 'reverse concert' position. The obvious problem arising from this layout is that the soloist is very much on the left-hand side of the tree and will be skewed towards the left in the image from the tree. The soloist's spot microphones can be used to help correct this by panning them left and right to place the soloist in the centre. However, if there is too much orchestral spill on the spot microphones, doing this might produce some orchestral imaging problems if they are faded up too far. The spot microphones also bring some more focus and clarity to the soloist's image, which will be quite diffuse on the tree alone. A pair rather than a single microphone is used to avoid the soloist coming across as a narrow point source in the recording image. If a single microphone is used, the more it is faded up, the smaller and narrower the instrument becomes, and the spot microphone begins to draw attention to itself. It is destructive to the beautiful illusion of the music arriving naturally and unmediated at the listener's ear if the listener becomes aware of the presence of an individual microphone.

The spot microphones are around 2.5–3 m (8′2″ to 9′10″) high, looking quite steeply downwards towards the player. Cardioids would be a good choice for the violin spot pair as there will be some degree of discrimination against the orchestral players, even though the microphones' rear axes are pointing towards the ceiling. The height of the microphones depends on the player; they can be lowered if the player is quieter, or brought lower and further in front if you need to reduce the HF in the player's tone; you should experiment to get the best that you can from the individual player.

The addition of a central ribbon microphone (Royer R-121 or Coles 4038) to the soloist's pair is another technique that can be helpful in two ways. Firstly, if the player has a thin or scratchy sound, the ribbon microphone's inbuilt HF roll-off will ameliorate this naturally. Secondly, the ribbon microphone's side rejection means that it can be placed lower without picking up too much orchestra. In turn, this means that the reduced spill on the ribbon microphone gives you more scope for panning it centrally without causing disruption to the orchestral image. Therefore, if you are having trouble with the orchestral image from the violin spaced pair being panned L-R and having to be faded up too high, or trouble getting the violin image to be central, adding a single central ribbon microphone can be used to lock the image more into place.

In Figure 11.10, the violinist stands in the middle of the orchestra in a space created in front of the second violins and violas. This position is central to the midline of the tree and hence avoids any problems with lateral imaging. In this scenario, it should be sufficient to use just the pair of cardioids as spot microphones without running into image problems, unless you choose to use the ribbon microphone for reasons of violin tone. In this case, you could use a pair of ribbon microphones instead of the cardioids. Some players like standing in the orchestra like this, and others do not find it comfortable. It is generally only used as an option for a solo violinist, and not for other instruments as they can feel rather hemmed in by strings, even if they are standing to play. A seated cellist would certainly feel a bit too embedded, and for a classical guitarist, the instrument is too quiet to be placed in this position.

11.6 Violin concerto: concert layout

In a concert setting, placing microphones to pick up the violin from the front at a good enough height on a stand will be too visually intrusive to be acceptable. If it is possible to sling the microphones above the player and thus do without the stand, then this would be the best solution. Where slinging the microphones is not possible, or where the position of the slung microphones is not ideal, the next thing to try would be a pair of spot microphones mounted lower down in front of the player, roughly at the height of the violin. Bearing in mind that positioning microphones on a violin any lower than the height of the bowing elbow will result in a significant loss of mid and high frequencies, this solution will not be tonally as good; the lower the microphones have to be, the less 'presence' the instrument will have. These microphones can be rigged very discreetly with the use of very slim 'Pavarotti stands' (Schoeps RC series; see Chapter 6), which are designed to be aesthetically acceptable in a concert situation. Another potential solution is to place the violin microphones high up and behind the player, although the tone will be affected in a similar way, with loss of higher frequencies. An ideal scenario would be to combine a slung, high-up pair and the lower microphones on the Pavarotti stands. The higher microphones would contribute most to the tonality of the sound, with the lower microphones adding some focus and a touch of closer perspective.

11.7 Wind concertos: studio and concert layouts

The oboe, clarinet, and bassoon can be treated in a similar way to the solo violin for concert and studio layouts, although it would not be usual to try placing the soloist within the central front strings for a studio session (as in Figure 11.10). With the player at the front to the conductor's left, the principles of microphone placement and panning would be the same as for the violin. The microphones, as before, would be a narrowly spaced cardioid pair, placed so that they look down on the instrument from high up, just in front of the player. For a concert layout, the player faces the audience, making slung microphones and/or a lower stand necessary, but for the recording layout the player can turn around, take a step backwards, and face the orchestra and conductor. If lower microphones have to be used, you should avoid them being placed directly in the firing line of the bell. (See also Chapter 7.)

The flute is the only woodwind that can be successfully recorded from behind the player when necessary, as the combination of its inherent radiation pattern and playing position mean that it radiates to front and back equally. All that is really lost in this position is the wind noise from the player blowing across the mouthpiece, and in a classical music scenario, flautists do not want this to be obtrusive (in a similar way to classical guitarists not wanting to hear any fret noises). This is useful in a live flute concerto situation where microphones can be placed behind, either by slinging them or by using a stand if it is permitted by the venue. The rear of the spot microphones will be towards the orchestra in this position and will afford the engineer some useful additional discrimination in favour of the flute. For a recording session situation, the flautist can turn and face the orchestra, with the microphone placed high up and in front.

In a concert situation, the player will be standing on the left and closer to the orchestra than for a recording session. Some thought should be given to how far to go in trying to centralise the soloist in the recorded image. Any microphones facing the front of the soloist will also be facing the orchestra and will be picking up significant orchestral spill from the violins. Initially, you should try to pan the soloist microphones fully left and right, but note the effect on the orchestral image as you fade them up. If the image is affected too much by the addition of hard panned soloist spot mics, you can adapt the panning of the soloist's microphones by bringing the right microphone inwards, and moving the soloist back off centre to nearer his or her natural position in the room. *A word of caution*: do not try to make this judgement when listening on headphones; they are not the best place to judge stereo imaging. See Chapter 2 for a discussion about pros and cons of monitoring on headphones.

11.8 Cello concerto: studio and concert layouts

For the recording session, the cello can be placed at the front of the orchestra, facing towards them, in the 'reverse concert' position. The distance from the orchestra will need to be enough to accommodate the cellist's microphones. The cello is less directional than the violin in the radiation of its middle and upper range fundamentals, and the instrument faces in a different direction because the player is seated. For both these reasons, the spot microphones do not need as much height in order to obtain a good balanced tone and can be placed in the plane perpendicular to the instrument's front in the region of 2.25 m (7′6″) high and 1.8–2.1 m (6′ to 7′) away (see also Chapter 7). Omnidirectional microphones should be considered for the pair as they have a smooth LF response that can capture the depth of the instrument's tone (the cello's lowest fundamental note is around 60 Hz). They will produce a tone that coheres well with the rest of the orchestra being picked up on the tree, so as long as the balance can be made to work, they would be the first choice. Wide cardioids and cardioids would be next, and alternatively, the technique of using a narrowly spaced pair of omnis plus a central directional microphone (introduced in the context of recording the violin) would also be applied to the cello. It is often considered to have a tone that is close to that of a singing voice, and using a vocal type microphone for the centre microphone could be an effective choice. An example would be to use a Neumann U87 in this role, where the slight presence edge and fatness to the sound would work well to enhance the cello's singing quality; the AKG C414 is another good alternative. The addition of a central directional microphone means that the omnis would not have to be faded up as high, so reducing any attendant imaging problems for the orchestra.

In a concert situation with the player facing out towards the audience and positioned on the conductor's left close to the orchestra, the cello can be discreetly recorded very well with a lower-down pair, looking slightly upwards at the instrument from a distance of about 1.5 m (5′) and height of 60–75 cm (2′ to 2′6″). Ribbon microphones can be useful in this scenario as the side axis can be used to discriminate against the orchestra. They can also be placed a bit closer in than high-end cardioids because their HF roll-off will soften any increase in noises and scrapes and avoid any oppressive closeness. Orchestral imaging should be carefully observed when deciding how the spot microphones can be panned, as detailed in section 11.7.

11.9 Guitar concerto: studio layout

The guitar and the cello have some similarities in that they cover essentially the same musical range, are of a similar size, and the players are seated. The approach to recording the guitar for a studio recording session is also similar, the main difference being that it is a very quiet instrument. When you have a quiet source, there are a few ways of getting more of the instrument on the microphone signal and less of everything else. Remember that orchestral spill onto the guitar microphones is not so much of a problem for the guitar as it is for the orchestral image if the guitar is sitting off-centre and the guitar microphones have to be faded up high enough to be useful.

- Firstly, distance between guitar and orchestra – moving the guitar back from the orchestra will help, but only within the bounds of keeping comfortable communication with the conductor. It is not feasible to put the guitarist right on the other side of the room (and if you did this, the orchestral spill would be so distant sounding as to have a detrimental effect on the orchestral sound).
- Secondly, distance between guitar and microphone – to help with the low acoustic levels, you could place microphones closer to the soloist, bearing in mind that the closer they are, the more localised sound and fret noise from the instrument they will collect, so microphone placement will become more critical. A closer spot microphone will also be drier, and it might need additional reverb when blending into the mix.
- Thirdly, you could use high-quality directional microphones and place them to discriminate against the orchestra whilst maintaining a good guitar sound. However, you should be careful using directional microphones too close, as the proximity effect will start to colour the sound.

Putting all these ideas together, the main danger to avoid is placing microphones so close to the soloist so that they give an artificial level of focus on the guitar and pick up a great deal of finger noise. The best approach would be to use some physical distance between the orchestra and guitar in the first instance and then bring in the other ideas if needed. The limit to placing the guitarist behind the podium will be around 1.2–1.5 m (4′ to 5′) before communication will start to suffer.

To find a good place for the microphones, it is suggested that you try sitting on the floor when the guitarist is playing and move around in an arc about 1 m (3′4″) away to see where it sounds best. A useful pair to start with would be some small omnis such as Neumann KM83s, which will produce a good tone and can be placed around 30–60 cm (1′ to 2′) back from the instrument without having to worry about colouration from the proximity effect. An additional central ribbon microphone can also be used to help with focus and soloist imaging (as for the violin/cello) and, as noted previously, the HF roll-off will soften any squeaks from fingers.

Whichever microphones are used, there will need to be a lot of fader riding during mixing to bring out the details in the guitar part and prevent the instrument from being swamped. In modern-day performances of the well-loved *Concierto de Aranjuez*, the guitarist is almost always amplified with a tiny clip-on microphone. When the piece was written, the orchestral forces were assumed to be smaller, and the venues more intimate than a modern orchestral concert hall. Bearing this in mind, if setting out to record a guitar concerto, a small chamber orchestra will make it easier to create a natural sounding result that sets the guitar and orchestra together in a believable acoustic space.

11.10 Brass concertos: studio and concert layouts

Brass instruments are loud, and the forward facing trumpets and trombones can be particularly hard to manage in a concerto recording because they spill onto the orchestral microphones.

The concert position with the player with their back to the orchestra is sometimes easier to handle; although the tree will pick up the soloists, the microphones are above and behind the instrument, so not on its front axis. Soloist's microphones will still be needed if only to give some focus to the image, and these should be placed above or below the bell to avoid an over-bright sound and large changes in tone as the player moves (see section 7.6 on brass and piano and Chapter 13). In a concert situation, below the bell is the most discreet option and could be managed with a pair of 'Pavarotti' microphones (Schoeps RC series; see Chapter 6). In this low position, they can be around 90 cm (3′) high and still avoid sight lines. They should not be too close, as you are only trying to collect enough of the soloist to provide a little focus. 'Too close' in this context means around 60 cm (2′); move the microphones as far back as you can get them given the constraints of the stage, which is likely to be in the range of 90–150 cm (3′ to 5′). Apart from Pavarotti microphones, the microphone choice should be one that doesn't make the actual placement distance feel even shorter (i.e. it shouldn't have any sort of HF lift or be too directional). A ribbon microphone will provide useful HF roll-off, and wide cardioids will sound less close than cardioids. Unless your placement has to be very close to the soloist, omnis will probably pick up too much in the way of other unwanted signals from all directions.

For a recording layout, placing the soloist behind the conductor (as for all the other soloists discussed in this chapter) is the usual option, but in this case the priority has to be how far away they can be positioned before contact with the conductor and orchestra becomes too compromised. The trombone or trumpet soloists will potentially be playing directly towards the orchestral microphones in this scenario, and it is a good idea to ask them to stand sideways on to the orchestra, so aiming sideways across the front, see Figure 11.11. It is also possible to use a tree of cardioids or wide cardioids instead of omnis to reduce soloist pickup.

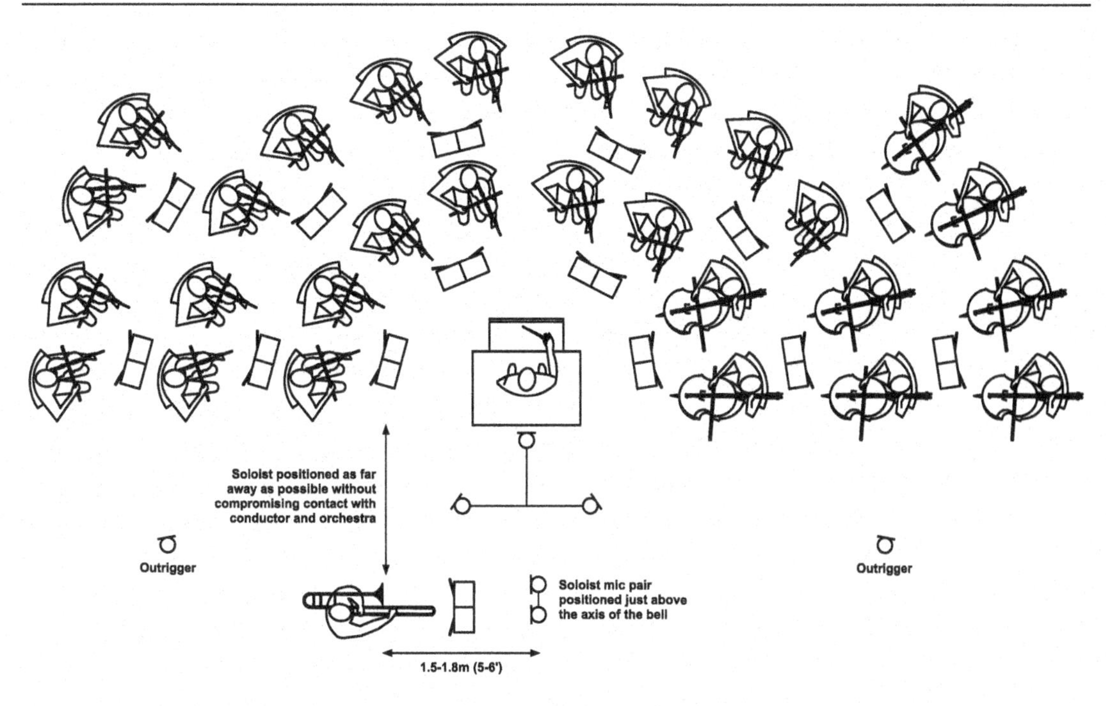

Figure 11.11 Showing placement of forward-facing brass for a studio concerto recording

The soloist's microphones should again be placed off the direct axis of the bell, and they could be cardioids or ribbon microphones as the extended LF is not required (the lowest note on the trombone is around 80 Hz). The main priority when placing the brass solo microphones is to get a good tone and enough focus on the soloist, and you should not be overly concerned about orchestral spill. For recording a standing trombone or trumpet soloist in this context, the microphones should be just above the bell and as far back as possible without losing the focus, which will be around 1.5–1.8 m (5′ to 6′) away. Cardioid and ribbon microphones will pick up a similar amount of reverberant sound at the same distance, but the ribbon microphone will be able to go in a little closer because it will not pick up so much HF. It goes without saying that you should not place a ribbon microphone in the direct stream of moving air from the bell of a trombone, as it may be physically damaged.

Horns and tubas are much easier to deal with as they can be placed at a similar distance in front of the orchestra and behind the conductor as the violin (see Figure 11.8 and section 11.5). Neither instrument is firing directly at the orchestral microphones, and the horn can be recorded from behind and the tuba from above, both off-axis to the bell to avoid the HF and wind noises. For the French horn, typical distances would be 1.2–1.5 m (4′ to 5′) back from the horn and about 30 m (1′) above the bell looking down and across it. For the tuba, the placement would be about 60 cm (2′) above the bell but off-axis; this is close enough to capture enough attack to give it some focus in the mix. If the microphones are much further back than this on a tuba, some of the important attack is lost and the signal will be less effective at bringing the tuba to the foreground.

11.11 Percussion concertos

There are concertos for marimba but also for groups of untuned percussion. These instruments can be very loud, and the difficulties when recording a live concerto are similar to those when recording the trumpet or trombone. The percussion will be situated directly under the tree for a live concert and will easily dominate the sound in terms of level and perspective. One solution to avoiding the dominance of the percussion on the main pickup would be to place the microphones for the orchestra closer in, including both the tree and sectional ancillary microphones. Alternatively, as with the forward facing brass, a tree of wide cardioids or cardioids could be used to reduce the percussion spill on the orchestral microphones.

In a studio session, it would be usual to place the percussion facing the orchestra, on the rear side of the tree, and screen the players off to some extent. Where there are no screens available, getting some distance between the percussion and the orchestral pickup will also help, but this has to be balanced against loss of communication with the conductor. As with the live concert, closer orchestral microphones and/or a tree of more directional microphones can also be used. Microphone placement on the percussion themselves would be the same sort of distances as noted for the percussion section in Chapter 9, as the aim is to pick up the transients and attack and use these to add clarity and definition to the sound of the instrument on the main pickup.

Notes

1 *PROKOFIEV/BARTÓK/Piano Concertos/OSM/Dutoit/Argerich EMI (1998) UPC 7243 5 56654 2 3*
2 *Example – MOZART/Piano Concertos 19 & 24/Ashkenazy/Philharmonia Orchestra DECCA (1980) LP SXL 6947*

Chapter 12

Chamber ensembles

Chamber ensembles come in many varieties, but they all perform music designed for a more intimate space and work without a conductor. Communication and eyelines between the players are therefore very important, and this needs to be taken into consideration when deciding on studio recording layouts. Small classical brass ensembles (of quintet size) have been included in Chapter 13.

We will consider some of the most common ensembles in turn.

12.1 String quartet in studio layout

The string quartet consists of two violins, a viola, and a cello; music has been written for this combination of string players from the 18th century to the present day. Whatever method is used to record the quartet, the aim is stability and focus in the recorded image. This is a greater priority than it would be for a large orchestra or choir where the overall blend of players is most important. Each player in a quartet has a different musical line and so needs to be heard as an individual as well as part of the whole.

The traditional seating layout from left to right is first violin, second violin, viola, cello. However, sometimes when recording, the cello and viola prefer to swap places, and this might affect both the internal acoustic balance and hence the recorded balance. One of the common difficulties with any quartet layout is the cello sounding a little recessed in perspective compared with the other players. The cello's radiation is less directional than that of the violins and violas, and from the point of view of microphones that are positioned looking down from a height, it is further away. Where this loss of contact with the cello is apparent, additional spot microphones might be used to help to bring it more into focus (see section 18.1.3 on using additional spot microphones).

There are a few approaches to recording the string quartet, depending on how close the players like to sit, how well acoustically balanced they are, and what sort of space you are working in. We will concentrate here on two primary techniques: the small Decca Tree and spaced omnis in conjunction with a near-coincident pair of cardioids.

12.1.1 Small Decca Tree and no spot microphones

The standard Decca Tree spacing is designed for orchestral use, and as such it is too wide to use on a string quartet; it would be wider than the ensemble and the imaging would be very unfocussed. A smaller Decca Tree can be used, with each of the three microphones spaced at around 50–56 cm (19.5″ to 22″) from the centre point of the tree and the centre microphone placed about 15 cm (6″) back from the front edges of the musicians' chairs. It allows the quartet to sit close together in a quite a tight, closed-in horseshoe, but it does require the quartet to have a really good internal balance; best results will be obtained from a quartet that prefers to sit close together. (This technique was used for the 1975 recordings of the Fitzwilliam Quartet, who liked to work in this way. The recordings included the Shostakovich string quartets series[1] and were made in All Saints' Church, Petersham, London.)

The microphones are placed at a lower height than for an orchestral tree, at about 2.5–2.7 m (8′2″ to 8′10″). If they are much lower than this, the instrumental noises become more intrusive and the difference in height positioning of the cello and the violins becomes more exaggerated. The cello is naturally situated about another 60 cm (2′) further from the microphones than the violins, and so keeping the distance to the microphones to over 2.5 m (8′2″) reduces the impact on perspective. The outer tree microphones will be pointing directly at the first violin and the cello, and the cellist should be able to turn a little towards the right-hand microphone to gain a little extra focus. The microphones will be panned hard left, centre, and hard right. See Figure 12.1a and 12.1b for plan and side views.

The imaging from a small Decca Tree will not be as sharp or stable as with other stereo techniques, but the overall tonality will be excellent, and with the smaller tree spacing and a tightly knit group, the imaging should be more than acceptable.

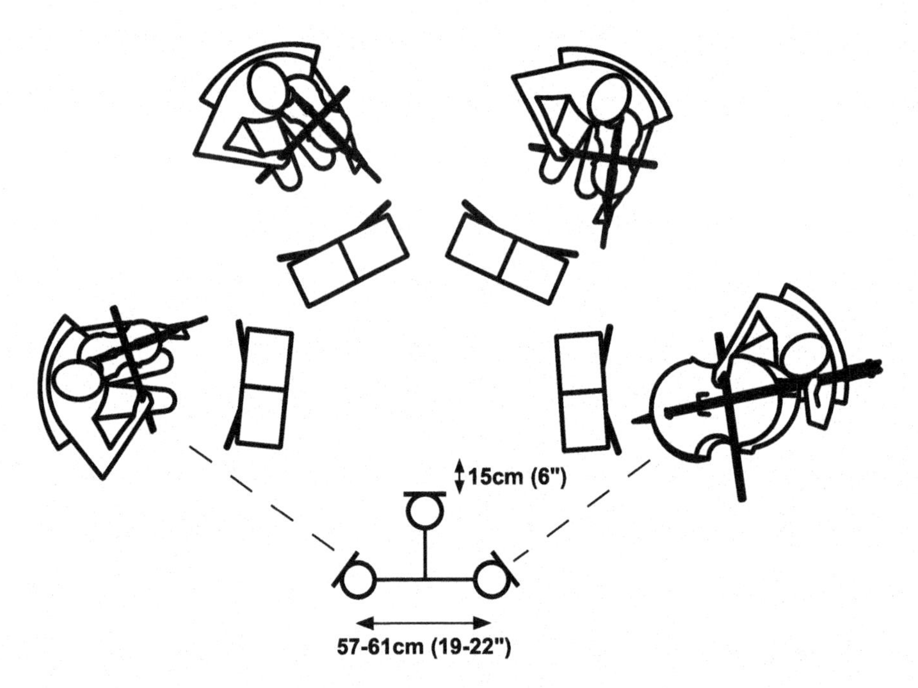

Figure 12.1a String quartet recording layout using a small Decca Tree – plan view

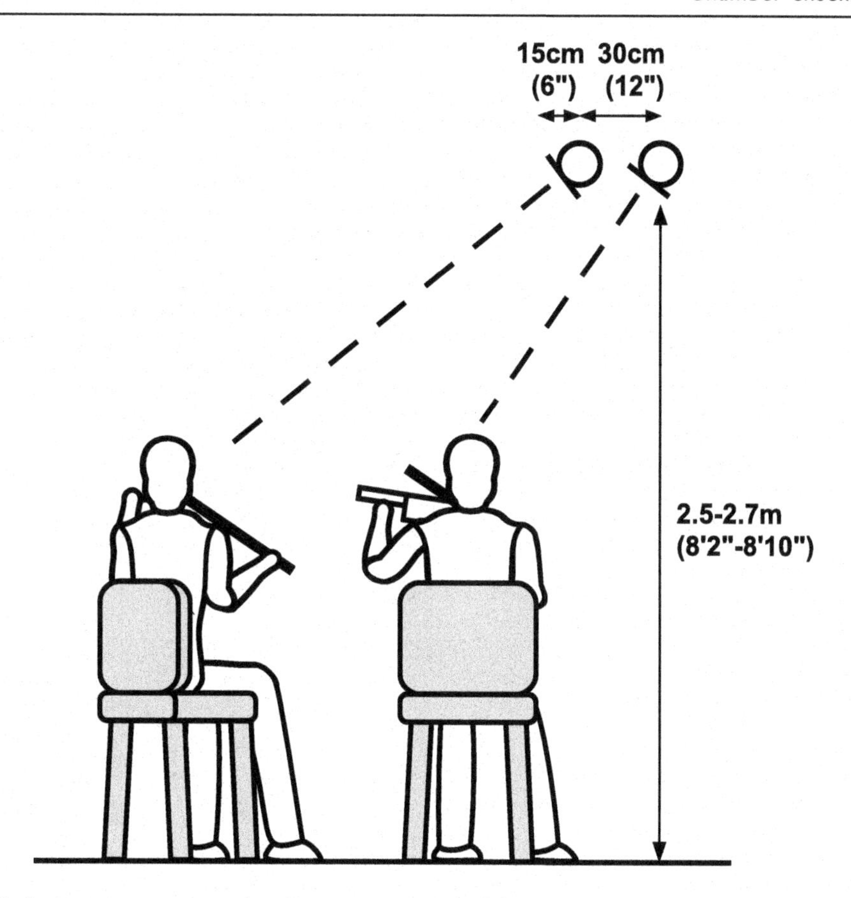

Figure 12.1b String quartet recording layout using a small Decca Tree – side view

12.1.2 Two-microphone tree (spaced omnis) plus ORTF/NOS-type pair

The success or otherwise of two spaced omnis used alone depends on the width of the ensemble, the distance from the ensemble, and the spacing of the microphones. (See Chapter 3.) The distance from the ensemble will be decided by the desired balance between the amount of close sound from the instruments and the amount of reverberation. If the microphones are too far apart on a string quartet (e.g. 1 m (3′4″)), the image will be rather unfocussed and unstable, and player movements can be exaggerated. At even greater extremes of spacing, the effect will be of two mono sources placed in the loudspeakers, with little occupying the centre of the image. If the microphones are too close together (closer than about 30 cm (12″) in this context), then the resulting image will be lacking in width. Therefore, the range of spacings and distances from the players in which this technique will work on its own for a string quartet is quite limited, and it is common practice to combine a pair of spaced omnis (for their quality of sound and extended LF) with a near co-incident pair such as ORTF to provide stability and focus to the image.

Figure 12.2a and 12.2b shows a typical set up of this kind. The rationale for placing the cardioids further back is that they pick up less reverb than omnis, and the pairs will sound similarly reverberant when the omnis are placed closer. (It is also possible to reverse the position of the pairs, and place the ORTF pair as the closer of the two, with the omnis further back where they will collect a more reverberant sound. In this situation, the ORTF pair will necessarily dominate the mix because the omnis will be too reverberant, and they will just be used to blend in a little

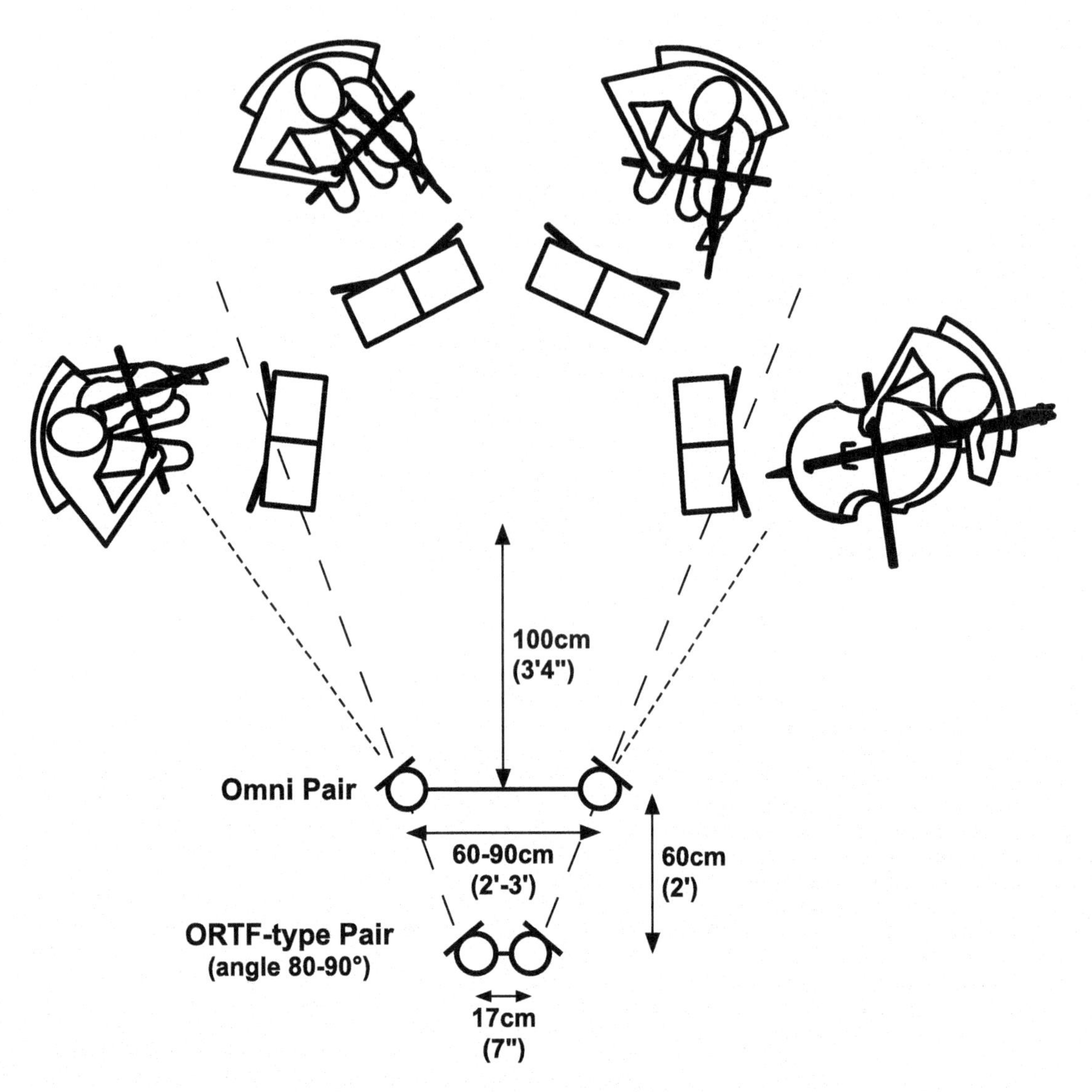

Figure 12.2a String quartet session layout using spaced omnis and ORTF pair – plan view

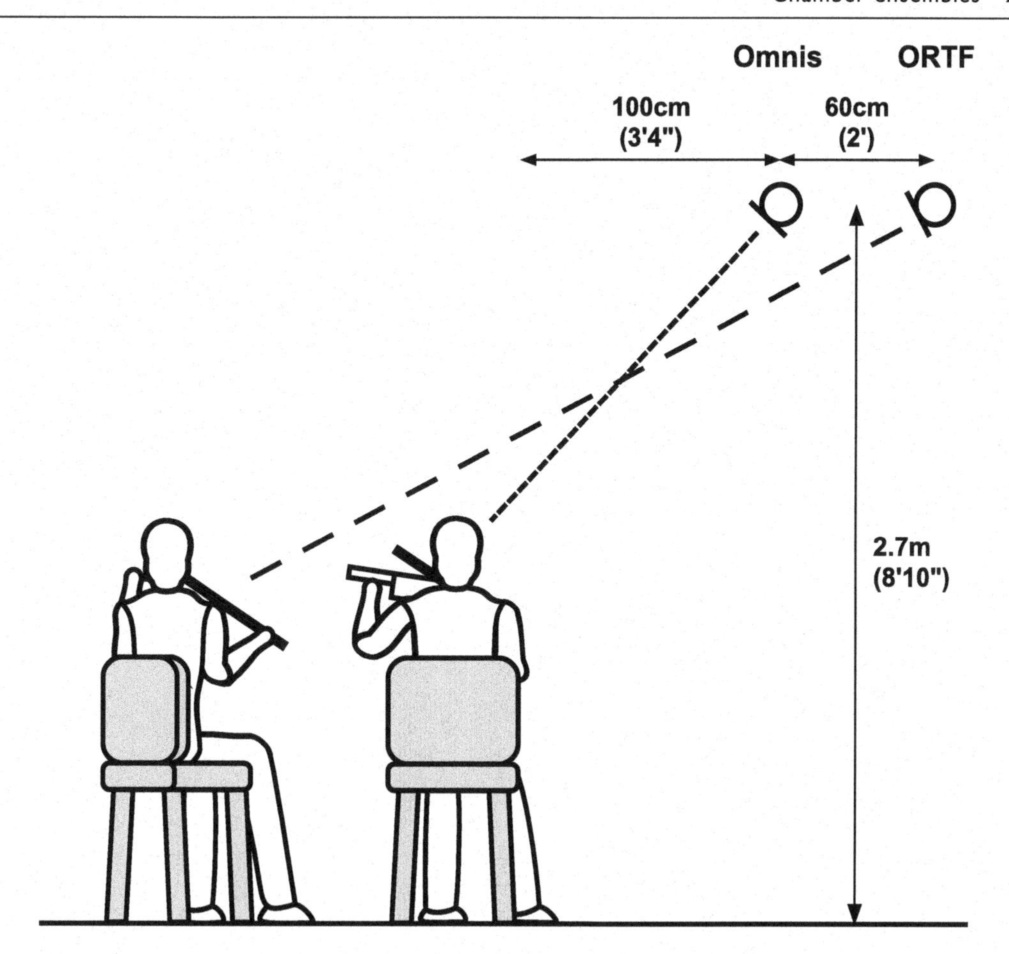

Figure 12.2b String quartet session layout using spaced omnis and ORTF pair – side view

more room sound. It means that the sound will be primarily based on cardioid microphones, so it will have reduced LF and a different tone quality to that based on omnis.)

12.1.2.1 The omni pair

The omni pair should be positioned at around 2.7 m (8′10″) in height and 30–100 cm (1′ to 3′4″) back from the instruments, depending the desired ratio of reverb to direct sound and the acoustic of the room. Height is used to gain some distance from the group whilst avoiding the front players appearing significantly closer. The spacing of these microphones is something that can be altered between about 60 cm and 90 cm (2′ and 3′). If it is left at its maximum of around 90 cm (3′), the image will be wider and less focussed, although the addition of the ORTF-type pair in mixing will narrow this to some extent, depending on how much is added. They can be closed in to around 60–70 cm (2′ to 2′4″) whereupon a little breadth will be lost, but you will be less reliant on the ORTF pair to both narrow the image and provide stability. There will be less potential image conflict in the recording if the image of the quartet on each pair is about the same width (see also 12.1.2.3 Mixing the pairs together). The addition of spheres to these microphones (as discussed in Chapter 8) will add some directionality and some HF lift if required.

12.1.2.2 The ORTF-type pair

This does not need to exactly conform to the ORTF standard (refer to Chapter 3 for a discussion of spaced and angled pairs), and the microphone angling or spacing will need to be modified to produce a suitable image width at the distance they are being used. They should be placed about another 60 cm (2′) back from the omnis, and the quartet is likely to fill much of the acceptance angle of a true ORTF pair at this distance, producing a wide image that stretches the full width between the loudspeakers. Generally, a string quartet would need to fill around 50% of the whole image width on a finished recording, so the image width will need to be reduced. Rather than panning inwards, it is preferable to reduce the angle between the microphones to nearer 80°–90° and retain full left and right panning. You will also find that this results in the microphones being pointed between the first and second violins and between the viola and cello. This will avoid the first violin and cello being directly on-axis to the cardioids and dominating the sound.

12.1.2.3 Mixing the pairs together

As with all 'mixed pair' techniques, one of the pairs should be chosen as the dominant pair in order to minimise any image confusion. This can occur where two stereo pairs are superimposed on top of one another and both contribute equally to the mix. The confusion can be particularly problematic if the source image width is somewhat different on each pair, but making one pair dominant and paying some attention to matching image widths will usually avoid the problem. With the string quartet, the suggestion would be to use the omnis as the dominant pair and bring in the ORTF-type pair just enough to focus the image. This will usually mean the omnis are around 6–8 dB higher in level. They will give a nice warm sound, extended LF, and the visceral appeal of being slightly closer to the players. As noted in section 12.1.2.1, if the image from the omni pair is slightly over-wide, the addition of some of the ORTF-type pair will pull it in a little. In terms of our lighting analogies, the omnis represent a pair of floodlights and the cardioids some spotlights placed further back. The microphone feeds from the two pairs can be easily captured on multi-track alongside the mix and will enable you to adjust the amount of focus and depth of field later on if needed.

12.1.3 Using spot microphones

It is very common to try and record the string quartet with a spot microphone on each player and a spaced pair of omnis placed further back. Because the spot microphones produce a fairly close-sounding perspective, this makes them hard to blend with the omnis other than at a very low level. However, in order to focus the image from the omnis, the spot microphones often end up being used at quite a high level, and this results in the quartet separating out into four mono sources. This effect can be avoided by using the technique outlined in 12.1.2; using an ORTF-type pair rather than four spot microphones to add focus will avoid isolating each instrument.

However, there are times when a tiny bit of help with balance and/or perspective might be needed, particularly on the cello and viola, and spot microphones can be added but used in small amounts. To avoid these microphones sounding too close, they should be placed at least 2 m (6′7″) high, and aiming them over the top of the players will reduce the amount of HF that is picked up. (The instrument will be situated just off the front axis of the microphone, and this will tone down the HF and any 'close' noises.) Spot microphones placed in this way can be added to either of the main techniques outlined in 12.1.1 and 12.1.2, but care should be taken not to overuse them. As noted earlier, occasionally on a recording session a quartet will sit with the cello seated more centrally and the viola seated on the right-hand side, thus projecting away from the main microphones. In this situation, spot microphones for adding a little more contact with the cello and viola might be needed.

12.2 String quartet in concert

The techniques outlined in section 12.1 could be used in concert, as there is little difference in the musicians' seating arrangements other than a tendency to open out the semicircle a little, but there will be a number of microphone stands in the way of the audience, and you will need to reduce these. To use the 'omnis plus ORTF' technique from 12.1.2, there are a few approaches that would be worth considering.

Firstly, mounting on stands. A single stand is far preferable in a live situation, and it is possible to mount both pairs on a single stand by using a central cross piece as shown in Figure 12.3. The attachment shown allows for adjustment of the front-to-back distance between the pairs. It is often possible to improvise a central cross piece using an additional wide stereo bar, although this will usually mean that the front-to-back distance between the pairs has to be fixed.

Where it is not possible to use an arrangement like the one in Figure 12.3, mounting all four microphones on a single wide bar would be a workable compromise. The omnis will be at the outer end of the bar at 60–70 cm (2′ to 2′4″) apart, with the ORTF-type pair in between them. The rig should be placed to get the best sound from the omni pair, even though this will place the ORTF pair a bit too close to the players. The width of the image on this pair will now be a little wider than before (because it is closer and the players take up more of its stereo recording angle). They could be angled a little further inwards to compensate.

Slinging microphones is always more difficult, but it is also more discreet, and the single wide stereo bar would be the easiest option if suspending microphones becomes necessary, as it only requires a single slinging line across the width of the stage (excluding a second safety line). See Chapter 2 for notes about slinging microphones.

12.3 Piano quintet: studio and concert techniques

The piano quintet can be considered as a combination of a string quartet with a piano, and as such, we can combine microphone techniques for the string quartet with those for the piano already outlined in Chapter 5. Please review these sections for more details.

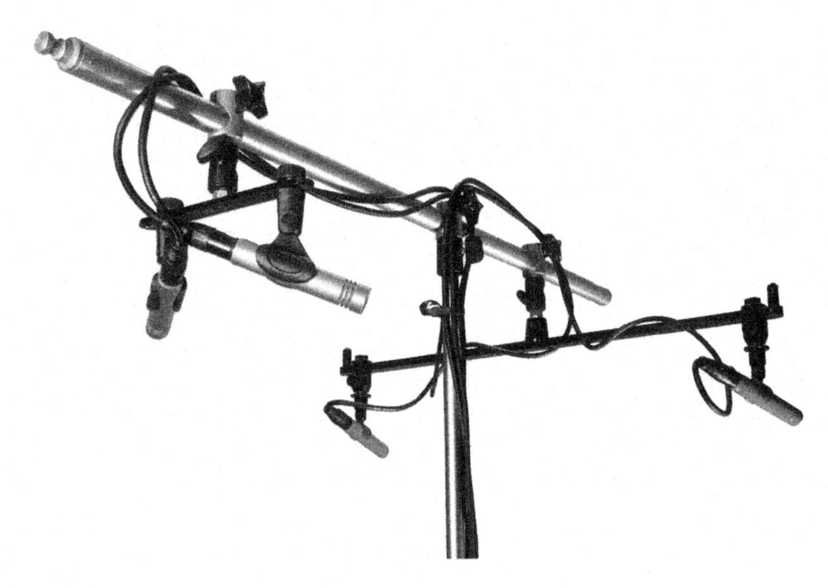

Figure 12.3 Mounting an omni pair and ORTF pair on a single stand

Photo: Mark Rogers.

For both live performance and studio work, the players can either sit in the well of the piano or further out and over towards the piano keyboard, as their preference dictates. Figure 12.4a and 12.4b shows these alternative placements. From the engineer's point of view, moving the players out of the well of the piano would always be preferred, as there will be lower levels of piano reflecting off the lid and into the string microphones. Therefore, the position in Figure 12.4b would be the favoured recording set-up if it can be arranged. (See section 12.4 for discussion of the difficulties of the reverse-seating position in this context.)

In both these positions, either of the main string quartet techniques outlined in section 12.1 can be used for the string players (a small Decca Tree with optional spot microphones if there is not enough detail or too much piano, or a combination of spaced omnis and ORTF-type pair). It should be noted that in the position in Figure 12.4a, an ORTF pair is likely to pick up a lot of piano from the lid, and this can be helped by lowering its height. If this is still a problem at a lower height, using spot microphones will be the best approach. In addition to the string microphones, a pair needs to be added to the piano to add some focus and to control the piano image placement. If one of the techniques from section 12.1 is used for the strings, there will already be some omnis contributing to the overall tonality of the piano sound. With the lowest LF taken care of by these omnis, the piano microphones themselves can be cardioid or wide cardioid to give a little more image focus and to reduce spill from the strings. In Figure 12.4b in particular, the piano is somewhat to the right on the string microphones, but it should be possible to anchor the piano image more centrally using the piano microphones and panning them fully Left and right around the centre of the image. The tree itself will give a nice full sense of stereo space to the string quartet and the wider room.

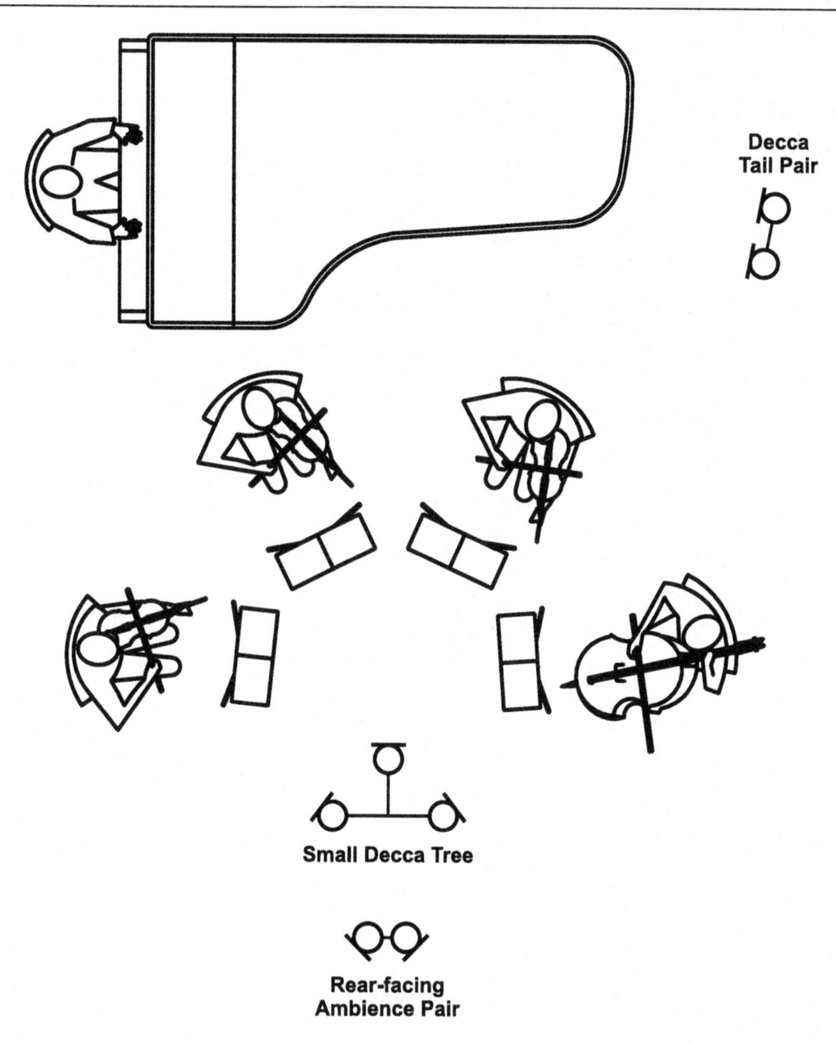

Figure 12.4a Piano quintet sessions layout with string players in the piano well. String spot microphones are optional if the tree does not give enough detail. The mixed omnis and ORTF microphone technique shown in Figure 12.4b can also be used in this layout, although the height of the ORTF pair might need to be reduced if it is picking up a lot of piano reflecting straight off the piano lid.

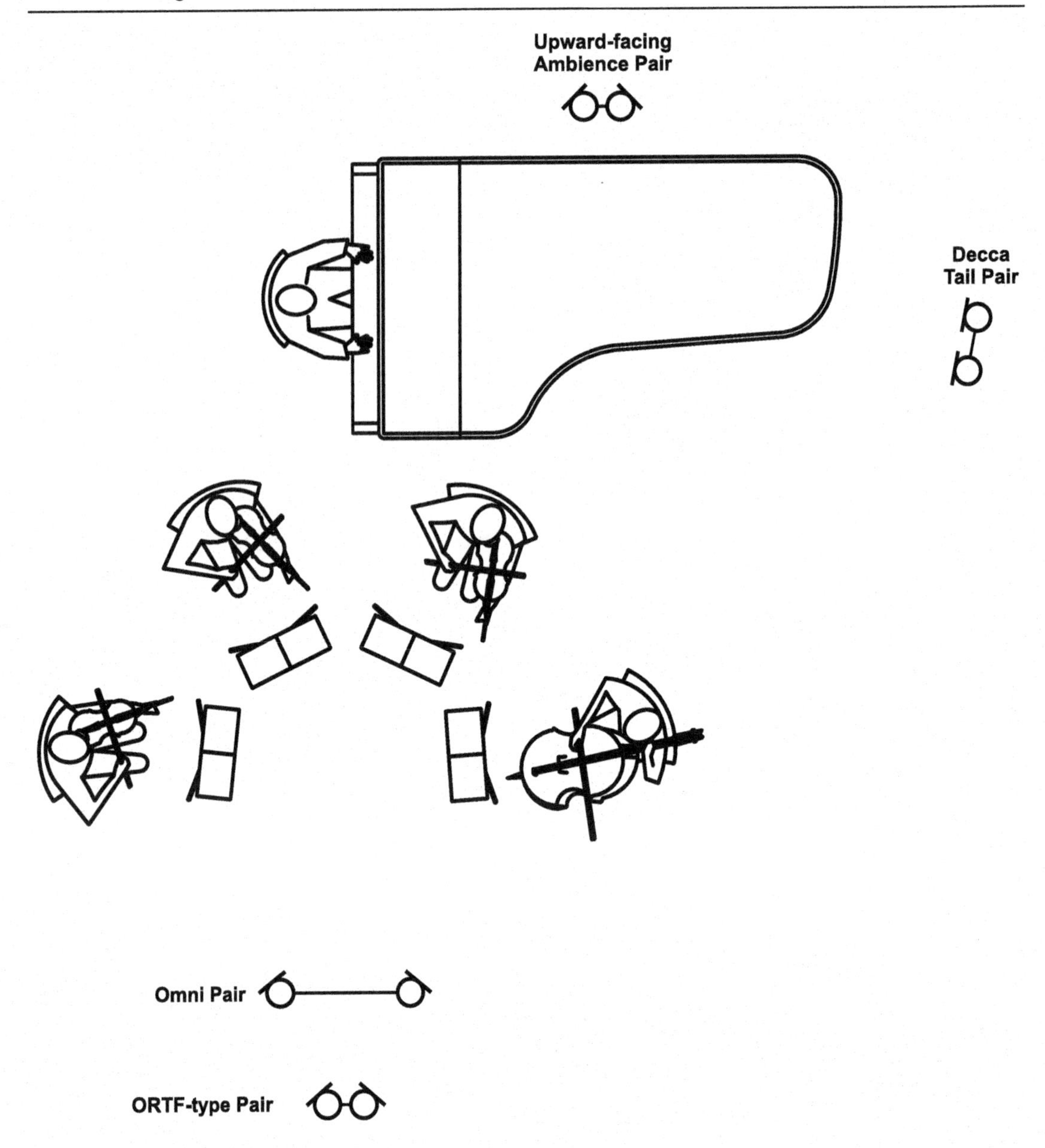

Figure 12.4b Piano quintet sessions layout with string players sitting away from the piano well, which would be preferred for recording as it reduces the level of piano on the string microphones. The small tree microphone technique shown in Figure 12.5a can also be used in this layout.

In order to pick up some more reverberant sound when using either of these positions or microphone techniques, an ambient pair can be rigged, either upwards-facing behind the piano (remembering that this is a good discreet position in a live situation) or behind the main array of tree or spaced omnis and ORTF. If placed behind the tree for a studio recording, they could be facing backwards instead of upwards without the danger of picking up a lot of audience noise. The aim of these microphones is to provide some natural room reverb that contains all the instruments to help bind them together in the same space.

12.4 Piano quintet: reverse-seated studio position

The string players will often prefer to face the other way around so that they can see and hear both each other and the pianist really well, as shown in Figure 12.5a and 12.5b. Many of the Ashkenazy, Perlman, Zukerman, and Harrell recordings[2] were done in this seating position using the microphone technique shown in Figure 12.5b with individual spot microphones.

Once the players are seated this way around, the techniques outlined in sections 12.1 and 12.3 will not be suitable because the string microphone array will be right in the middle of the players and very close to the piano. If there is enough distance between the quartet and the piano (Figure 12.5a), it might be possible to fit in a pair of ORTF-type cardioids to pick up the strings, but any omnis in this position will pick up too much of the piano. Given that the players prefer to sit this way in order to hear and see each other well, the likelihood is that they will be quite close to the piano, and it will not be possible to fit in a pair at an appropriate distance from them. In this situation, the only solution will be to give the string players a microphone each (Figure 12.5b). As noted in section 12.1.3, a great deal of care needs to be taken when relying on spot microphones to avoid the recording sounding like a collection of individual mono sources. The spot microphones should not be too close – at least 2 m (6′7″) high – and some additional ambience microphones might be essential in providing a blended sound of all the players together to soften the effect of the spot microphones. These can be placed either behind the piano or behind and above the string quartet in order to collect a more distant sound from all the players.

Ultimately, the decision about the seating arrangement will be in the hands of the players, and if you are not sure how they are going to want to sit, you should allow yourself enough microphones, stands, and channels to be able to give them a spot microphone each, if that becomes necessary. More experienced players tend to prefer the reverse-seated recording position, but younger players might prefer the outward facing 'concert' position as it is familiar. Most players prefer to avoid sitting in the well of the piano, as the sound is loud and overwhelming.

12.5 Piano trio: studio and concert techniques

The piano trio (piano, violin, cello) will normally be seated for performance, as shown in Figure 12.6a, and the techniques outlined for the string quartet and piano quintet can be adapted for use here: a small Decca Tree with optional spot mics, or ORTF and spaced omnis centred on the string players, plus a piano tail pair and an ambient pair. Again, if there is too much piano on the ORTF pair, try reducing its height, and if this doesn't work, move to using spot microphones.

Figure 12.6b shows an adapted layout for recording with the players moved over to be nearer the piano keyboard for closer communication with the pianist. This includes a small Decca Tree

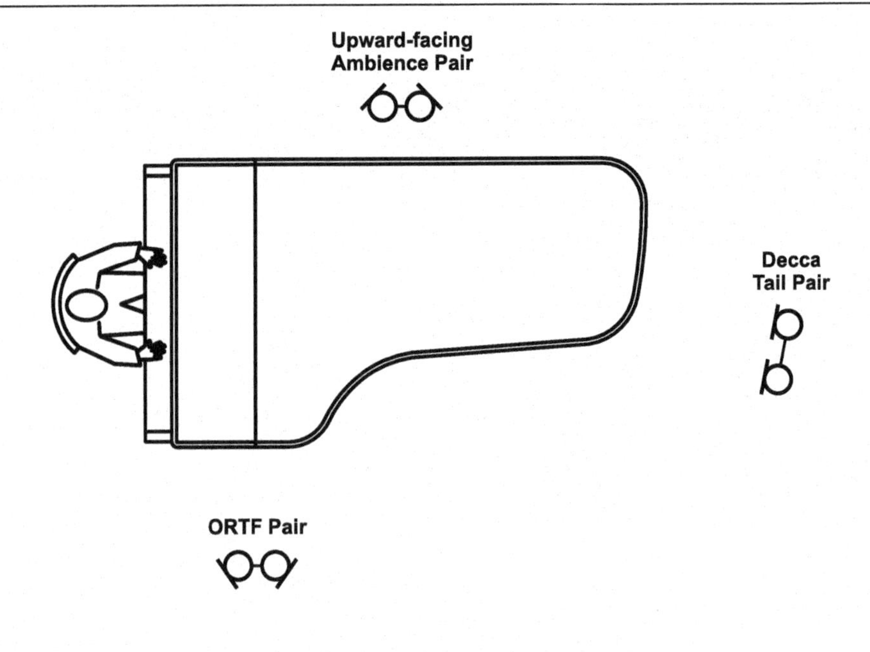

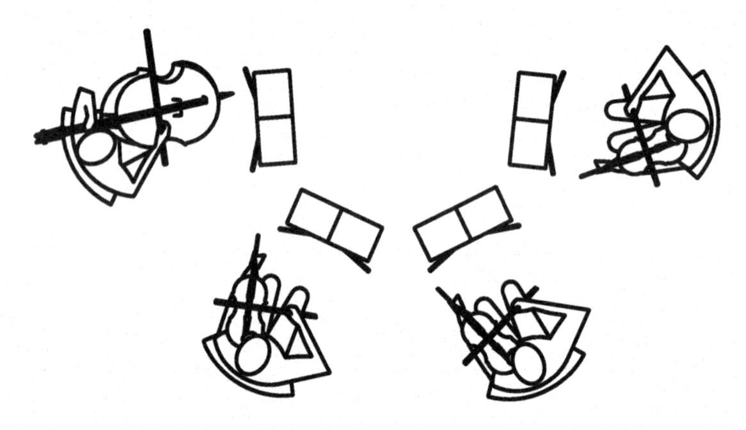

Figure 12.5a Piano quintet reverse concert layout for recording. This shows some distance between strings and piano, enabling use of an ORTF pair on the strings.

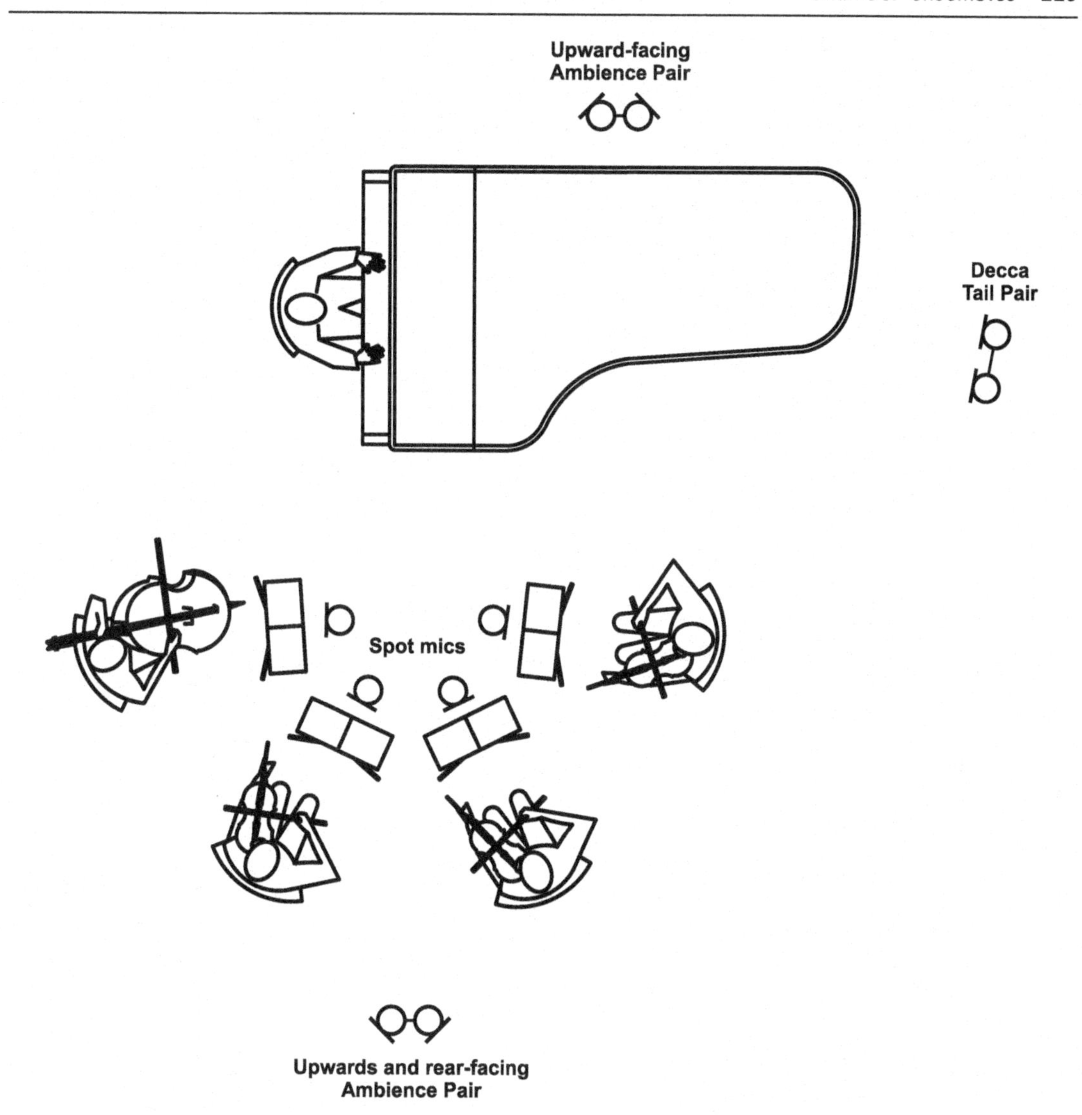

Figure 12.5b Piano quintet reverse concert layout for recording. This shows the players grouped more closely than in Figure 12.6a, in which case spot microphones plus an ambience pair will have to be used.

centred on the string players to place them centrally in the recording image and to provide a good sense of space and reverb around the strings, a microphone each on the violin and cello for a little focus, a piano tail pair for the main quality of the piano sound, and a vertically pointing ambience pair to add a little reverb to the piano, as the tree will provide more reverb to the strings than to the piano. The piano pair will be panned L-R in mixing in order to anchor the piano image centrally – the piano will be appear to the right on the string image from the Decca Tree.

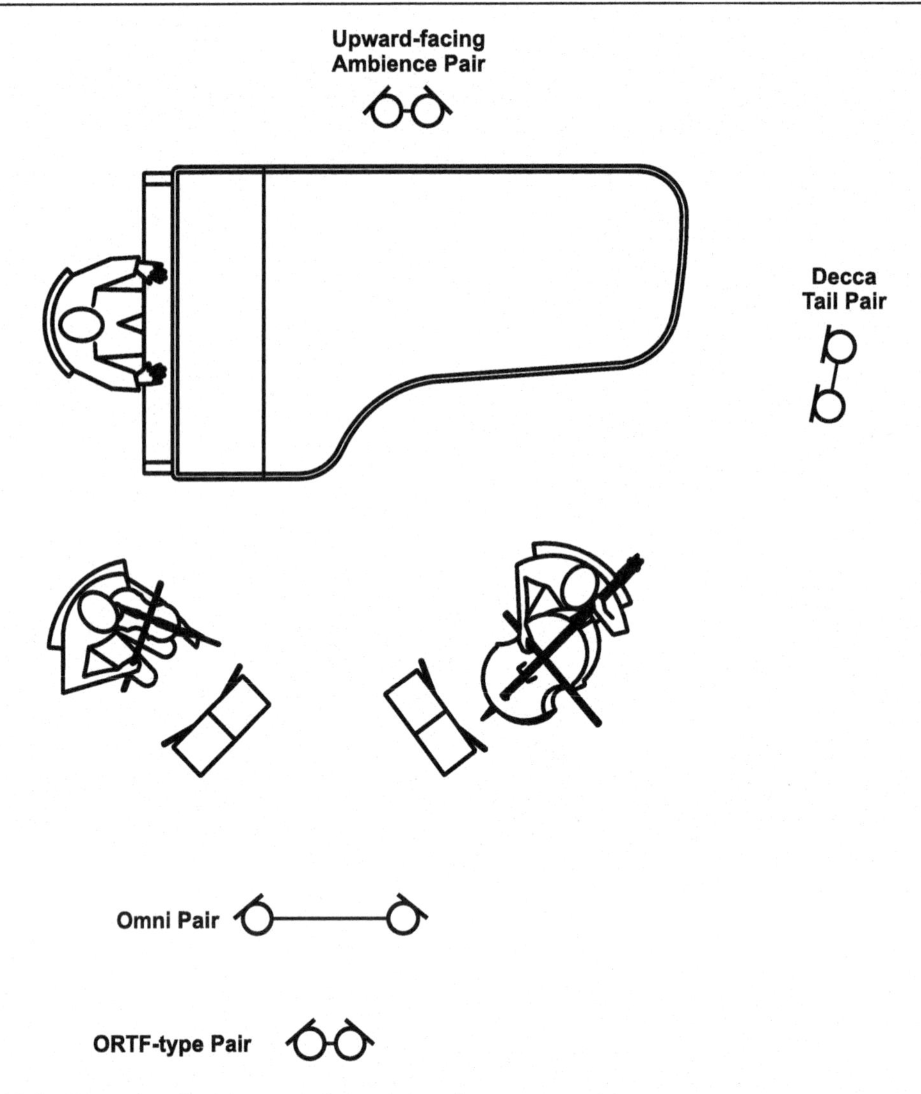

Figure 12.6a Piano trio facing outwards in a typical concert position

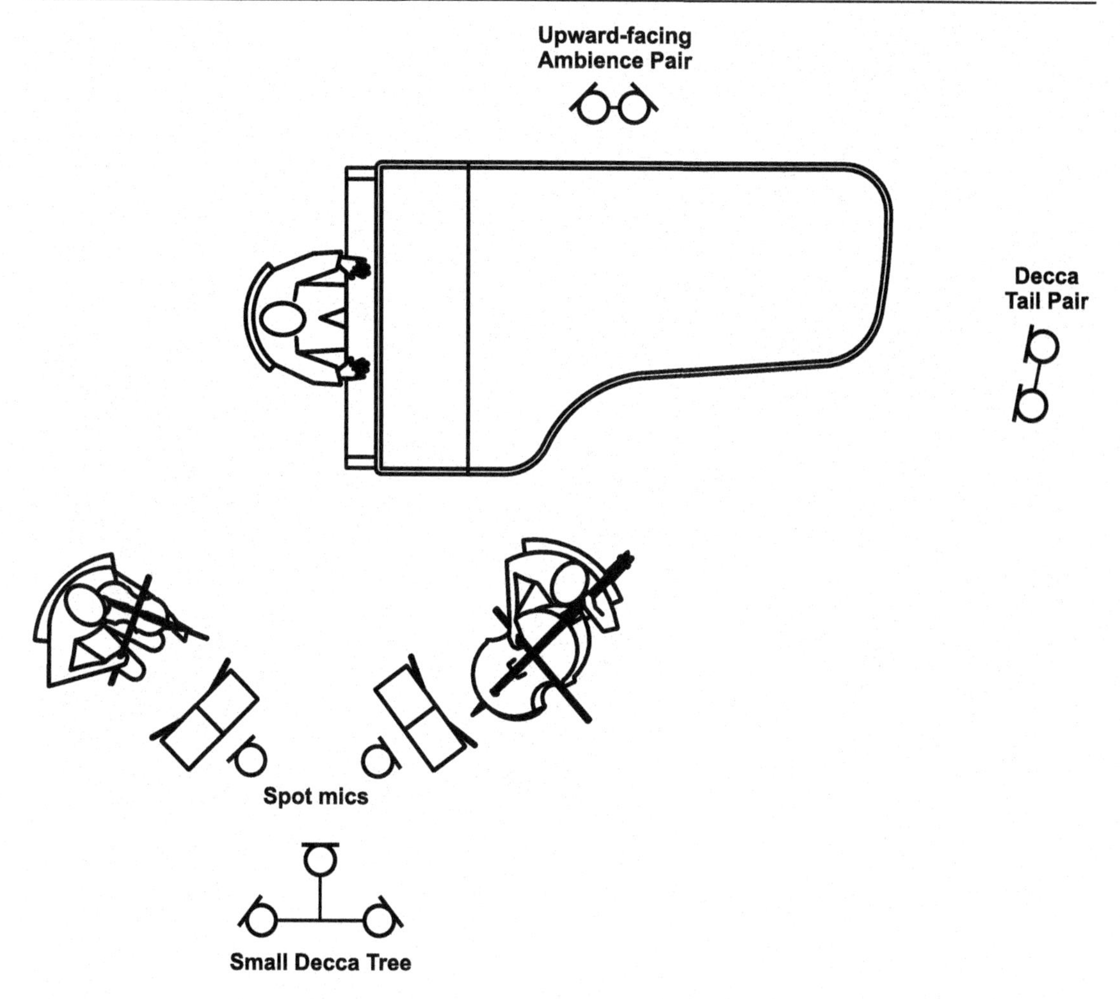

Figure 12.6b Piano trio facing outwards in preferred recording position

As with the piano quintet, if the string players prefer to turn and face the piano for recording, you will probably have to rely on using individual spot microphones placed at least 2 m (6′7″) high; see section 12.4 for notes about the piano quintet. Unless the string players leave a very large gap, there will not be enough room to place a Decca Tree between them and the piano without it being too close to the piano.

12.6 Small wind ensembles: studio or concert layout

This section is aimed at chamber wind ensembles of up to about 12 players. For information about how to record a larger, wind band–sized ensemble, please see Chapter 13.

Figure 12.7 Photograph from Abbey Road string trio session set-up similar to Figure 12.6b

Photo: Carlos Lellis, Programme Director, Abbey Road Institute.

There can be a subtle difference in approach when recording wind ensembles as opposed to string ensembles. Strings have a homogeneity of sounds that easily blend together, whereas the woodwinds have their own more distinctive and individual sounds that need greater punctuation and clarity. A certain amount of blended sound is desirable when recording a wind ensemble, but the need for some subtle additional detail means that spot microphones are more likely to be needed. The mixing needs to be more dynamic, relying on small fader moves to bring out individual lines and produce a really musical recording.

For small wind ensembles with piano, the techniques outlined for piano quintets and trios above can be used. Remember that getting too close to the woodwinds is to be avoided because of key noise; the section on wind soloists with piano in Chapter 7 would be useful reading at this point for advice on distances and microphone types.

12.6.1 Wind ensembles without piano: recording or concert

Wind ensembles without piano might range in size from 4 to 12 players, and where possible, it is best to arrange them seated in a single-row semicircle or horseshoe to help the communication between the musicians. All the elaborate microphone techniques in the world will be wasted if the players cannot communicate well, and sight lines between them are of primary importance. Figure 12.8 shows the layout used for recording the Amadeus Winds with Christopher Hogwood in 1987 performing Mozart's 'Gran Partita' K. 361 for 12 winds[3] (two oboes, two clarinets, two basset horns, two bassoons, and four horns) plus a double bass or contrabassoon. See Appendix 3 for the set-up sheets for this session.

This is based on a Decca Tree for the overall sense of the ensemble and the space, with an additional ancillary microphone for each pair of instruments and one for all four horns, applied in a similar way to the ancillary microphones used for orchestral recording (see Chapter 9). Also shown are an upwards-facing pair of cardioids for capturing more indirect sound.

The tree used is a full-sized tree rather than the smaller version used for the string quartet, although it would also be possible to use the spaced omnis plus ORTF-type pair technique also discussed in the context of the string quartet. In this case, the omnis would need to be spaced at

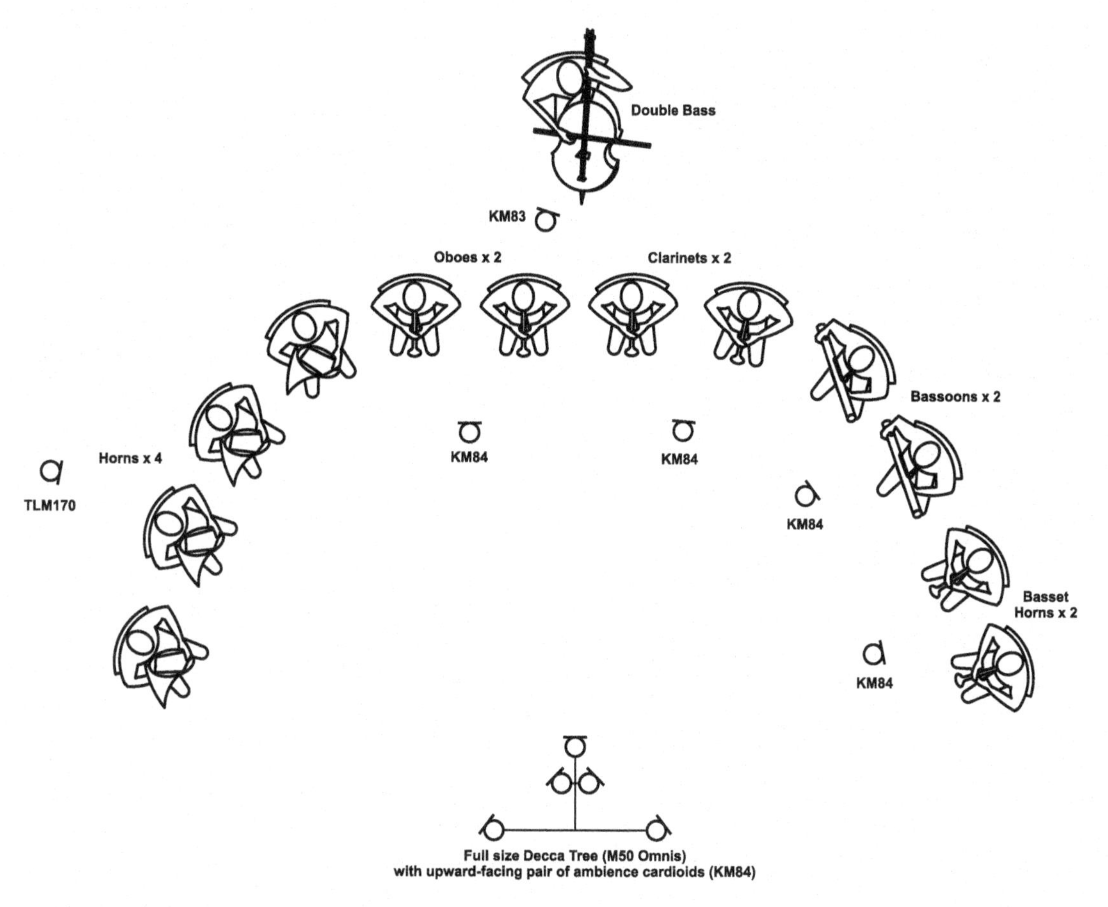

Figure 12.8 Wind ensemble layout for Mozart Sonata No. 10 for winds 'Gran Partita'.

the wider end of the range given the increased size of the ensemble. This ensemble of 13 players would need to occupy about 70% of the full image width, so the spaced omnis should be panned fully and their spacing adjusted to produce an appropriate image size. Take the same approach with the ORTF pair, and adjust the angle between them to produce an appropriate image width when fully panned. The spot microphones are used in conjunction with either main pickup method to bring a little focus to each pair of players; avoid using microphones with HF lift as spots to reduce the impact of key noise. The spot microphones should be panned as they appear on the image from the main pickup to avoid lateral image movement when they are altered in level during mixing. For example, if spots are panned so instruments appear closer together than on the main image, individual instruments will move inwards as their spot microphones are increased in level. If the panning is well matched, different musical lines can be brought out during mixing using the spot microphones, depending on what is musically important at the time, but the image will remain laterally stable.

If this were a concert recording, and rigging a tree was considered too intrusive, using the spaced omnis plus ORTF discussed for the live string quartet in section 12.2 would be more discreet. It would still be important to retain some imaging control with ancillary microphones for each pair of players, but having them rigged on stands in front of the players at 2 m (6′7″) high is likely to be unacceptably cluttered. These could be boomed in from behind the players to avoid having to place them very low and in front, but where this is also not possible, lower spot microphones in front of the players will be the final resort. In this case, make sure the microphones are high enough to be above the bell of clarinet and oboe-type woodwinds, and keep as far away as you can to reduce key noise and a localised sound.

Figure 12.9 shows the Wiener Oktet recording in the Sofiensaal, Vienna. A small Decca Tree is shown with some additional spot microphones, including one placed behind the horn, which has acoustic screens placed behind it.

12.6.2 Wind quartets

Where there is a smaller wind group, such as a wind quartet or flute, oboe, clarinet, and bassoon, the overall techniques (small tree, or spaced omnis and ORTF pair) outlined for string quartet earlier can be used as a good basis for the sound. Spot microphones might not be needed if you can get your overall pickup in the right place and the players are well balanced, but if it is a live concert, you should allow for using them just in case.

For the Amsterdam Loeki Stardust recorder quartet recordings of Baroque recorder music in Abbey Road Studio 1 in March 1987,[4] an AKG C426 stereo microphone was used with some spaced M50s further back at about 2.75–3 m (9′ to 10′) high. Because of the intermodulation effects that are a feature of recorder ensembles, it is not a good idea to place the microphones too close. Because this group liked to sit close together, there was no room to fit in spot microphones at a suitable distance, and so the AKG stereo microphone was used to provide the detail and the spaced omnis for the overall tonality and sense of space. A recorder group has a homogeneity of sound that is more like the string quartet than a typical wind quartet of flute, oboe, clarinet, and bassoon, and if they are well in tune, well balanced and in time together, the blended sound they produce can sound almost like a pipe organ.

Figure 12.9 Wiener Oktet in Sofiensaal, Vienna

Photo: Courtesy Decca Music Group Ltd.

Notes

1 *SHOSTAKOVICH/String Quartets 1–15/Fitzwilliam Quartet/DECCA (1980) LP D188 D7 (1998) CD 455 776–2*
2 *Example – DEBUSSY/RAVEL/Vln and Cello Sonatas/Trio/Ashkenazy/Perlman/Harrell/DECCA (1995) UPC 028944431827*
3 *MOZART/Serenade K. 361 'Gran Partita'/Amadeus Winds/Christopher Hogwood/L'OISEAU LYRE (1989) UPC 028942143722*
4 *BAROQUE RECORDER MUSIC/Amsterdam Loeki Stardust Quartet/L'OISEAU LYRE UPC 028942113022*

Chapter 13

Wind, brass, and percussion bands

The focus in this chapter will be on larger groups of players, typically requiring a conductor, although small classical brass ensembles have also been included. For smaller groups of 4 to 12 string or wind players, please refer to Chapter 12.

13.1 A note about dynamic range and ear protection

It perhaps goes without saying that brass and percussion ensembles in particular can get very loud, and an adequately sized room is needed for performance and recording or the room will saturate acoustically. (See Chapter 1.) When recording and setting up microphones around any very loud ensemble, you should consider the benefits of ear plugs or ear defenders for when you find yourself rigging microphones whilst close to an instrument's bell or percussion section. You may also need to make use of the pads in your condenser microphones to avoid clipping in the head amplifier.

Percussion groups and brass bands can have very high level peaks on the transients. When a brass band is playing legato, the sustained sound has a very smooth envelope, but if the whole band comes in together with a sforzando, there can be a very large peak in level at the start of the note. Managing and preserving the excitement of this dynamic range is an important consideration when recording them. This leads us to the question of whether compression should ever be used when recording classical music, and while different engineers will have slightly different approaches, all would agree that the music should never be audibly compressed. The dynamic range (between noise floor and peak level) of modern recording media is such that compression is not needed from any technical point of view, and the spirit of classical recording is to preserve the natural dynamic range. The norm, therefore, is not to use any automatic compression. However, broadcasters' requirements mean that they often need to reduce the dynamic range of natural playing. If you need to reduce the dynamic range of a performance for any other reason, then you can produce better results (i.e. inaudible) by manually altering the levels throughout. This is best left to the mastering stage, if it needs to be done at all (see Chapter 19 for more details on this).

13.2 Large wind ensembles

Wind ensembles come in many shapes and sizes, from concert wind bands to flute choirs, but they have no standardised line-up equivalent to that of the standard brass band or orchestra. The concert wind band might also include drum kit and bass guitar, and possibly a PA system for a singer. At this point, it starts to become a non-classical ensemble and the recording approach will be that used for a big band of similar line-up.

The wind instrument family contains a wide variety of tone colours, and the overall sound from a wind ensemble lacks the homogeneity that is characteristic of a string orchestra or brass band. In addition to an overall similarity of tone quality between all the parts, the exclusively string or brass ensemble can also play very soft, smooth attacks which if played together gives a feeling of a single, coherent unit. Wind instruments (the double reeds in particular) have more definite transients that punctuate the playing more clearly even when the group is playing perfectly together. Capturing some of the detail and individual tone colours of the woodwind lines is an important part of the recording approach.

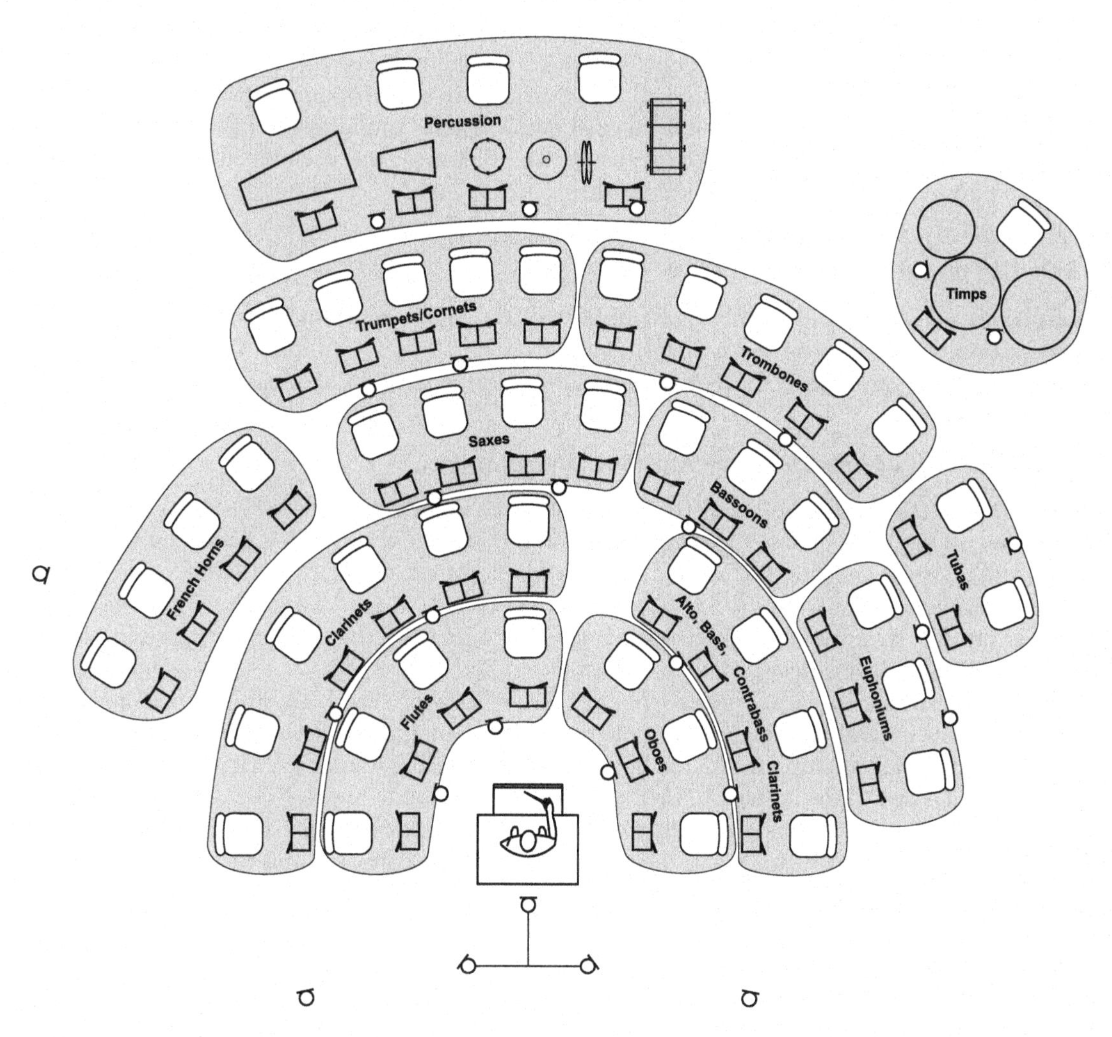

Figure 13.1 Large wind band in a typical layout

For wind ensembles of up to about 20 players, the semicircle layout outlined in section 12.6.1 can be adapted by placing the players in two rows instead of a single row and using an overall pickup with some spot microphones to capture more detail.

Figure 13.1 shows a layout for a typical larger wind/concert band where there are enough individual players to form distinct sections. Overall coverage is given by a five-microphone Decca Tree with ancillary microphones on each section. The ancillary microphones would be placed above and looking down, at a height of around 2.5 m (8′2″), and ideally about 1.2 m (4′) back from the instruments, although this is likely to be limited by chair spacing.

Large wind ensembles that include PA are really outside the remit of this book, however, there are a few principles to remember. Once you are working with a PA, spill from the PA loudspeakers onto the microphones has to be reduced to imperceptible levels to avoid feedback. Therefore, the direct sound from each instrument must be at a high level compared with any background reverb or PA sound, and this means placing microphones closer to the instruments. As we will see in Chapter 15, the closer microphones have to be, the more of them are needed to obtain even coverage of all the players; this will mean at least one microphone between two players, and possibly one each. More EQ will be needed, as a close-up sound will not give a good overall representation of the natural sound of the instrument. Balancing an ensemble in this way is much closer to pop recording than to classical work.

13.3 Classical brass ensemble

Instruments used for classical brass ensembles will usually come from the standard orchestral brass section, consisting of French horns, trumpets, trombones, and tubas. The individual tone colours are more varied than those that make up a traditional brass band (see section 13.4).

13.3.1 General approach to microphone placement

There are two features of brass instruments that make recording them different to strings and woodwinds. One is their huge dynamic range and ability to play sudden fortissimos, and the other is their strong and simple directionality of radiation from the bell. The whole sound from a brass ensemble needs to combine a good overall tone (acquired at a distance to allow the fortissimos to interact with the room) with a flavour of the attack and immediacy of the sound heard just in front of the instruments.

For most previous scenarios in this chapter, the balance of reverb to direct sound has been arranged by the careful placement of the main pickup at an appropriate distance from the players. Additional spot microphones have been added in small amounts to bring some focus and detail where needed. Brass come into their own when heard at an appropriate distance in a good room, so the overall pickup is placed further away, at a height of at least 3 m (9′10″) to capture a slightly more spacious blend that is better suited to the instruments and to avoid a feeling of congestion and saturation, which can arise with the microphones too close when the players play loud.

With the main microphones at a distance and the players facing one another in a semicircle and not projecting any HF towards them, the attack of the instruments will be lost. These transients would be easily picked up on a closer and more on-axis microphone, but this can sound very bright and does not make a well-balanced brass sound taken in isolation. It is important, however, that it is included as part of the mix as it represents an essential part of the instrument's character and provides detail and punctuation to the playing. To this end, spot microphones are used to capture the immediacy of the sound that is heard directly in front of the instruments, although you should avoid them being placed directly on-axis to the instrument's bell. This means that any change in tonality or level that comes from the player moving the instrument is reduced, and air noises will be minimised.

In many ways the approach to recording brass chamber music reflects orchestral brass techniques; they are heard from quite some distance on any orchestral main pickup, but spot microphones are used to add the essential bite and attack. When the players really let rip, the more distant microphones will add a wonderful bloom and can fully convey the excitement of just how big the sound is.

In the brass recording suggestions that follow, the balance between reverberant and direct sound is created in this more artificial way. A relatively distant main pickup that provides the main tonal character of the sound is blended with a more substantial level of spot microphones than would be usual for strings or woodwinds. The spot microphones and overall pickup will usually be contributing about equal amounts to the mix in this scenario, with the spot microphones placed a little closer than is usual in a classical context to really capture the attacks.

Figure 13.2 shows a set-up for a brass quintet comprising a spaced pair and some spot microphones for each instrument or pair of players. Figure 13.3 shows a set up for brass octet/nonet

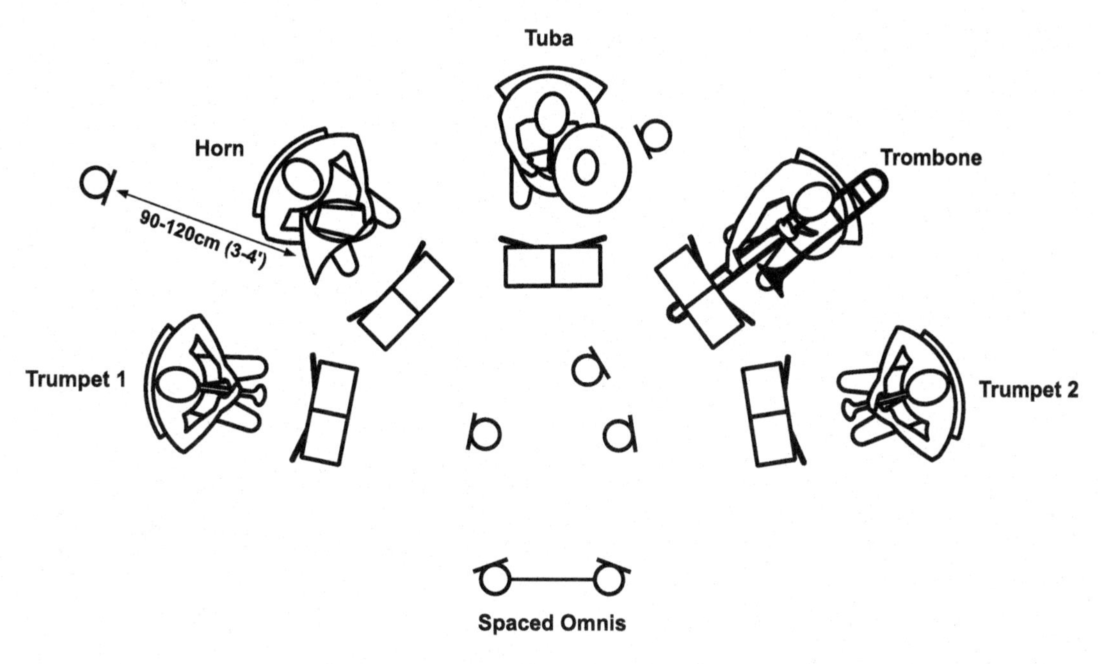

Figure 13.2 Brass quintet – this shows one commonly used layout, but some players prefer to sit: trumpet 1, trumpet 2, horn, trombone, tuba.

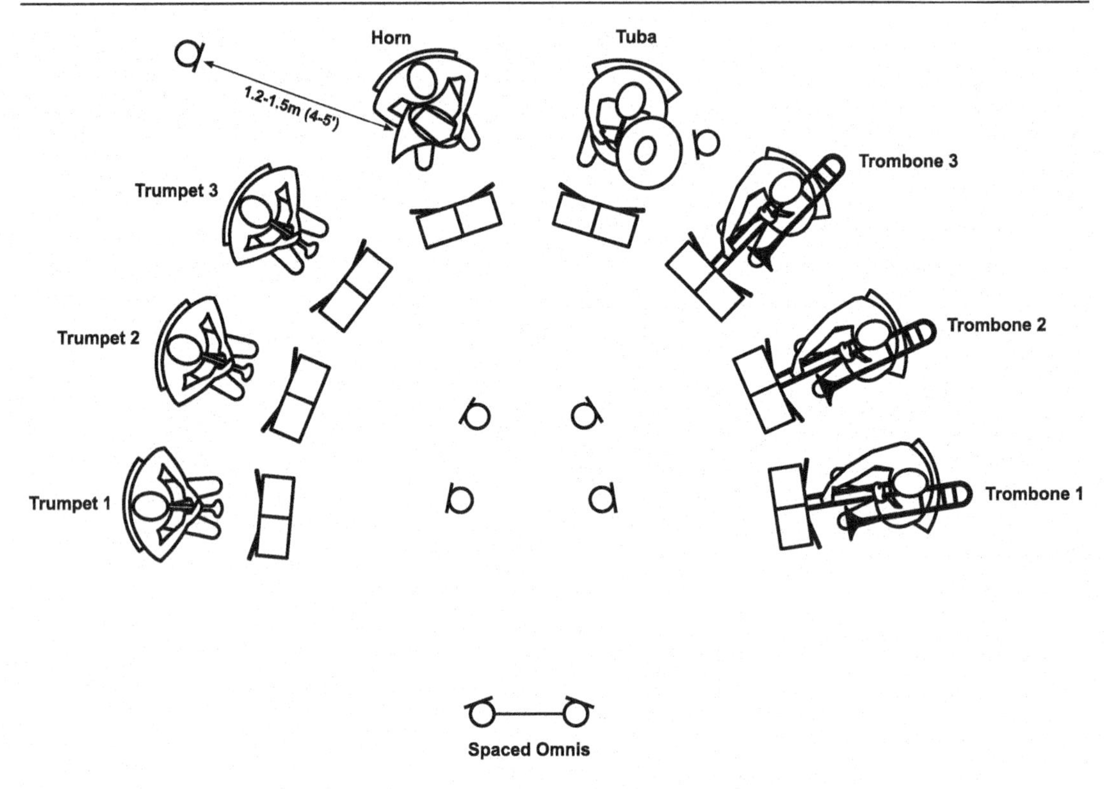

Figure 13.3 Brass octet/nonet showing a layout with the tuba in the centre. The tuba will sometimes be placed at the right-hand end (see Philip Jones ensemble in Figure 13.6).

following similar principles. The layouts shown have a single row of players, but if there is a second row of players (for a larger classical brass group), the microphones can be raised up to look at both rows, and if the back row can be offset a little, the microphone can be aimed through the gaps in the front row (Figure 13.5).

For the smaller quintet group, the microphones will be a little closer, as there is less space between the players as they are grouped around in a semicircle, so they will be about 60–90 cm (2′ to 3′) back and about 1.2 m (4′) high, again just above the bell. See Figure 13.4 for a sideways view.

For the larger nonet layout, the spot microphones on the forward-facing players are a little further away and slightly higher at 1–1.2 m (3′4″ to 4′) back from each individual player and about 1.5 m (5′) high, above the bell and looking down across it. The additional height is particularly important if there is a back row of players as well; if the back row can be offset a little, the microphone can be arranged to aim between the front row of players.

The upwards-facing tuba(s) will need a microphone 60–90 cm (2′ to 3′) above, but not on-axis to the bell (to avoid the blasts of air that are a particular problem with the tuba), and the French horn will need a microphone from behind. If you are working at the slightly greater distances

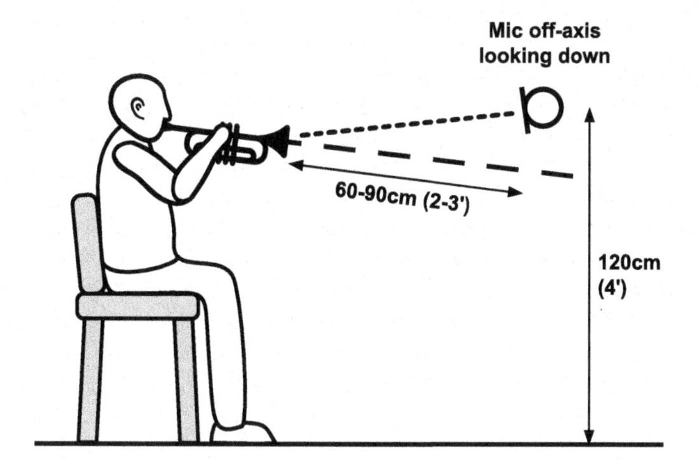

Figure 13.4 Sideways view of microphone placement on single row of trumpet players

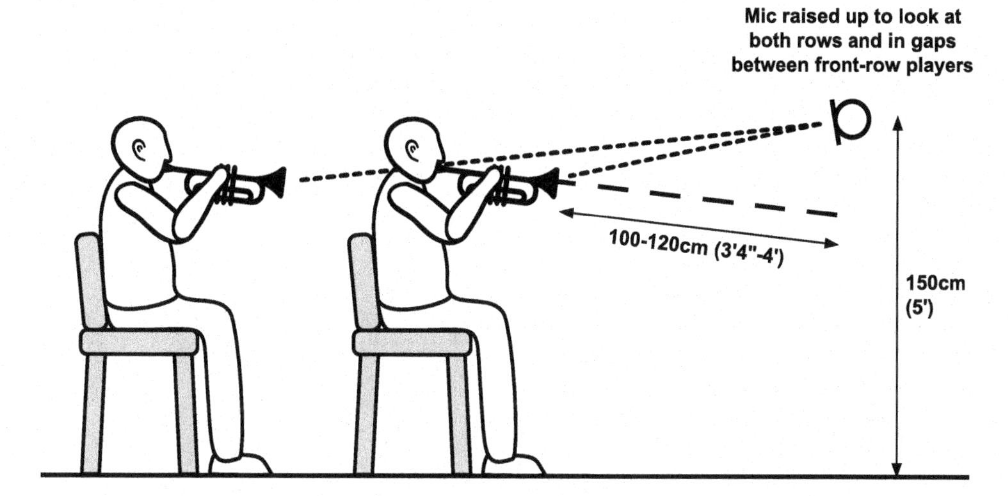

Figure 13.5 Sideways view of capturing a double row of trumpet players, with players offset on the back row to look through to the front

used for the nonet, the horn microphone will be at about 1.2–1.5 m (4′ to 5′) away, but if you are using the closer placement used in the quintet, the horn mic can be about 90–120 cm (3′ to 4′) away. In both cases, it should be high enough to be just looking down across the bell. When the players are sitting in a fairly close position, the spot microphones on the forward-facing players form a tight back-to-back horseshoe in the middle of the ensemble.

When the players are seated in a horseshoe shape, the choice of microphones will be limited to cardioids. If a fig of 8 is used, it will pick up any player situated behind it as much as the player in front, and the independent control over the level of the two instruments will be lost. The inverted phase of the rear lobe signal will also potentially result in phase cancellations. If the players are in a line, or the rear lobe is not pointing at any other player, ribbon microphones (Royer R-121 or Coles 4038) can sound wonderful on brass, as long as they are kept out of any wind flow to avoid damaging the ribbon by stretching it beyond its normal limits.

Figure 13.6 A photograph from a Philip Jones Brass Ensemble session showing the tight horseshoe of central microphones

Photo: Mike Evans; courtesy Decca Music Group Ltd.

13.3.2 Blending the microphones

When trying to find a suitable balance for the classical brass ensemble, it can be helpful to listen to the spot microphones alone first, making sure they are positioned well and are producing the sound you want. They will be used at a higher level than the spot microphones on a woodwind or string recording and will be contributing a similar level to the mix as the overall pickup, so spending some time getting them well placed is important.

As noted earlier, the microphones on the trumpets and trombones will be physically very close to one another in the middle of the ensemble and will produce quite good image localisation because of their proximity to one another. They should be panned individually between about three-quarters left around to three-quarters right, depending on the number of players, and this will give you a secure basis for the sound. Once you are happy with the sound and balance between the spot microphones, the more distant and warmer sound from the spaced omnis can be introduced to glue the separate spot microphone sources together and place them in the room. Because the image localisation from the omnis will be fairly unfocussed, the spot microphones can be panned as you prefer them and the addition of the omni pair should not cause any imaging confusion.

The act of balancing the ensemble starting with the spot microphones is the opposite to most other situations, including brass band (see section 13.4); the usual approach is to listen to the overall pair first and bring up the spot microphones, having adjusted their panning to match the main image. It is of course possible to balance a classical brass ensemble in the more conventional way by starting from the main pickup, but because the spot microphones are playing such an important part in the image and tonality of the sounds, it makes more sense to check on their contribution in isolation first.

13.4 Brass band

The brass band is potentially a much larger ensemble than the smaller classical brass group, and it is best approached as a 'brass orchestra' in terms of recording and balancing. The instruments used are different to those in the standard orchestral brass section, and include cornets (instead of trumpets), tenor horns, baritone horns, euphoniums and tubas, a trombone section, and a variety of percussion. The cornets have a darker and warmer tone than trumpets which arises from their having conical rather than cylindrical bores. The whole family of instruments taken together produces a homogeneity of sound that blends very well, and as the parts are written in a similar way to those of a string orchestra, it can help to think of them as 'strings in brass form' when approaching a recording. This is quite a different experience to recording a wind or orchestral brass group with their mix of very distinctive tonalities that don't blend in the same way.

The sound of a brass band should be mellow and smooth, and the attacks can be very soft when required, particularly when playing more lyrical repertoire. None of the instruments has its bell pointing directly at the audience; the forward-firing cornets and trombones are seated in the place of the violins and double basses and face the conductor and the centre of the ensemble. Figure 13.7 shows a typical layout with suggested microphone set-up.

Here a three-microphone tree with outriggers has been used, along with ancillary microphones for individual sections. As a general principle with this approach, there is no need to provide a microphone for each player; one microphone between two players would be plenty, the aim

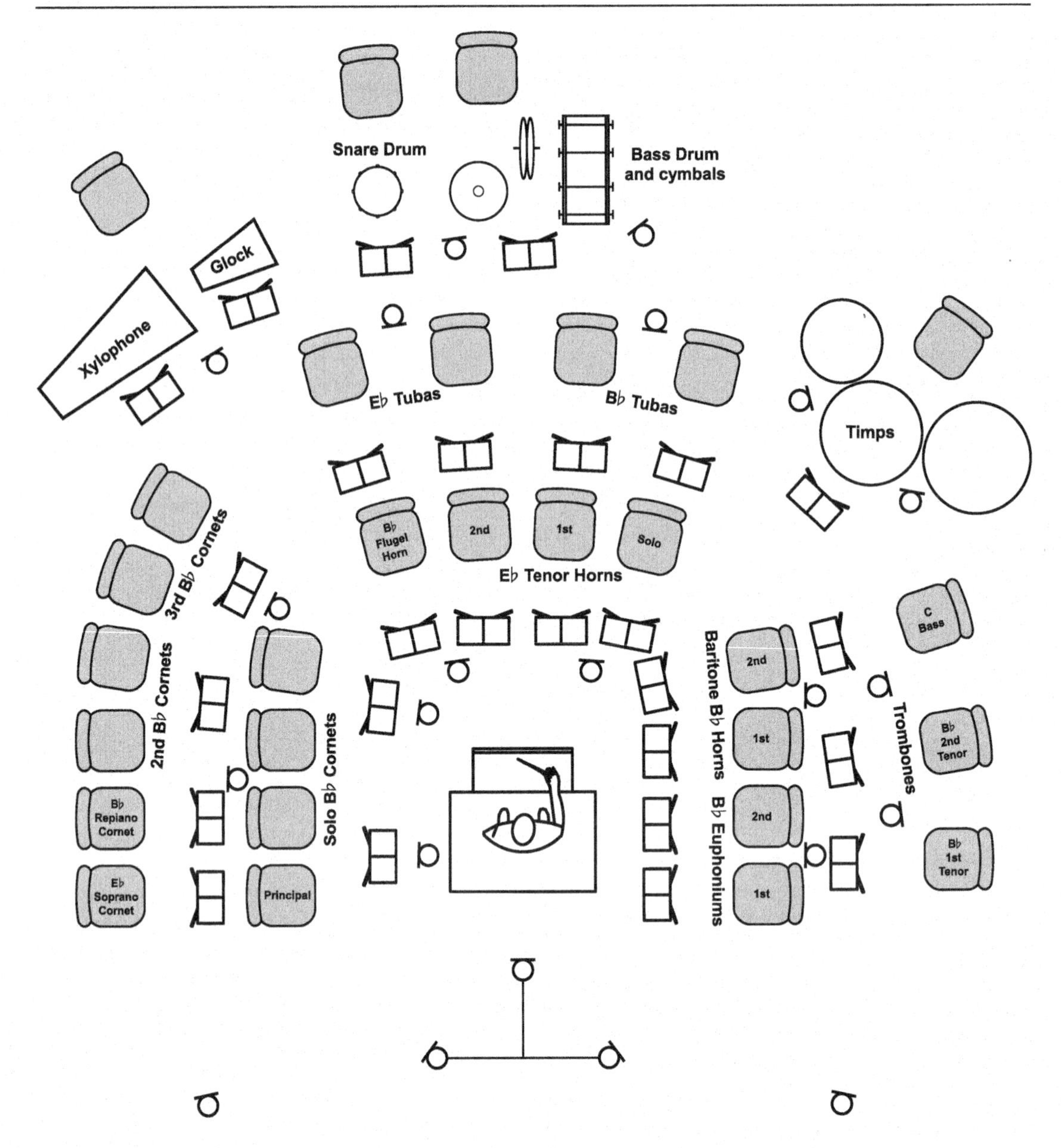

Figure 13.7 Brass band layout

being to obtain even coverage of each section as we would for an orchestral string section. At all times, remember the rule that brass instruments are best not recorded on-axis to the bell if a natural sound is required. The spot microphones will be used to provide more attack sound and are particularly useful for preventing the lower end of the ensemble (E flat and B flat tubas) from lacking any definition and becoming 'woofy'.

Percussion will be arranged along the back, as for an orchestra, and spot microphones can be added in the same way, including panning slightly wider than they occur in the natural image from the tree (see Chapter 9). The timpani can be covered with two microphones, and panned to about half right to three-quarters right. If you pan them too close together, it rather negates the effort of having rigged two microphones on them, and it will give them a very narrow image.

The forward-facing cornets and trombones are recorded from about 1–1.2 m (3′4″ to 4′) in front and just above bell height, looking slightly down. The larger, upwards-facing instruments (tubas, euphoniums, and baritones) are recorded from above but off-axis, with one microphone between two players, at about 60–90 cm (2′ to 3′) above the bell (about 2.6 m (8′6″) from the floor). The tenor horn bell faces more diagonally upwards and forwards when played and can be recorded from a position about 90 cm (3′) above the bell, 90 cm (3′) in front, and looking down. The trombones can be recorded from the front and above, but where there is difficulty erecting a stand between the trombones and euphoniums (they often prefer to sit fairly close together), a boom or booms can be lowered from behind the trombone section in order to manoeuvre the microphones into position.

Figure 13.8 shows a typical brass band set-up from Abbey Road studios.

Figure 13.8 Photograph from Abbey Road Studio 2 showing the University of Chichester Brass Band, conducted by Emma Button

Photo: Carlos Lellis, Programme Director, Abbey Road Institute.

When mixing the brass band, treat the brass band as an orchestral string section: start with the overall tree pickup and then blend in the spot microphones until there is sufficient detail. You should aim to retain some natural perspective (i.e. the tubas and timpani are further back) as you would with an orchestra.

13.5 Percussion ensembles

Percussion ensembles are all completely different in their composition. To record one, follow the principles outlined previously and use an overall pickup to provide the sense of space, depth, and overall character of the sound, and then add spot microphones for clarifying the attacks and stabilising the image.

The very percussive content will not sit well in a small, live room. Early reflections from nearby walls or surfaces will make all the transients very prominent and uncomfortable, and so the choice of venue for recording is important. If you decide to avoid prominent early reflections by recording in a large venue where the walls are some distance away, you run the risk of finding you are in a space with too long a reverb time. In order to manage the longer reverb in this situation, you would need to use a lot of spot microphones fairly close to the instruments to capture more direct sound and less of the overall pickup. If you cannot get access to a good-sounding hall, it might be a better approach to record somewhere fairly dry and add in artificial reverb afterwards, although this will be a less enjoyable playing experience for your musicians.

13.5.1 A note on small percussion ensembles

The method outlined in section 13.3 for the small classical brass ensemble would also work for a percussion ensemble of similar size. That is, placing the players in a horseshoe, relying more on slightly closer spot microphones for a large component of the sound to make sure the attacks are clean, and then adding some good overall room sound to set the group together into the space.

13.5.2 The steel band

The steel band is again a very loud ensemble, with immensely complex overtones best suited to outdoor or large venue performance. As with brass bands, one of the biggest concerns will be the acoustic saturation of the space that you are recording in. The space needs to be big enough to allow the acoustic energy to dissipate somewhat before rebounding off the walls, and you should avoid trying to record them in a village hall with a low ceiling or a similar environment.

Chapter 14

Organ

14.1 Brief introduction to the pipe organ

When recording the pipe organ, one of the first problems that presents itself is that of the organ's location, or rather the lack of a single location within the building. Because of the number and size of pipes, the ranks of pipes are commonly split up over at least two or more locations set within the architecture of the church or concert hall. It is worth taking a brief moment to present a simple overview of how the organ works so we can see how it comes to be split up in such a way.

The organ consists of at least one keyboard (or manual), usually of five octaves, and a set of foot pedals, usually of two and a half octaves. The number of manuals varies, but larger instruments might have four or five. The pipes are physically arranged in ranks of a single timbre and pitch type, which are grouped into divisions, each of which is activated by one manual. The player selects which pipes they want to use by means of stops (selecting a single rank of pipes) and mixtures (selecting a pre-set mix of pipes). The primary manual division is called the 'Great' (in the UK/USA), and it forms the main tone and character of the instrument. Other divisions include the 'Swell' (whose pipes are enclosed in a box with movable opening louvres controlled by a pedal, enabling a degree of volume control) and the 'Choir' or 'Positiv', which is a set of smaller pipes traditionally placed behind the organist's back, and therefore closest to the congregation.

Pipes can be made of metal (circular cross section) or wood (square cross section) and are of two basic types: flue pipes and reed pipes. Flue pipes produce their sound by means of a fipple, such as found on a recorder mouthpiece. They include the principal or diapason (producing the basic organ sound, often very visible and decorated at the front of the casing), flute pipes (wide bored, producing mainly fundamental tones) and string pipes (narrow bored, producing little fundamental tones and a lot of overtones). Reed pipes produce their sound by passing air through a single reed, such as found on a clarinet. These include trumpet and clarion stops, intended to be used as part of a mix, and solo reeds, designed to imitate orchestral reed instruments. Pipes are organised by pitch, where an eight-foot pipe (8′) is the reference unison pitch (i.e. the pitch of the same keys on a piano); 4′, 2′, and 1′ represent one, two, and three octaves higher, respectively, and 16′, 32′, and 64′ represent one, two and three octaves lower, respectively. The ranks of pipes are placed where convenient within the building and are often split across separate sites, especially in a large cathedral.

14.2 Doing a venue reconnoitre

A venue reconnoitre is uniquely valuable if you are planning an organ recording, as the instrument and the building are intimately bound together. This visit can be used to plan microphone layouts, listen for any acoustic quirks of the building, assess background noise levels, and befriend the venue's technical manager, verger, cathedral staff, or college porters. These are the people you will need on your side at every stage – for trouble-free parking while you unload, for permissions for slinging microphones, and to ensure quiet while you are recording (see section 2.2).

Sitting quietly in the venue will give you a good idea of where the noise problems are (e.g. traffic, heating pipes, organ blowers, pigeons or bats), and asking the organist to play will reveal the degree of isolation that you have between your control room and the main church building. Your control room is likely to be poorly acoustically isolated from the instrument, whether you set up in a vestry, storeroom, or crypt, and will sometimes be located directly under the organ, making monitoring your recording accurately quite a challenge.

While the organist plays something that uses a large range of pipes, take a walk around the building to identify where the nodes of various standing waves occur. You are looking for places where certain notes really become prominent and seem to hang around in the air longer. These places should usually be avoided when placing your microphones, or an uneven coverage of the instrument will result. However, in unusual circumstances, you might find that you only really get very low bass in one particular part of the building, and you might choose to place microphones here in order to capture this. (An example of this is Temple Church, London, where the 32′ stops are only really audible near the rounded end of the building and have to be captured with microphones at this position, which are then delayed to account for the great distance to the main pair.) It is a good idea to always bring some spare microphones to use on important sections of pipework that you might have overlooked.

14.3 Microphone choices, stands, and cable runs

To capture the lovely extended LF that the organ produces (the lowest C of a 32′ stop has a fundamental frequency of 16 Hz), omnidirectional microphones are essential to form the bedrock of the sound. It can be tempting to try and use cardioids to capture more detail and discriminate against some of the reverberant sound, but the instrument will not carry enough weight and hence will sound like a small organ if the extended LF is not recorded. Additional pair(s) of cardioids can be added to pick out more detail if necessary by positioning them high enough to capture the attack as the air hits the lip of the pipe at its opening. All the microphones used to record the organ (the omnis for overall sound, and cardioids for detail of each section of pipes) should be mounted at least as high as the front opening of the pipes (the part that looks like a recorder mouthpiece). If they are too low down, some of the attack will be missed, and if they cannot be raised high enough, they should at least be pointed upwards.

The most pressing practical problem here is getting stands that will go high enough (at least 6 m (20′) in most reasonably sized buildings) while remaining safe and stable. If you are fortunate enough to be able to record the organ in King's College, Cambridge, Westminster Abbey, or Winchester or Coventry cathedrals, you will need stands that go to around 12 m (40′) as the organs are all very high. You should use stands that are designed for this purpose and that make use of weights or sandbags on the base of each stand. Apart from this being a safety-conscious course of action, you will also come across as serious and professional, and this is another important factor in gaining the confidence of the venue staff. Record companies that make a lot of organ recordings often make use of telescopic lighting stands or even aerial masts to mount microphones. Decca adapted Clarkson aerial masts for use as microphone stands (these were originally used by the BBC and the military for radio transmissions). You should never make any sort of improvised arrangement using 'stands on top of stands'; your own safety and that of others has to be the highest priority, followed by the fabric of the church building, and finally, the safety of the microphones.

Slinging microphones rather than using stands is another option to be considered, but you will need to make sure that you can do this safely (see Chapter 2) and that you have the correct permissions from the venue. The staff of old churches and cathedrals can justifiably be very difficult to negotiate with when it comes to attaching ropes or cables to parts of an ancient building, and you must be sure that you can avoid damage to stonework and wooden structures. It is worth noting that cable runs within a large church can very quickly run into hundreds rather than tens of metres, and you should include this in your planning.

14.4 Basic technique: straightforward organ layout

Figure 14.1 shows a nice symmetrical organ arrangement as a place to start our discussion, with part of the casing on either side of the central choir/positiv division.

This diagram shows some useful principles which will have to be adapted to your own situation. There is a two microphone tree of omnis spaced at around 1 m (3′4″) apart, at a distance of at least 3–4.5 m (10′ to 15′) away and probably more, which is aimed at capturing a good overall sound from the organ. The exact distance will depend on the instrument itself, but the aim is to avoid as much blower noise and action noise as possible. If these sounds are quite dominant, your microphones are too close. Because the casing in this particular example is spread out over quite a wide area, a single additional omni is placed on each side, rather like orchestral outriggers. If the casing was more self-contained, as might be the case for a smaller organ, there would be no need for these. (If you do not have access to enough tall microphone stands, one suggestion is to rig these wide outriggers on a long piece of aluminium slotted angle, screwed to the top of a single microphone stand rather like a very wide stereo bar (in the order of 3 m (10′) wide). Care must be taken to ensure this is absolutely symmetrical and safely attached.)

The cardioids in Figure 14.1 are shown centred on the choir/positiv organ (which is usually sited lower down nearer to the player). A pair of cardioids can also be used for more detail from the main organ without having to get any closer. They will discriminate against some of the reverberation from the building and so sound clearer, catching the note attacks before they are swallowed up by the building. If you were to move the omnis in closer instead of using cardioids, you are likely to find that the wind and blower noise becomes intrusive. Therefore, the combination of more distant omnis for a good tone (without blower noise) and some cardioids to pick out some detail from the reverb should enable you to find a good balance of all the desired attributes of the sound.

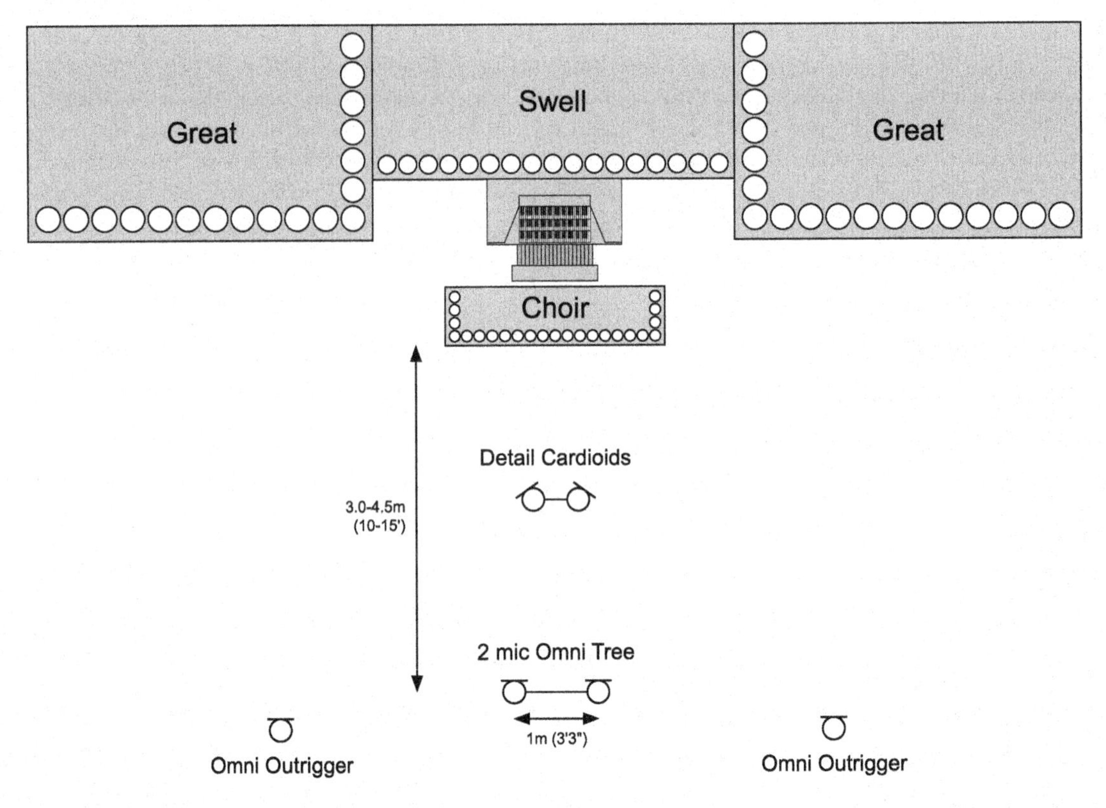

Figure 14.1 Simple organ layout

14.5 Organs with pipe divisions in several locations

Where the case is divided into two or more widely separated sections within the building, we can develop the principles from the straightforward organ in Figure 11.1 further. An overall pair of omnis spaced at around 1 m (3′4″) should be used as ambience microphones to tie all the contributing sections of pipework together in the same space. Additionally, each section of casing and pipework should have its own pair of microphones (cardioids or omnis) which are panned so that they overlap but do not completely overlay one another. For example, two casing sections could be panned fully left and half right, and half left and fully right, respectively. This will give some sense that different parts of the organ are spatially separated without trying to reproduce exactly the experience of standing in the building. Overlaying the different parts in this way should not cause any stereo imaging conflict because they are separated by relatively large distances within the building. This means there is insufficient correlation between the separate microphone pairs to confuse the stereo imaging. Coventry Cathedral has widely separated sets of pipes, and for the 1994–1995 recording of Janáček's Glagolitic Mass with Thomas Trotter,[1] completely separate organ booms were placed on either side of the nave.

14.5.1 Very distant pipes and use of delays when mixing

There might be some very distant, specialised ranks of pipes, such as a fanfare rank mounted over the west door (as in St Paul's Cathedral, London). It is always worth talking to the organist and finding out when, or even if, they will be using various ranks of pipes during a piece. It is a good idea to fade out very distant microphones if they don't need to be in use, as they might cause problems with imaging and delayed sound compared with the main pickups. (See also Chapter 17 for discussion about the use of delays in general.)

Figure 14.2 includes a very distant set of fanfare pipes. There is always some degree of timing difference when a main pickup and additional spot or ancillary microphones are used. The sound of the distant instrument or organ pipes will arrive at the nearby spot microphones earlier than it arrives at the main pickup. There are many different opinions on the effectiveness or necessity of using an electronic delay to retard the signal in the spot microphones so that it is time-aligned with the same instrument on the main pickup. With a complicated multi-microphone set-up over large distances such as a cathedral organ, the question of which signals you delay with respect to which other signals does not always have an obvious answer. In order for delays to work at all, there must be a single pair of microphones that are the dominant pair in the mix to which all the others are delayed. In the illustration in Figure 14.2, the dominant pair will be the omnis that are capturing the overall organ sound, labelled 'Main Pair'.

It is possible to get into quite a muddle if you start introducing electronic delays at the time of recording, so it is recommended that you do not try to address this until the mixing stage. If at

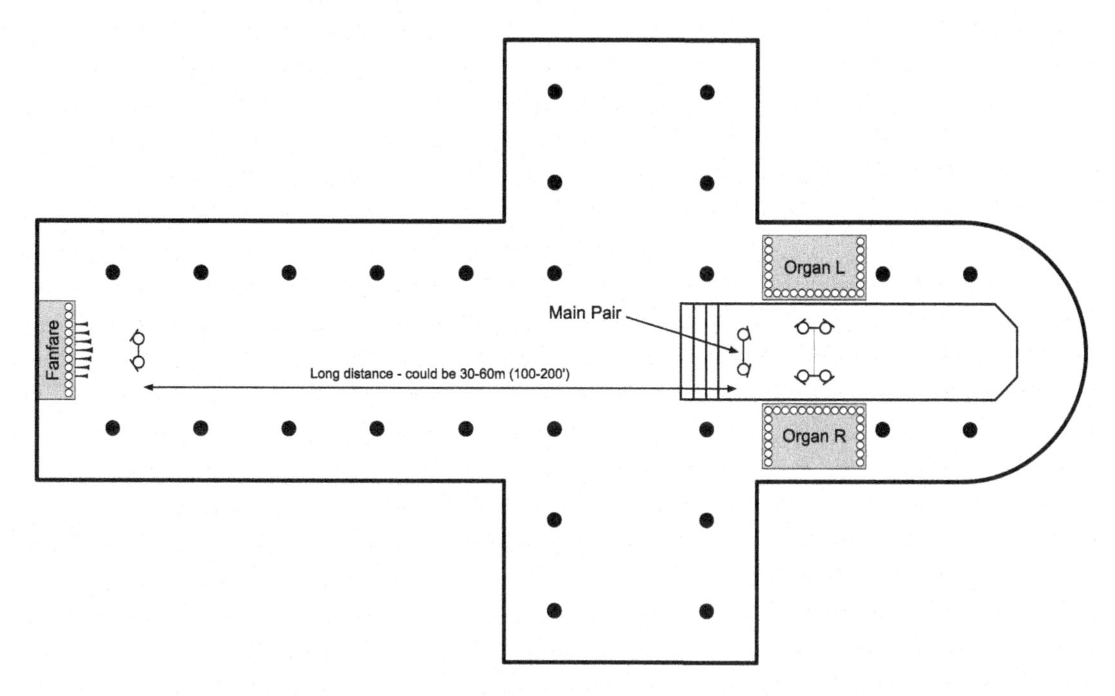

Figure 14.2 Organ with ranks of pipes spaced around the building

all possible, it will be better to capture a multi-track recording with no delays and mix later on, away from the live instrument. This is not to discourage the skill of mixing straight to stereo at the time, but because of the difficulties of monitoring accurately in a church (see section 14.7). You should also measure the distances between the sets of microphones and note these so that if you do decide to try out electronic delays while mixing, you will have a good idea of the sort of delays that might need to be involved. On a practical note, measuring long distances within a venue is most easily accomplished by use of a laser measuring device. If you are going to use one of these, make sure you notify the venue staff beforehand so that you don't cause any alarm by waving a laser around the building.

Each additional metre causes approximately 3 ms delay, and once the distance is more than about 10 m (35′), or 30 ms delay, it is likely that you will hear the time difference between sources as a discrete signal (i.e. as an echo). Below this threshold, your ear will tend to fix on the signal that arrives earliest (which will be the spot microphones) as being the timing of that rank of pipes. A later arriving signal will not be perceived as a discrete echo but just as part of the original sound. Although there is a grey area where you might perceive some blurring of note attacks, this depends on the relative levels of the two signals, and it is more likely to occur with more transient sounds. Therefore, with moderate distances, not using any delays might well be perfectly satisfactory.

Another potential source of delay with a mechanical organ is the time that different ranks of pipes might take to speak, with the greatest problems caused by the largest distances. This is something that is best discussed with the organist to see how much of a problem it is in that particular venue, whether they are able to compensate for this at all in their playing, and whether it is another factor of which you need to be aware.

14.6 Interesting acoustic effects and other awkward corners

Figure 14.3 shows the set up used at the Temple Church, London, whose organ was used on the *Interstellar* film sound track. This church has a curved back wall at the end of the nave, and this has an interesting acoustic effect on the lower frequencies of the 32′ stop such that they are best picked up by placing some microphones near to the curved wall. The signal from these microphones is then delayed to match the main pair. This method has been developed by practical experience of this particular church, and if anything is to be taken away from this chapter on organs, it is that every organ and every situation create different problems to be solved, and you should always take some time to walk around the building and see how the organ's sound develops in different areas.

An arrangement of pipes that is frequently found in the UK is to have the organ mounted all together in a corner of the church, with pipes visible on two sides at right angles to one another. Using the principles that we have looked at, the approach to an arrangement like this is to put a pair (a two-microphone tree, or a pair on a normal-sized stereo bar, but splayed outwards to increase their spacing) on each section of pipework, and then use an overall, more ambient pair to join them all together. In this case, the overall pair would be panned fully left and right, and then the pairs for each section could be panned to overlap (e.g. fully left and half right, and half left and fully right) as previously discussed. A corner organ is illustrated in Figure 15.4.

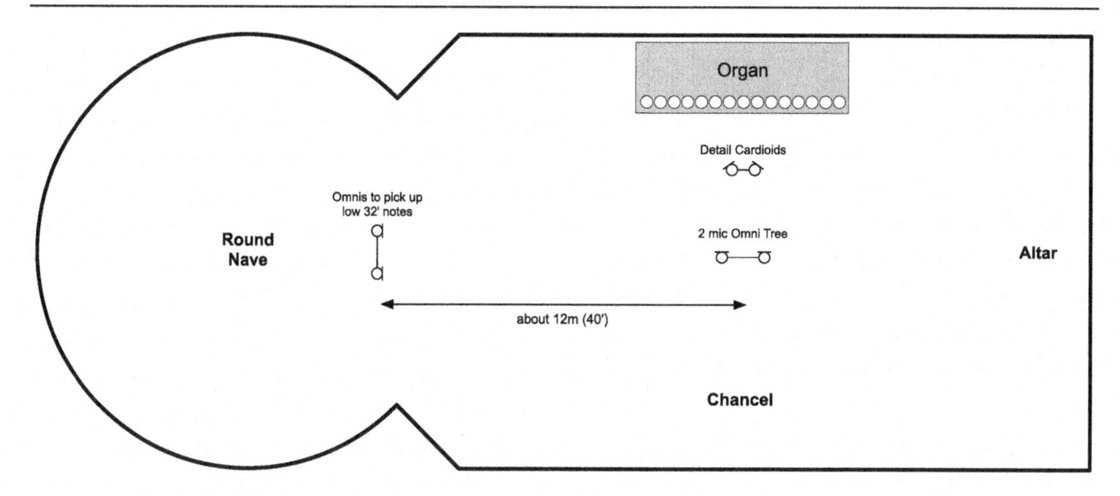

Figure 14.3 Microphone set-up for Temple Church

14.7 Monitoring limitations

In addition to the organ sound being audible in every far corner of the building, there are other monitoring limitations that mitigate against trying to mix at the time of recording (as discussed in Chapter 2). You are likely to find that you end up using headphones because of the spill of the live instrument into your monitoring area, and it is important to recognise the limitations of headphones when it comes to panning, imaging, and reverb amounts. Small near field monitors will help you with these aspects of the recording but will not give you any sense of how the low frequencies are sounding. Your headphones will be better at the LF reproduction, but in the absence of some really large studio monitors and a good isolated control room, you should leave as many mixing decisions as possible until after the event.

14.8 Communication: talk microphone

Communication with the player is very important as always, and in addition to rigging the normal talkback speaker near the player so you can speak to them, you will need to rig a dedicated organist's talk microphone in order for the player to be able to communicate with you. A talk microphone is only needed when a player is not sitting close to the microphones that have been placed on their instrument; the organ is an excellent example of this problem given that the player maybe sitting a long way from the pipes, and it is easy to forget that the player won't be audible unless they have a microphone near to where they are sitting. There is a very natural tendency for musicians and conductors to speak towards the talkback loudspeaker as this is the source of the producer's voice, but this of course only provides one-way communication. Remember to fade down the talk microphone during takes, unless you want to capture humming, singing along, mumbling, and other chit-chat from the player.

14.9 Noise

Churches and cathedrals are likely to have high levels of broadband ambient background noise from rumble, traffic, and sounds from the central heating system. The organ itself also contributes to the background noise with its blower, and minimising picking up of the blower noise is part of the challenge, as noted earlier. To remove rumble, our instincts might be to reach for a high-pass filter of some sort, but this will inevitably take the LF of the organ away with it. (See also note about limited monitoring in section 14.7). A better solution is to use software that can analyse the rumble and remove it whilst leaving the LF notes intact. (See Chapter 19 for software suggestions.)

Other common noise sources include birds from inside or outside the building. Pigeons or bats can be nesting in the roof area, and the cooing of pigeons from high up will reverberate around the whole space. Some engineers carry a starter pistol to scare them away when recording is to start, which is at least a non-lethal solution. Outside, songbirds such as blackbirds have been known to imitate tunes being recorded (which is so endearing that you might choose to leave some of them on the recording). St Jude's Church in Hampstead Garden Suburb, London, is a frequently used recording venue and is situated in the middle of a leafy residential area. Here, lawn mowers and nearby school playing fields can be a problem, and persuading the neighbours and children to keep quiet is another task for the diplomatic engineer.

Recording overnight is a common solution to bird noise and traffic noise in particular, and you should consider if this might be a possibility for your project.

14.10 Electronic organs

In this context, this refers to electronic organs that are mounted within a church, using loudspeakers where the pipes and casing would normally be sited. This has become more common as electronic organs have improved, and the cost of replacing old instruments or installing a new one from scratch has become insurmountable for many churches. The wonderful Victorian Gothic Roman Catholic cathedral in Norwich, for example, has a beautiful acoustic but has never had a real pipe organ, and whilst fund-raising continues, it has an electronic organ for now.

The first thing to say is that these instruments need the sound of the building in which they are installed to bring them to life, and they should not be DI'ed (directly injected, or plugged straight into the audio interface or mixing desk) or closely recorded, as would be appropriate for a Hammond-type organ in a jazz context. The core approach is to place the microphones at the same distance from the loudspeakers as you would for ranks of pipes in order to capture the effect of the church acoustics as well as the direct sound. Treat them as you would an acoustic organ in terms of panning and overlaying different sections of the instrument if they are split up within the building. The two main differences from the sound of a real, live organ are that you may notice are a lack of 'edge' and upper harmonics, and the pedal fundamentals are not usually reproduced as well because of the loudspeaker limitations.

Portable electronic organs are sometimes used for concert performances with other instruments if the organ in the venue is at a pitch that doesn't match with the performance practice of the players. The modern standard pitch is notionally A = 440 Hz, and authentic early music performers are likely to be using lower values such as A = 430 Hz as a reference instead. The organ within the building might not match any of these, which leads us onto making recordings with organ and orchestra.

14.11 Organ with orchestra: overdubbing or simultaneous recording

This is a combination that can be hard to balance at the time of recording because the organ and orchestra have some conflicting requirements. The orchestra is best suited to a reverb of around 2 seconds, and the organ tends to be in a building where the reverb time is around 5 seconds or more. It is very common to record the organ and orchestra separately, and it is often essential to do so to make a success of it. The basic workflow when recording separately would be to record the orchestral session first, edit it in the usual way (see Chapter 18), and use this as a track onto which the organ can be added. For cases where the organ pitch does not match the orchestral pitch, see section 14.12.

There are a two primary reasons for choosing to record the organ as a later overdub:

1. *The type of organ.* The organ in the recording venue will have different registrations and possible combinations of sounds depending on when and where it was built, but the repertoire being recorded might have been written for an organ that produces a particular sound that works with the orchestration. A good example are the renowned Cavaillé-Coll organs found in the Gothic cathedrals and churches of northern France (such as the church of St. Ouen in Rouen), for which much of the big romantic French repertoire was written. An organ of a different type to this might need to have many stops out simultaneously to produce the required tone, at which point it can easily be so loud as to overpower the orchestra.
2. *Acoustic saturation of the recording venue.* Even with the right organ in the building, if it is recorded together with the orchestra, it can take up so much of the acoustic space at low frequencies that the room will saturate too quickly. Recording simultaneously for a large scale work is only likely to be successful in somewhere as huge in volume as the Royal Albert Hall.

A good example of a complex project with overdubbed organ is the OSM (Orchestre Symphonique de Montréal) recording of Fauré's Requiem,[2] which was recorded in St Eustache Church, Montreal, with the organ of Gloucester Cathedral added afterwards.

14.11.1 Dealing with two different acoustics

Recording the orchestra and organ separately in different venues can then easily lead to a problem with mismatched acoustics, with the organ recording having a longer reverb characteristic than the orchestral recording.

In order to help this problem when mixing, it is possible to artificially shorten (by fading out) the reverb tails left at the ends of organ phrases and especially at the end of the piece so that the orchestra and organ reverbs appear to finish at the same time. However, this requires a great deal of skill and painstaking attention to detail, and the producer needs to be aware of the potential for expensive remedial post-production work when deciding where and how to record the organ overdub.

Rather than opting for the grandest and most enormous instrument available, it might be a better match to record a smaller instrument in a smaller acoustic to avoid a lot of salvage work later on. It is very important that the orchestra and organ can be made to feel as if they are in the same space. A very fine example of a successful combination of orchestra and organ is the OSM/Dutoit recording of Holst's *The Planets*,[3] where the orchestra was recorded at St Eustache, Montreal, and the organ was recorded at King's College, Cambridge. UK churches that are known for having a big sounding organ but a shorter reverb time include St Paul's Church, Knightsbridge, and Temple Church, London.

14.12 Organ pitch

Organs are tuned to a variety of pitches, depending on the time of construction and local choral traditions, and are often found to be not tuned to A = 440. This is something else that needs to be researched in advance and factored into the recording and production process. Where the organ that needs to be overdubbed is not at the same pitch as the orchestra, the normal process would be to pitch shift the edited orchestral recording to match the organ, overdub the organ, and then pitch shift the organ back again, and put it back into sync with the original orchestral track. The OSM's recording of *The Planets* was managed in this way: the OSM play at around A = 443 Hz, and the King's College organ is at A = 440 Hz. The orchestra was pitch shifted down to A = 440 for playback to the organist at King's, and then the organ overdub itself was pitch shifted up to match the original orchestral recording and the two resynced together.

Pitch shifting can be done as purely a pitch shift or as a vari-speed (where the duration and tempo will be altered along with the pitch). Care should be taken to understand exactly which you are doing, as well as making sure that your audio files are time-stamped and can easily be synced up again.

It is also worth taking a moment to think about audio quality during post-production; it is not a good idea to process something twice in order to get back where you started because there will be some small loss of quality or increase in noise. If the original orchestral file (let's call it A) is pitch shifted in one direction to make a new file (B), and then file B is pitch shifted back up again to make a third file at the original pitch (C), the audio quality of C when compared with A will have suffered a small amount of degradation. The original orchestral recording at pitch (A) should be used in the final master, matched up with the pitch shifted organ.

14.13 Sampled organs

This leads us to one obvious solution to pitch problems, which is to use a sampled organ that can be played back at any pitch. Where a part is a subsidiary one and not a solo, using a sampled organ would be a very good alternative to recording one live. The best known sampled organ system is Hauptwerk (www.hauptwerk.com), which makes available complete sets of organ samples of every note from a variety of organs all over the world, and enables them to be played using MIDI controllers. When you load in a set of samples from a particular church or cathedral, the available stops can appear on the computer screen and you can control them as you would on the real instrument. Now that technology has moved on, these samples and other systems far surpass the quality of the early Allen-type organs from 30 to 40 years ago.

Notes

1 *JANÁČEK/Glagolitic Mass/Orchestre Symphonique de Montréal/Dutoit/Troitskaya/Kaludov/Leiferkus/ Randová Organ: Thomas Trotter DECCA (2006 release) CD 436211–2*
2 *FAURÉ: Requiem Orchestre Symphonique de Montréal/Dutoit/Kiri Te Kanawa/Sherrill Milnes Organ: Chris Hazell DECCA (1988) CD 421 440–2*
3 *HOLST/The Planets/OSM/Dutoit/Decca (1987) CD 417 553–2*

Chapter 15

Choirs

Choirs and singing groups come in many sizes, from the small vocal quartet through the chamber choir and church choir to the large opera chorus or choral society of 80–150 singers. They could be performing in a concert hall or a church, and they could be positioned on the floor, on risers, on raked seating, or in the traditional choir stalls in a church. Each situation will need a different approach, but for a choir rather than a small vocal group, the aim of the recording is to present the overall blend of singers and avoid highlighting individuals. Achieving a blended sound that is well balanced between the parts is the responsibility of the choir and conductor, but the recording engineer also needs to bear this aim in mind when choosing and placing microphones.

15.1 General notes on microphone choice and placement

One overall principle to note when recording choirs is that it is not essential to use omnidirectional microphones for their extended LF frequency response, as no one is singing low enough for this to be necessary (unless you happen to be recording a Russian male voice choir where contrabass singing is a feature). You can of course choose to use omnis for other reasons (organ or piano accompaniment), but some good cardioids will cover the frequency range of the singers perfectly well. The lack of extreme low bass also means that the application of an appropriate HPF to choir microphones to alleviate rumble should leave the musical content unaffected.

15.2 Choir spacing

When planning choir recordings, estimating the space required by a choir of a given size can be a challenge, and you may find the following guidelines useful:

- Each singer and their chair will be approximately 75 cm (2′6″) wide.
- Each singer and their chair will need a depth of at least 90 cm (3′), although 120 cm (4′) is more comfortable – giving them enough room both to sit and to stand whilst holding their music in front of them.
- The height rise for each level of most staging is at least 20 cm (8″).

15.3 Small choir

A small choir means a choir of around 20 singers, and because of its small size, it will work best to rely on overall microphone techniques rather than the additional use of section microphones as would be used for the choral society (see section 15.4). Trying to use SATB section microphones on a small choir is likely to make individuals prominent, as the microphones may have to be too close to obtain an even and well-blended coverage.

15.3.1 Small choir: unaccompanied

A straightforward overall approach is to start with an ORTF pair (see section 3.3) at about 3 m (10′) high. Start with it placed in the room further back than you think you will need, and listen to it. Move it forwards until the balance of reverb to direct sound is satisfactory and you are getting enough diction detail from the singers without any of them starting to be too singled out. If obtaining sufficient diction means you are not picking up enough reverb, you can add in a spaced pair of omnis at a similar height and about 70–100 cm (2′4″ to 3′4″) apart to collect more of the room sound.

Next, consider the image width. The choir should fill the stereo recording angle of the ORTF pair and so fill the space between the loudspeakers in the image. (If it is a very small chamber group of individual singers, it will work better if individuals are not located exclusively in either loudspeaker.) If the image is not filling the space when you have the distance from the choir adjusted, you can change the angling of the microphones a little to alter its width. Decreasing the angle will decrease the image width and vice versa. (See section 3.3 for general discussion of spaced and angled cardioids.) If you end up with two pairs, the ORTF and the spaced omnis, remember that as a general rule you will need to choose one of them to form the main component of the sound; do not mix them in equal parts, as this would maximise any comb filtering effects. In this scenario, the ORTF pair will be your main pickup, and the omnis can be used to add in a little extra room sound and spaciousness if needed. The image width of the choir on the spaced omnis can be adjusted to match that of the ORTF pair by altering the spacing between the omni microphones.

15.3.2 Small choir with piano accompaniment: concert

The ideal situation would be to have a grand piano placed centrally so that it is straightforward to produce a recording with the piano central to the image. (In practice, the piano is much more likely to be placed over to the left-hand side for a concert; see following discussion and Figure 15.2.)

Figure 15.1 shows a concert layout designed with the recording engineer in mind. It includes an ORTF pair of good condenser cardioids (fully panned) as the main coverage on the choir at 3 m (10′) high to produce a good overall image and to discriminate against the piano. Also included are spaced omnis to add some room sound and a pair of forward-facing cardioids on the piano itself (using techniques from Chapter 5). In a concert, the piano is likely to be on a short stick if it has been placed centre front (to make sure the choir can be seen) and the piano pair will have to be arranged to ‘look’ into the small opening that is left when a piano is open on a short stick.

When this ensemble performs live, it is usual practice to place the piano (whether it is an upright or a grand piano) off to one side in order that the choir can take centre stage and be more visible, as shown in Figure 15.2. From the recording point of view, a more central piano image would be preferred, and it will be possible to use the panning of the piano microphones to pull the piano image partially towards the centre. Bear in mind, however, that the piano image on the main choir image will be on the far left, and to try and manipulate this too far towards the middle could result in a confusing or unstable image.

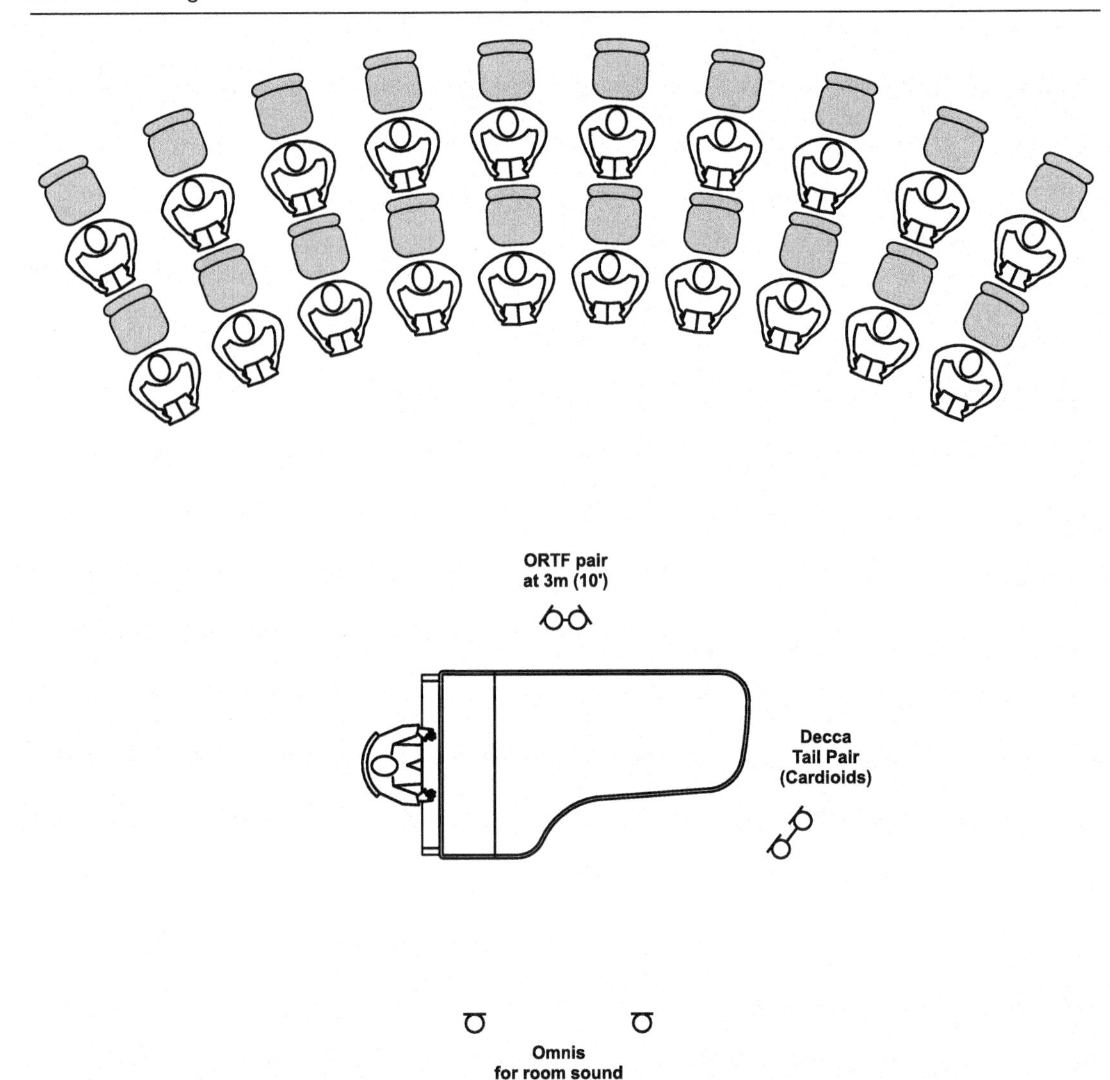

Figure 15.1 Small choir concert layout – ideal for recording with piano in the centre

One method of piano recording that usually allows for greater range in artificially moving the piano image towards the centre is to use the LCR three-microphone technique (see Chapter 5). Because these microphones are closer to the piano than the usual piano pair, you will need only a small amount of them for focus, and because they contain less choir, they can be panned more centrally without image confusion.

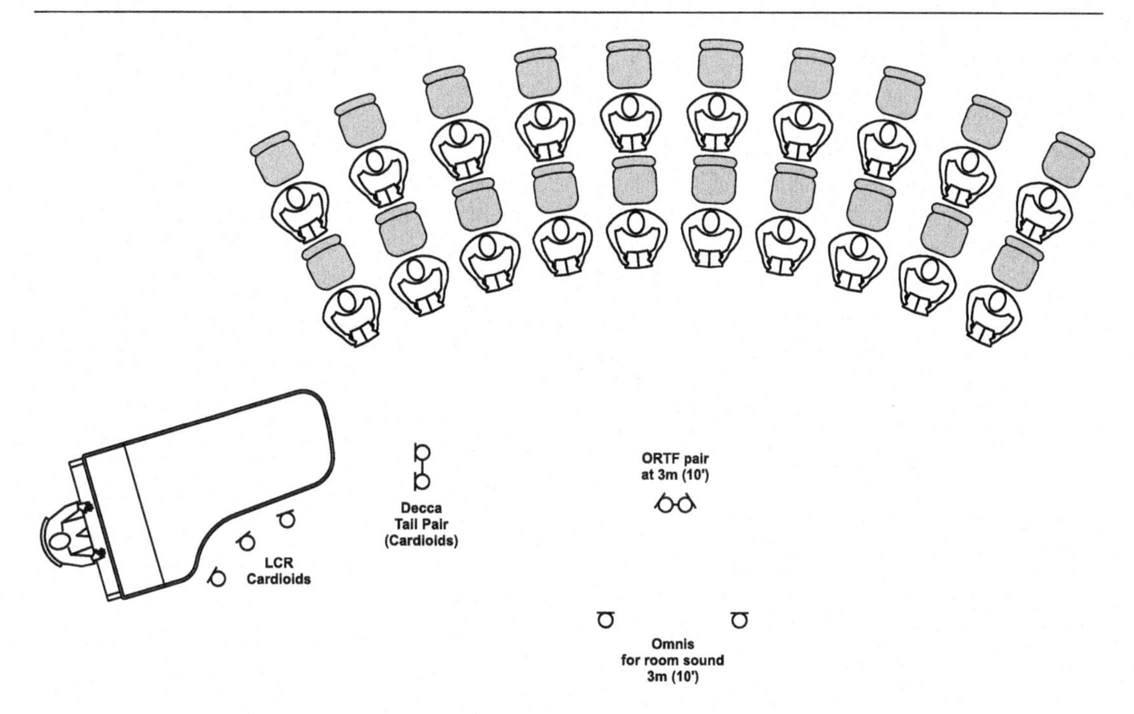

Figure 15.2 Small choir concert layout with piano off to one side and alternative microphone techniques on the piano

Figure 15.2 shows a grand piano to one side of the stage with both a normal piano pair and the LCR microphones in place. Only one set of these microphones needs to be used.

15.3.3 Small choir with upright piano accompaniment

An upright piano might be used because it is what is available, but good results can still be obtained. However, because an upright piano is often used in pop sessions, or live concerts of more pop-orientated repertoire, there are a lot of suggestions for microphone techniques that are *not* suitable in this classical choral context. This includes using two widely separated spot microphones placed close to the opening at each end of the open lid. These will behave as two individual mono sources when rigged this close to the piano and will not create a believable stereo image with a sense of real space around it.

An approach that works well is to open up the lid at the top and place the microphones pointing down into the instrument from a height of about 60 cm (2′) above the top of the instrument. This is more or less the equivalent to placing microphones at the tail end of the grand piano that look along the length of the soundboard and strings, and (as we would for a grand piano) using an ORTF-type or narrowly spaced pair will give good stability of image with the sense of spaciousness that comes from non-coincident microphones. On an upright piano, a pair of cardioids spaced at around 15–20 cm (6″ to 8″), angled at about 80°–90° apart and widely panned, will give an appropriate piano image width. In a choral accompaniment context, the piano image could be a little narrower than would be usual for a solo piano recording; it needs to be in keeping with the scale of the choir. Co-incident cardioids at 90° can also work quite well in this scenario; they will give a naturally narrow and centre-heavy image when fully panned, which would form an acceptable piano image when blended in with the overall choir pickup. (See Chapter 3 for discussion of use of spaced and angled cardioids to produce an appropriate image width without panning inwards too much.)

15.3.4 Small choir with piano accompaniment: recording session

For a recording session with a small choir and grand piano, the piano can be placed centrally, on full stick, and turned around to face the choir. This means they can hear the piano, communication is better, and the piano microphones and choir microphones can be used to discriminate against each other. (See Figure 15.3.)

15.3.5 Small choir with organ accompaniment

This scenario is a common one for church choirs, youth choirs, and music societies using a church as a performance space, and capturing a recording of an event might be used for a fund-raising CD.

The choir is very often placed on the chancel or altar steps, rather than standing in the choir stalls. The most common questions concern how to capture the choir sound whilst also dealing with the organ, which will almost inevitably be placed off to one side of the choir position or in several places around the church. It is not possible to deal with every conceivable scenario given the variety of church layouts and organs that might be encountered, but we can suggest some principles to be adapted.

As we saw in Chapters 14 and 7, it is possible to overlay and overlap the images of different instruments generated by different pairs of microphones, if the pairs are a sufficient distance apart so that their signals lack any significant correlation with one another. Where distances are large, in the order of 10 m (35′) or more, the images become effectively completely separate, but there is still sufficient decorrelation at smaller distances for stereo pairs to be overlaid. The choir image should always be prioritised and should be panned fully about the centre of the image. If there are any problems with image conflict (e.g. a high level of choir spill onto one side of an organ pair), then the organ panning should be adapted so as not to disturb the choir image. So if there is significant choir spill onto only the right microphone of an organ pair, then that organ pair will have to be panned part right (or centre) and fully left, placing that section of pipework spread around left of centre rather than centrally.

The organ microphones should be omnis in order to pick up the very low frequencies, and using an HPF should be avoided on any organ microphones that will be picking up the lowest pipes. The choir microphones can be cardioids, and they can usefully have a high-pass filter applied if it helps with LF church noise. Any overall ambient pairs can also be filtered if necessary, as the organ LF will still be present on the organ microphones.

See Figure 15.4 for a general layout, with the choir using the chancel or altar steps instead of risers and an organ in the corner as an example.

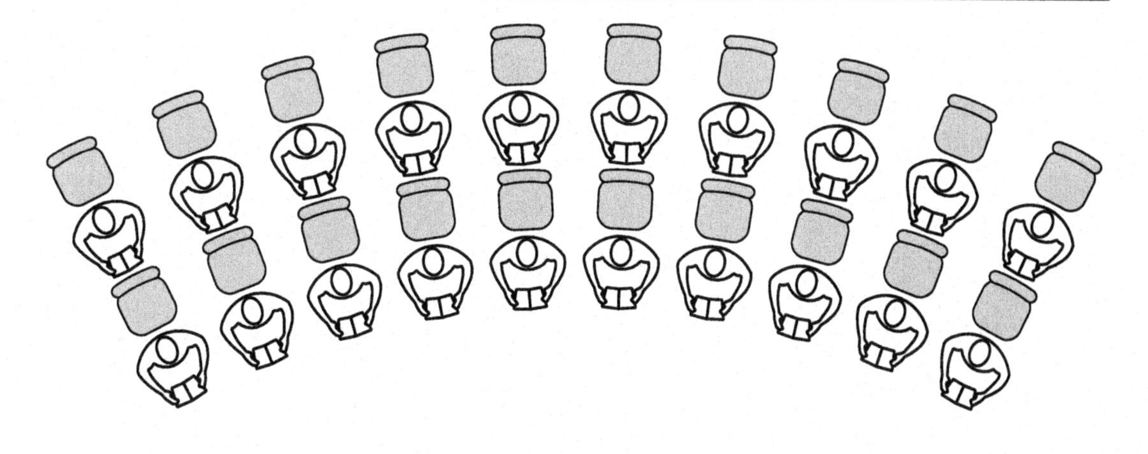

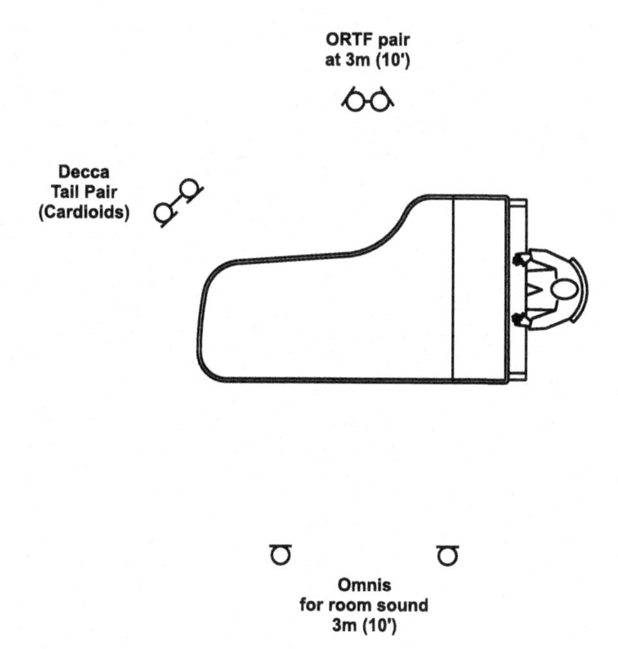

Figure 15.3 Small choir with the piano facing the choir for recording session

15.3.6 Children's choirs

This section will concentrate on the general school choir; professional choristers will be dealt with in section 15.6.

Of course, approaching recording a children's choir presents the same scenario as recording a normal choir of any size. However, an additional challenge, and something to bear in mind if

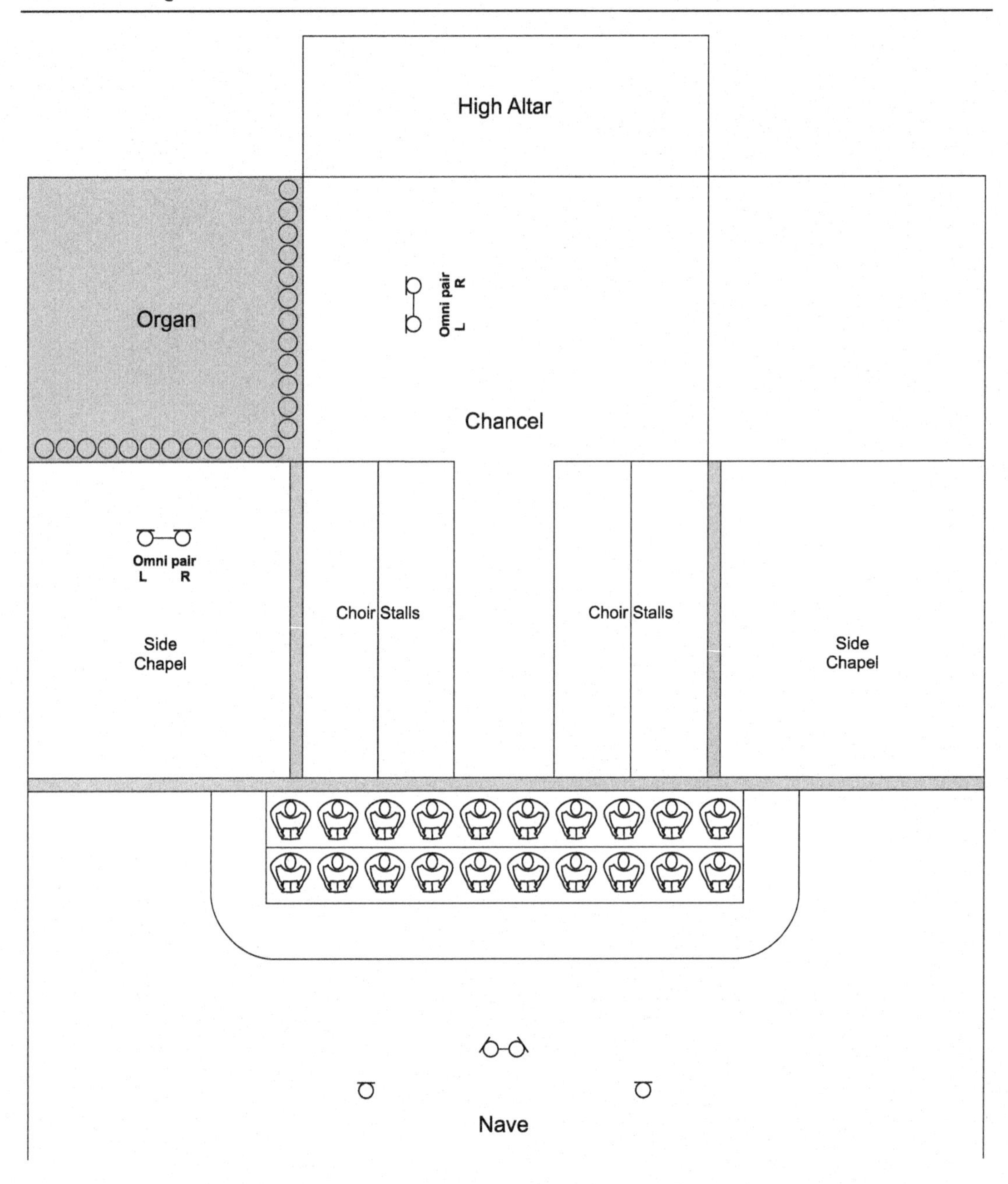

Figure 15.4 Organ and choir concert with choir on the chancel steps. Organ is sited in the corner with pipes on two sides.

you are called upon to record a school choir, is that there might be a group of singers who cannot really sing well, or especially cannot sing in tune. It would be a good idea to talk to the teacher beforehand and find out if it is a choir that represents the whole school, or whether it is a selective choir. For the sake of the recording, and everybody's sense of happiness and achievement and pride in the result, it is worth discussing confidentially with the teacher the possibility of placing any 'non-singers' together at one end, so that section microphones (as used for the choral society; see section 15.4.2) can be used to discriminate against them. You can then rig an overall pickup and some section microphones along the front, only fade up the section microphones covering sections of good singers, and pan these section microphones across the full image width.

Another alternative is to place the 'non-singers' at the back, or even in the front row with directional section microphones looking over their heads to the singers behind. Whichever method you use, the whole choir will be picked up by the main pair, but the section microphones should be predominantly the good singers. Your aim is to give the impression that all are participating equally during the recording session, while hopefully producing a recording that is more listenable.

One useful attribute of all children's choirs is that there is no LF to speak of, and therefore a high-pass filter can be used a little higher up in the frequency range than usual if needed.

15.4 The choral society

The choral society is usually a large choir of up to about 200 amateur singers, and singing standards can be very high. Three rows of singers is a common layout, but a choir approaching 200 singers will need to be arranged in five rows to avoid being excessively wide. It is highly desirable for the rear rows to be arranged on risers for recording and performing; this makes the rear singers more audible in performance and makes it easier to keep the perspective and level consistent between front and rear singers during a recording. It is most common to find the singers arranged in blocks containing singers of one type (soprano, alto, tenor, and bass). From a performance perspective, this allows weaker singers to be supported by standing with others who are singing the same notes, and when recording, it can make it easier to rebalance the parts by use of section microphones if that turns out to be necessary.

A common problem with amateur and semi-professional choirs is an imbalance between the sections due to different numbers of singers, and different levels of strength in singing technique. There will very often be more women than men, which usually results in sopranos being too loud and the bass section underpowered. The blocks of singers might be arranged from left to right as SATB, but choirs often have their own preferences depending on section strength, repertoire and what works best for them. If the soprano and alto sections are stronger than the tenors and basses, a common performance layout is STBA, with the women on either side, and the men projecting directly down the centre of the room. It can also help to put the women at the back and the men in front in this situation. Figure 15.5a and 15.5b shows two common layouts, both using risers. When it is not possible to stand the choir on risers, it is usual to put the women in front (because of their shorter average height) and offset the singers so that the back rows are singing through the gaps in the front rows and can be 'seen' by the microphones (see Figure 15.5c). Because of the potential for the choral society choir to be unbalanced in the room, it is usually

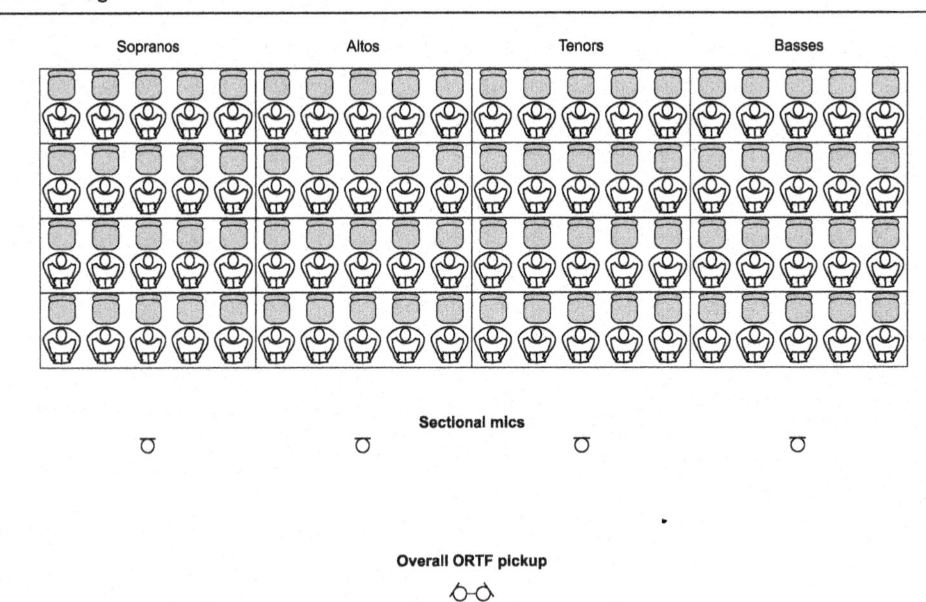

Figure 15.5a Large choral society arranged in blocks with risers SATB left to right

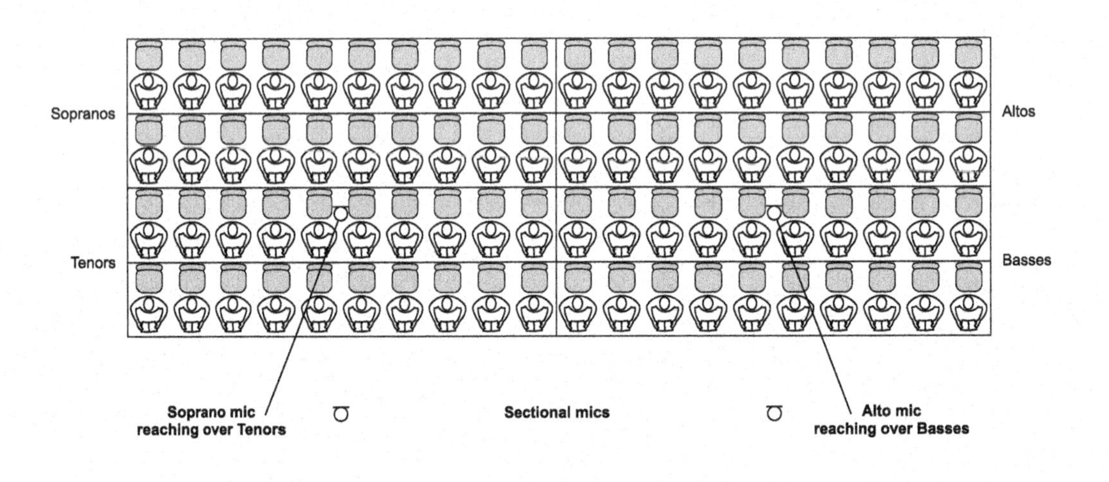

Figure 15.5b Large choral society arranged in blocks with SA at back on risers and TB in front

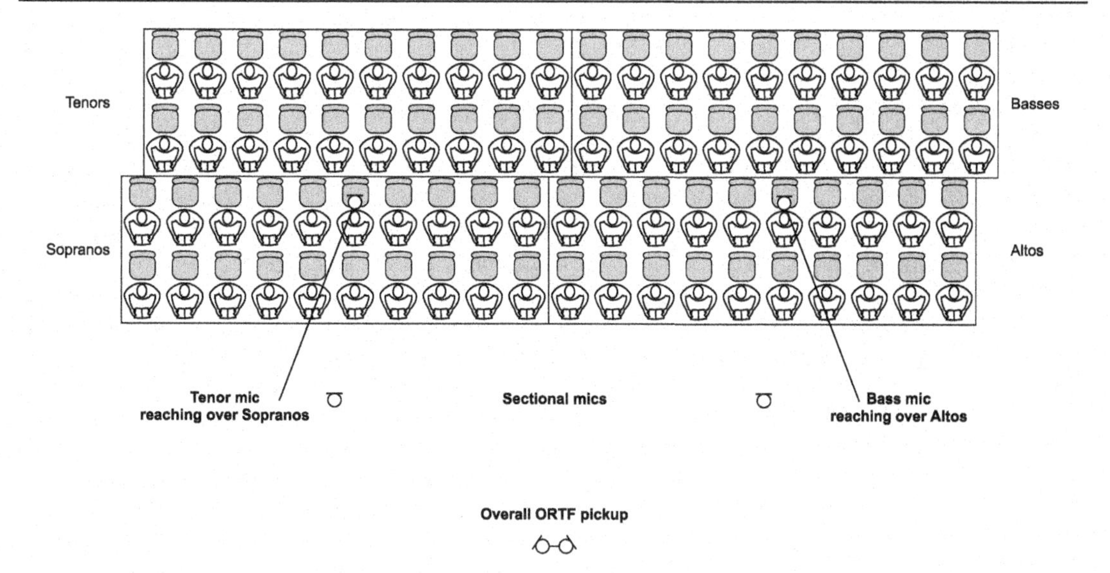

Figure 15.5c Large choral society with SA in front and TB at back without risers, with singers offset along the back rows

necessary to put a microphone on each section, which gives the engineer more control over the levels and internal balance.

A second potential problem with an amateur choir is the individuals who are particularly prominent due to lack of skill at blending into the overall sound of the choir. If you discover this problem, have a listen in the room to work out who they are, and see if you can avoid the culprit(s) to any extent by judicious positioning of section ancillary microphones. It should also be possible to have a discreet word with the conductor about a problem individual to see if any rearranging of seating plan is possible. The conductor is likely to be all too aware of the problem, but it might be politically difficult to address if the singers have their own particular positions that they have been sitting in for years.

15.4.1 Main microphone placement

The core approach will be some sort of overall pickup to provide the basis for the main image, with the addition of section microphones to support detail and diction and to give some control over the musical balance.

The main pickup should be placed at a height of around 3 m (10′) (possibly higher if the risers are more like steeply raked seating), and the distance from the choir will depend on the acoustics and the microphone technique chosen. The standard ORTF pair (see Chapter 3) is useful in that the overall pickup angle is quite wide (at around 100°), enabling them to be placed close enough to the choir to hear the words clearly without the choir falling significantly outside this angle. As noted earlier, cardioids can be used on choirs without loss of LF, but they should be of good quality with a smooth off-axis response for the ORTF techniques to work well, as almost all the singers will be singing into the side of the microphones. The image from an ORTF pair is well defined and sounds closer than spaced omnis at a similar distance, and the image width can be tailored to a certain extent by modifying the angle between the microphones. If your choir is physically very wide, and you cannot move the ORTF pair further away (perhaps due to venue restrictions, or acoustics), you might find that the choir presents too large a subtended angle at the microphones, and the outer regions of the choir will appear to come solely from each loudspeaker. This can be helped by narrowing the angle between the microphones a little until the image has a more even spread. Another idea is to use an ORTF pair with a pair of spaced omnis mounted on either side, approximately 70–80 cm (2′4″ to 2′8″) apart. A blend of the two can be used to produce the desired overall sense of distance and space.

15.4.2 Section microphones

Positioning a section microphone to obtain an even coverage across three rows of singers is similar to trying to 'light up' an orchestral woodwind section (see Chapter 9). You should avoid section microphones being too close, as individual singers will be picked out despite the choir being well blended in the room.

There are a couple of approaches to getting the even coverage that is required. One is to point a directional microphone over the heads of the front row, aiming at the back row. If the back rows are on risers, the microphones will need to be about 2.7–3.0 m (8′10″ to 9′10″) high for this. The front singers will have their level reduced by virtue of the microphone's directionality, and this will help to compensate for them being closer to the microphone (see Figure 15.6a).

Alternatively, you could take the microphones higher still and look down at the section, foreshortening the distance to the back row and making it more equal to the distance to the front row (see Figure 15.6b). With increasing height will come a loss of HF, as this content is not projected upwards into the microphones; this is even more the case if the singers are looking slightly down at their scores. Plosives such as 'p' and 'b' emerge forwards and downwards from the mouth, and the HF that gives shape to 't', 'd', and 's' sounds is also projected forwards. Therefore, some clarity of diction will be lost from any singers where the microphones are raised above the on-axis line of projection from their mouths. This may or may not be a desired effect depending on your context, but it is worth remembering as a rule of thumb that higher microphones on a choir will pick up less HF. The arrangement in Figure 15.6a should retain more diction from the choir than that shown in Figure 15.6b.

It should also be noted that if the choir is very wide (assuming SATB layout left to right), you might need an additional centre section microphone to avoid a 'hole in the middle' appearing because the distances between the section microphones become too large for the signals to have enough correlation. There would then be five section ancillary microphones 'lighting up' the choir, and they would be panned approximately left, half left, centre, half right, right. Where only four section microphones are used, it is a good idea to avoid panning the outer microphones fully to prevent a whole choir section being placed exclusively into the loudspeaker. Therefore, panning in this case might be approximately three-quarters left, half left, half right, three quarters right.

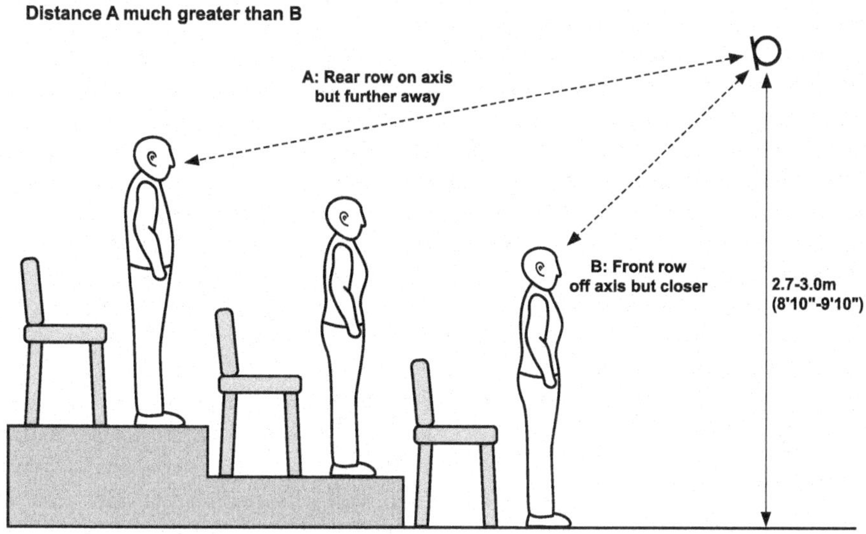

Distance A much greater than B
A: Rear row on axis
but further away
B: Front row
off axis but closer
2.7-3.0m
(8'10"-9'10")

Figure 15.6a A sideways view of a directional section microphone on a choir block with risers. Directivity is used to reduce the level of the front row.

Distance A still greater than B,
but less than Fig 15.6a
A
B
Mic raised higher
e.g. 3.5m (11'6")
Optional
diction mic
for front row

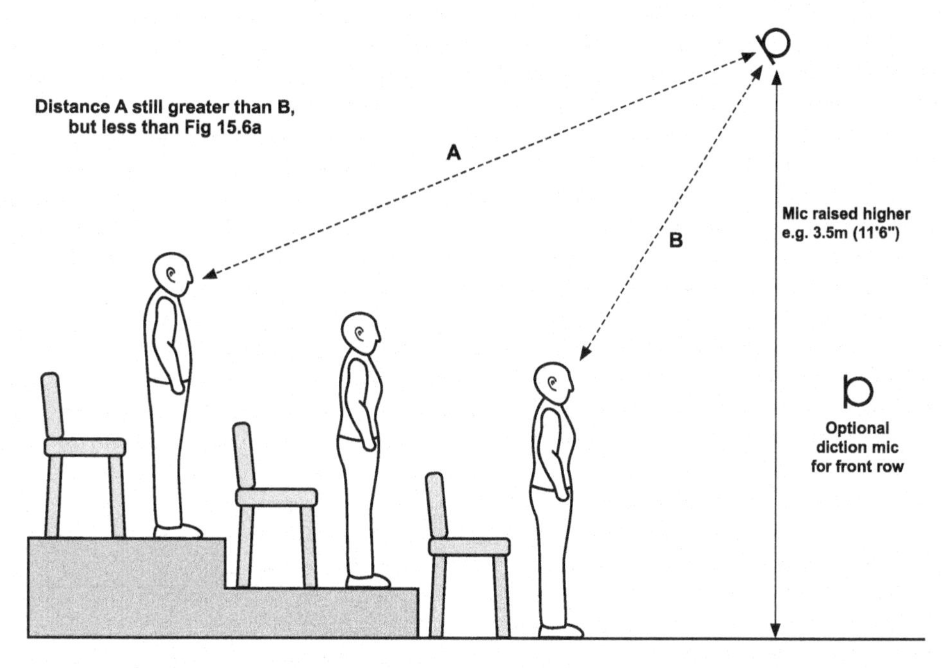

Figure 15.6b A sideways view of section microphones with raised height to equalise the distances of the front and back rows from the microphone. This will result in a reduced level of HF from the choir. Optional 'diction' microphone also shown.

The larger the choir, the higher and further back the section microphones will need to be to avoid discriminating in favour of a particular area of the choir, but this will eventually result in some loss of diction from all rows, not just the front. If the overall sound is working well in terms of choir image and control over section balance, a useful method of overcoming the loss of word clarity is to add in microphones whose primary role is to pick up diction sounds. These would be pointing at the front row at head height, and the resulting signal containing all the hard consonants and sibilant sounds can be blended into the overall mix at a low level. These additional microphones can also be useful if you need to pick up a semi-chorus within the choir.

Where the choir is laid out such that the back rows and front rows are not part of the same section (as in Figure 15.5b), the section microphones need to be aimed at covering the correct rows. It is sometimes possible to boom in microphones from the back or side to cover the rear block sections (see Figure 15.7).

Microphone choice for the section ancillary microphones would be cardioids or wide cardioids for their ability to discriminate against the front rows of singers, as noted earlier. Wide cardioids would be the preferred choice because the amount of overlap and spill between the signals at each microphone will create a nicely blended sound. There will be some altos and basses on the tenor microphone, for example, but this spill should not be feared, as it helps to avoid the creation of very clinically separated parts when the section microphones are faded up. Cardioids can also be used, but they have reduced levels of spill; another option to try is wide cardioids on the outer section microphones and cardioids on the inner pair.

Switchable directivity pattern microphones have particular benefits, as the effect of the section microphones can be fine-tuned simply by changing the directivity at the time of recording

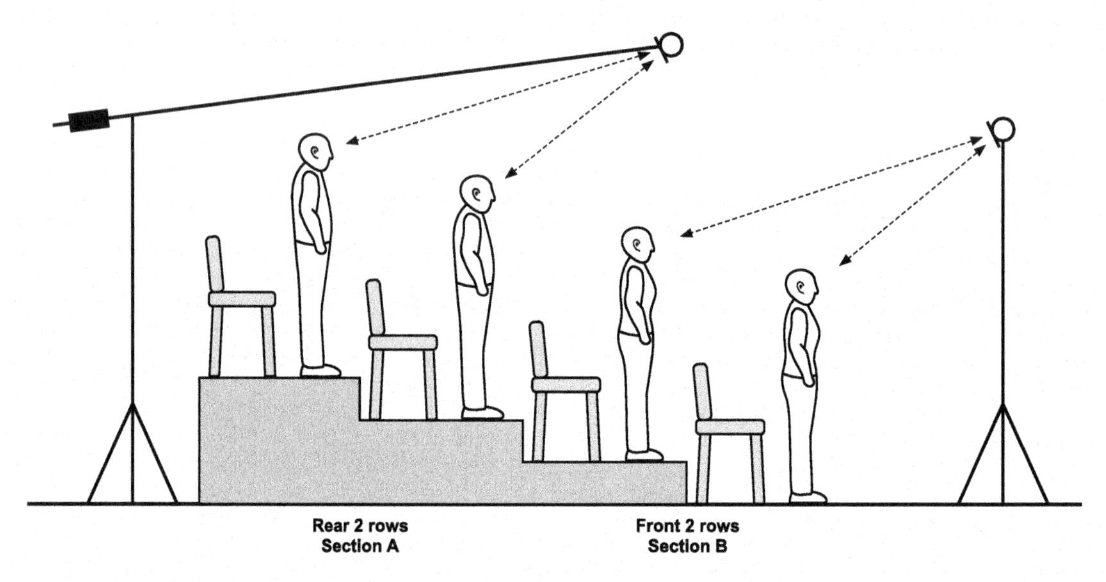

Figure 15.7 Sideways view of section microphones where back rows are from one section (e.g. basses) and the front rows from another (e.g. altos)

and hence changing the amounts of spill. In a concert situation where microphone placement is restricted, they also enable the user to alter the amount of reverberant sound being picked up from the room without moving the microphones.

15.4.3 Mixing overall pickup with section microphones

The main part of the sound and image is created by the ORTF or spaced omni pair, fully panned, and the section microphones blended in to achieve enough focus but without making their presence obvious. The amount used will depend on the acoustic, and more will be needed if this is very reverberant. If you can hear the section spots as little mono pools of 'closeness' in the image, you have faded them up too far. If you find you are not getting enough detail when you pull them down in level so that they no longer stick out, you then need to move the microphones a little closer. Section microphones should be evenly panned across the image, and avoid hard panning the outer microphones if it places too much of the choir directly into the loudspeakers. (See note on four- and five-microphone set-ups in section 15.3.2.)

The section microphones will be used on a professional choir for very occasional and subtle rebalancing of parts, but if you are dealing with a less experienced or non-professional choir, the sectional microphones will be of some use to you in correcting inter-part balance where the choir are not able to do it for themselves.

15.5 Choral society with soloists and organ in concert

Oratorios such as Handel's *Messiah* are the core concert repertoire that requires choir, soloists, and organ. A concert will often combine an amateur choral society with professional SATB soloists and organist. Figure 15.8 shows the layout of the choir and soloists, and the organ could be anywhere, so it is not shown. In a concert situation, consideration needs to be given to being unobtrusive and also to being aware that your favourite microphone technique might not be the best one if you are unable to position your microphones where you would like to. If you can, you should include some omnis somewhere in your set-up (as main pickup or organ pickup) to capture the LF of the organ.

15.5.1 Choir microphones

The choir coverage is achieved in the same way as was detailed in sections 15.4.1 and 15.4.2. The main pair could be spaced omnis at 80–100 cm (2′8″ to 3′4″) apart, cardioids in an ORTF-type arrangement, or a blend of the two if you need some flexibility in mixing after the event. The choice of ORTF or omnis will depend on the acoustic, where you are able to place the microphones, and personal preference. If you are in a reverberant space, and you have to place your microphones a little further back than you would like, the ORTF pair will work better as it will discriminate against some of the reverb.

Shown in Figure 15.9 is another option to give you some control over the amount of reverb when the microphones cannot be moved. This is an ORTF-type pair with an additional rear-facing ORTF-type pair which can be mounted on two stereo bars joined together to form an X on a vertical stand. The rear-facing ORTF pair is to collect reverb rather than direct choir sound;

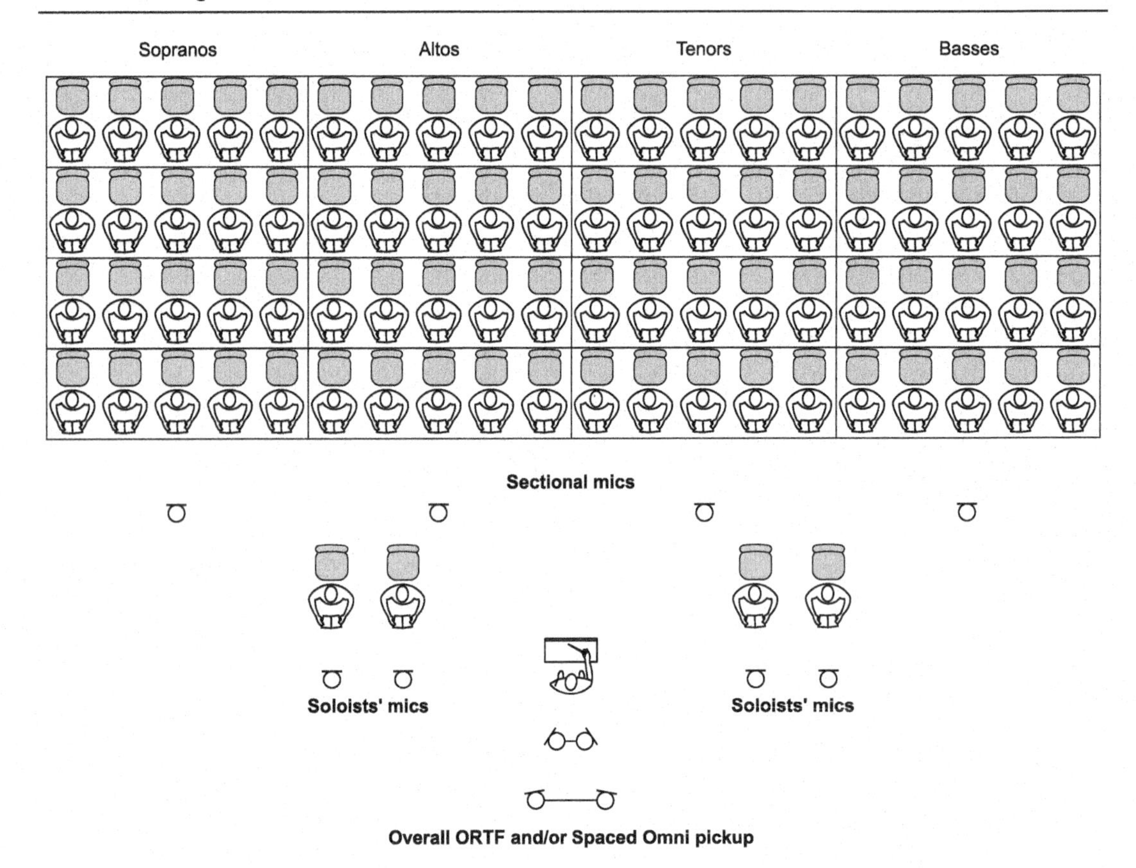

Figure 15.8 Choral society and soloists in front

this is a very tidy and unobtrusive rig, so it is good for concert use. Figure 15.9 shows how to mount two small stereo bars together.

15.5.2 Soloists' microphones

Soloists are usually placed in front of the choir and are sometimes seated within the front row. Each individual soloist is not continuously involved during a large choral work, so when they are not singing a solo, their microphone should not be left faded up. If you leave the soloist microphones open all the time, you will have a close-up on individuals in the front row of the chorus that will ruin your attempts at a good blended image. If you plan to mix to stereo at the time of recording, you will either need to know the work well or have a musical score to follow, or someone to follow it on your behalf.

Figure 15.9 Mounting two small stereo bars in a X with an ORTF pair on each

Photo: Mark Rogers

A single directional microphone can be used for each soloist. Although this seems to contradict the advice given previously about placing two microphones on a soloist to avoid presenting a strongly mono point source in the image (as discussed in Chapters 4, 6, and 7), you will find that each soloist's image will be given enough feeling of width by the singer's voice being picked up on the choir microphones. Provided the soloist's microphone is not too close or faded up too high, there should not be a problem of their image collapsing to a patch of dry, mono sound. Care should be taken in mixing to listen for this.

In a concert setting, the soloist microphones will cause the most problems with visual obstruction in terms of stands, unless you are able to suspend them from above. Microphone placement would ideally be as for a solo voice (see Chapter 6).

For oratorio performances, the soloists will usually be holding their music, which can get in the way when using alternative, lower microphone positions. Placing the microphone over to one side but pointing towards the singer can help with the obstacle presented by the score. See also section 16.2.1 for concert oratorio with orchestra.

15.5.3 Organ microphones

The organ should be approached as for an organ recording as much as possible (see Chapter 14) and using an omnidirectional pair in order to pick up LF where needed. If you are lucky enough to be in a dedicated concert hall which has an organ, the main pipes are likely to be placed around the central axis of the building, which means that your choir pair will also pick up a good central image of the organ, even if it sounds a little distant. You can then modify this with the addition of organ microphones.

If you are in a church and the organ is to one side or ranks of pipes are dotted about the building, you should pan your organ pairs as close as possible to how you would have done for solo organ (see Chapter 14). This means overlapping stereo images from the various organ pipe ranks, although not necessarily overlaying them all evenly spread about the centre.

However, if there is significant choir spill that affects only one side of a pair of organ microphones, you will have to pan the organ partly to one side to avoid disturbing the choir image. See section 15.3.5 for discussion of conflict between choir image and organ placement.

15.6 Antiphonal church choir

In the Anglican choral tradition, the church or cathedral choir is split into two sections, arranged facing each other across the aisle of the church at the eastern end of the building. The two sides are 'Dec' ('decani', the side on which the dean sits, the south side) and 'Can' ('cantoris' the side on which the cantor sits, the north side). Each side contains voices of all four parts (trebles, altos, tenors, basses), and so this layout can be used for straightforward four-part writing, antiphonal arrangements, or eight-part writing.

For a service, the choir will be situated in the choir stalls, but if recording a church/cathedral/chapel choir as a recording session, it is more usual to bring the choir out of the stalls onto the chancel steps, as this makes stereo imaging much easier.

15.6.1 Recording during a service

Because the choir from the congregation's point of view is seen sideways on and separated to the left and right, the listening perspective of someone seated in the pews is not one that will translate directly into a satisfactory recording. Some thought has to be given about how to capture and then pan the parts to fill the centre of the recorded image.

See Figure 15.10a and 15.10b for a church choir layout set in the choir stalls using microphones on stands or suspended.

This figure shows four 'floodlighting' microphones covering the choir on each side, which will form the main basis of the sound, and a vertically mounted ambient pair of microphones pointing upwards above the choir at a height of about 3–3.3 m (9′10″ to 10′10″). These can be used if the organ ambient microphones are not picking up enough ambient choir, which is often the case if

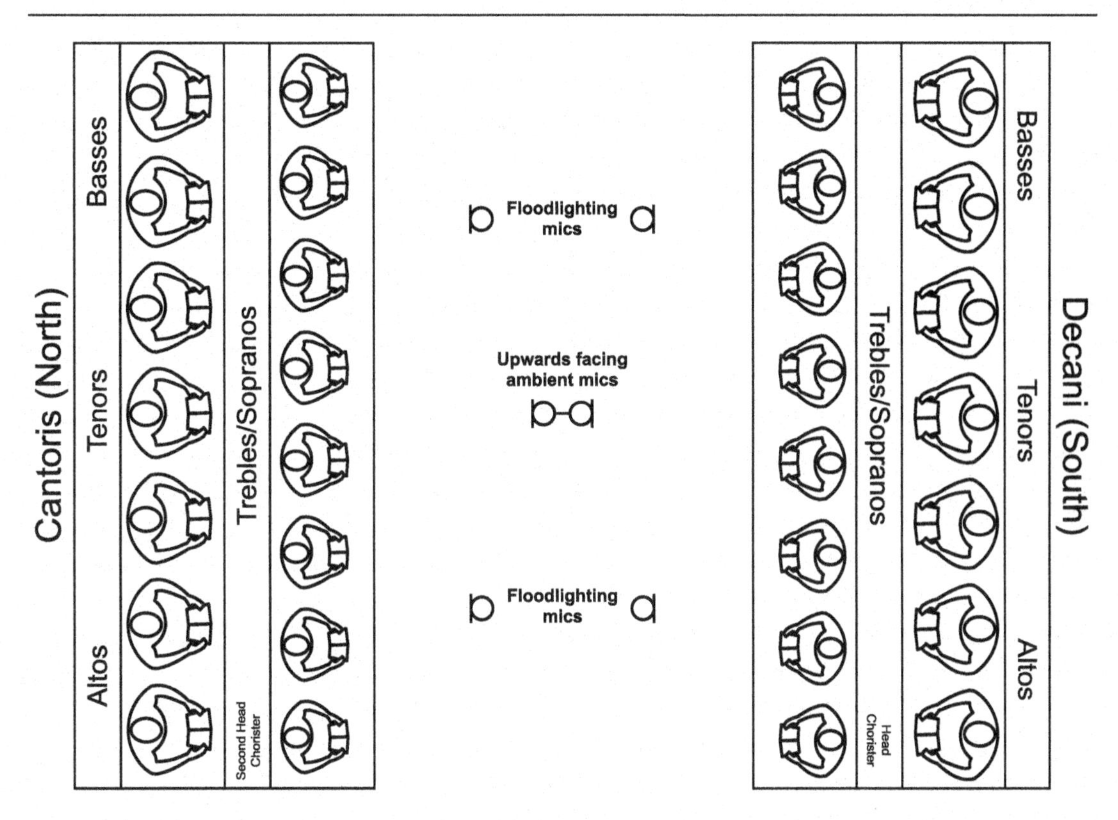

Figure 15.10a Microphone layout for church choir in choir stalls – plan view

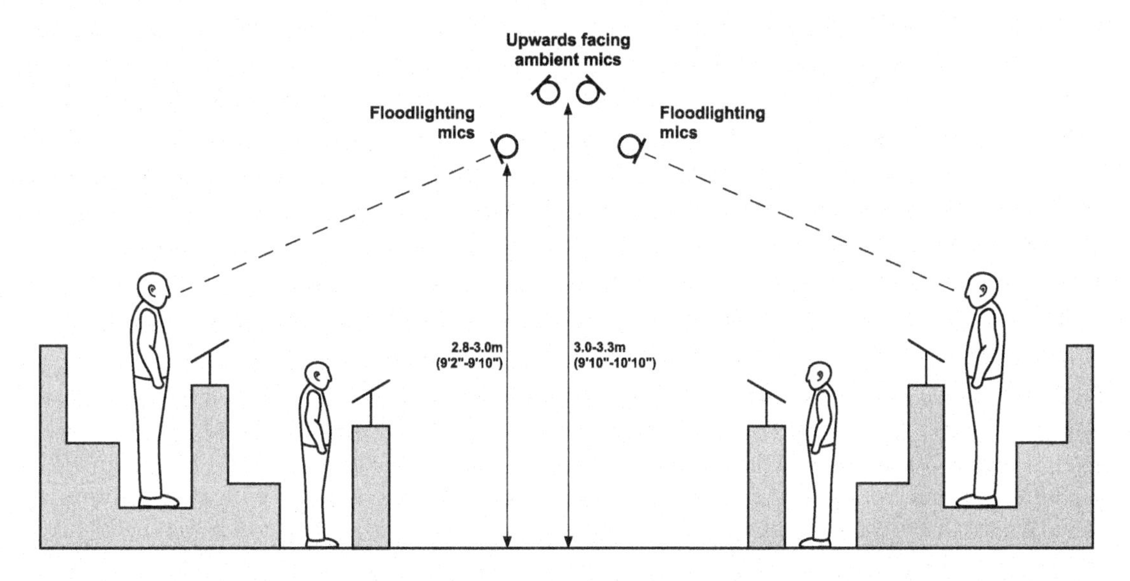

Figure 15.10b Microphone layout for church choir in choir stalls – side view

the organ is playing very loudly. The four floodlighting microphones are at a height of 2.8–3 m (9′2″ to 9′10″) and are aiming more towards the men in the back row, as the boys and/or girls in the front row usually come across very clearly. If this is not the case, then the microphones can be angled differently to pick up the front and back rows more equally. These four microphones can be panned to spread the two choirs to fill the width of the recorded image, rather than presenting them all split between left and right with an aisle up the middle. The panning would be along these lines:

Cantoris Alto end – left
Cantoris Bass end – half left
Decani Bass end – half right
Decani Alto end – right

Therefore, if the choirs perform something that is written antiphonally, the listener to the recording will still get this effect, with one choir spread between the far left and centre and the other spread from centre to far right. The person seated in the congregation will not experience the antiphonal effects as coming from very hard left and right, as the reverb within the building muddies any extreme localisation.

The disadvantage of the arrangement in shown in Figure 15.10a and 15.10b is that if you are unable to suspend microphones and have to use stands, these will be very obstructive to any movements up and down the aisle during the service unless the aisle is unusually wide. An alternative to this layout is shown in Figure 15.11. This shows an ORTF pair at each end of the choir stalls looking along the length of the aisle where the width is not so restricted, although slinging these would still be a better option if possible. They will capture two images of the choirs – one with the altos in the foreground and one with the basses in the foreground – which can be overlaid. The pairs should be panned L-R and R-L, as the image of one with respect to the other will be left/right reversed because they are looking at the choirs from opposite ends. If you are only able to rig a single pair due to restrictions, it will usually be preferable to choose the west end (furthest from the altar) because that is where the head chorister and second head chorister are seated, in the decani and cantoris, respectively. You will inevitably lose something of the basses' end of the choir to a more distant perspective and lower level with this arrangement. If there is room on the back row on one side or the other, it might be possible to rig an ambient pair of ceiling-facing cardioids on a vertical stand at least 3 m (10′) high to help with adding in some more reverberant sound.

Where the choir is unaccompanied, then a high-pass filter of appropriate turnover frequency can be used to help with any very low frequency church rumble. If organ is also used, the organ microphones can be left unfiltered while still filtering the choir microphones.

When organ is in use, the main question that arises is how to combine the choir in the choir stalls and various sets of organ pipes into a coherent and believable image. As discussed in section 15.3.5 and Chapter 14, we can use the principle that if pairs of microphones are sufficiently separated from each other, the lack of correlation between them means that the images from the separate pairs can be overlaid onto one another, either directly on top or with partial overlapping. Where there is any image conflict, the choir should take precedence and be panned left to right across the centre of the image, and the organ panning should be adjusted to work with this.

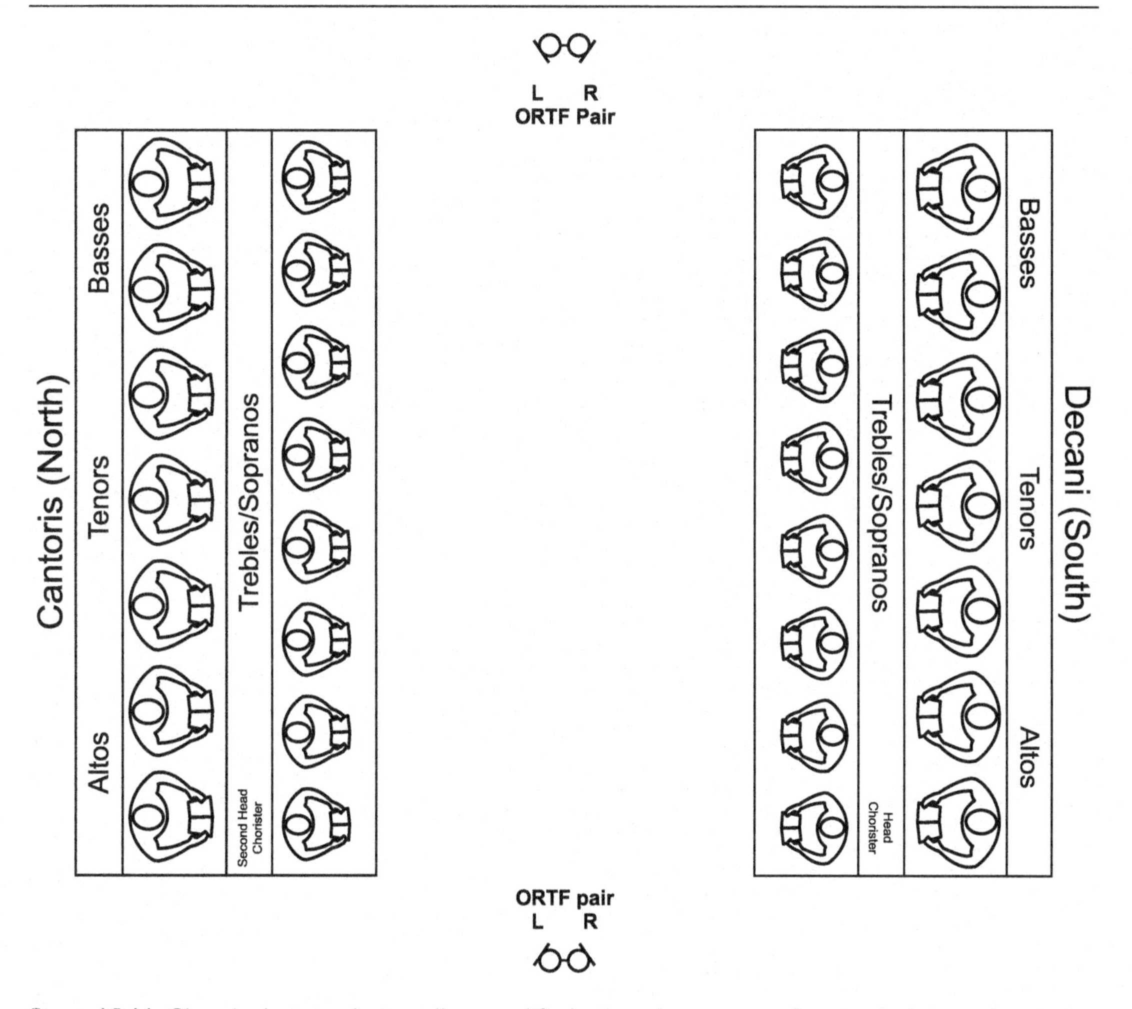

Figure 15.11 Church choir in choir stalls – modified microphone set-up for use during service

In the layout shown in Figure 15.12, the choir are in the stalls and the organ is placed in the corner behind the cantoris. This will certainly result in some choir spill onto organ microphone 3 in particular, and this part of the organ will have to be panned partly to the right (panning microphone 4 to right and microphone 3 to centre or partially left) in order not to disturb the choir image.

15.6.2 Antiphonal church choir recording session

Unless it is a service that is being recorded, choosing to record in the choir stalls simply makes the process more difficult. For a recording session, there is no need to have the choir seated in the

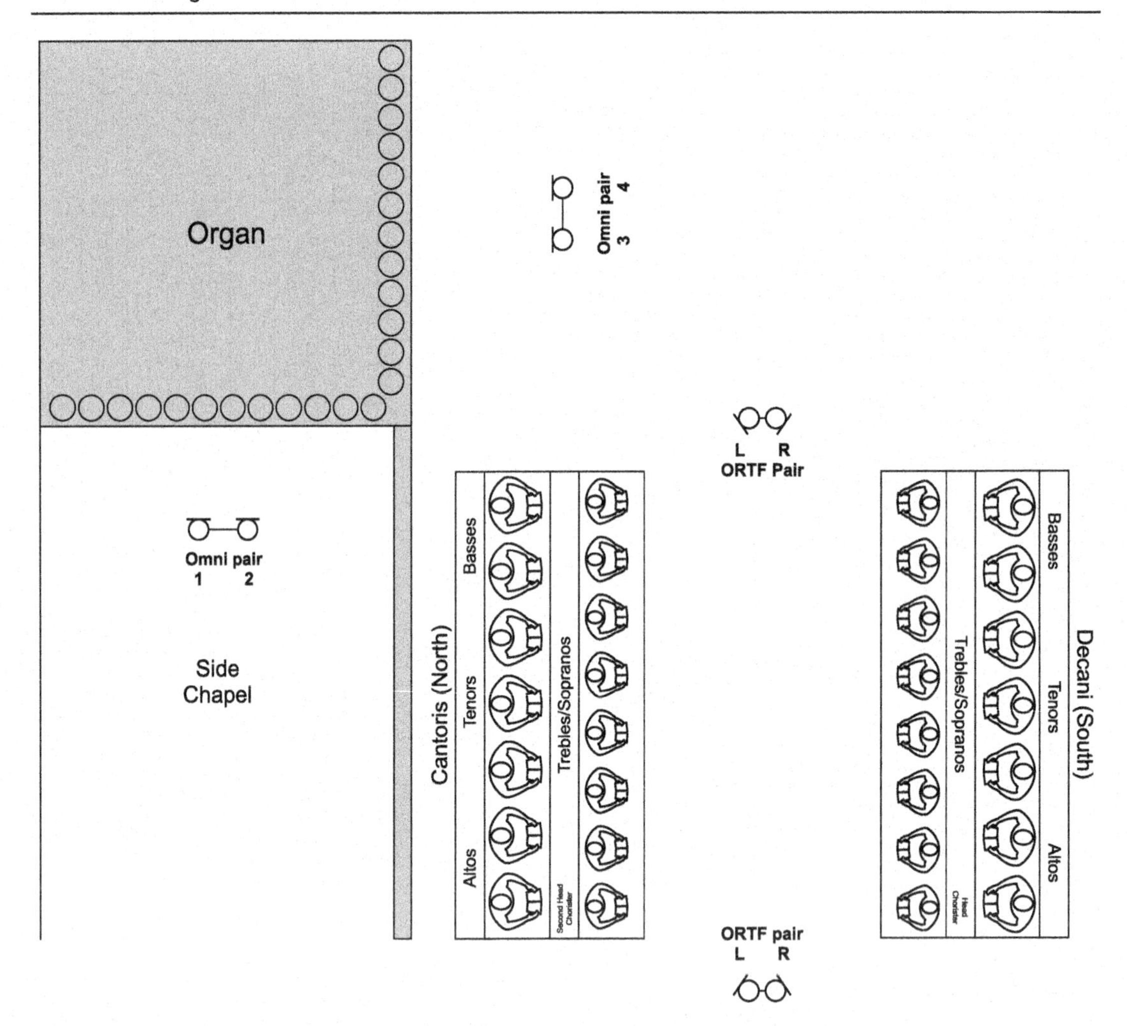

Figure 15.12 Separate pairs on choir and organ

choir stalls, and they can be placed wherever the best sound might be achieved in the building. This is very often in front of the choir screen and chancel steps, as for the many Decca sessions with King's College Choir, Cambridge. For these, the choir were arranged in a wide, semicircular layout in front of the choir screen. If the separation between cantoris and decani is to be preserved, then the layout will commonly be to have the trebles in the front row and the altos, tenors, and basses in the back row, with the cantoris to the left and the decani to the right (see Figure 15.13). This will work best if the choir has good internal balance; blocking the sections by part (as for the choral society) will usually prove helpful if there is an imbalance between parts (e.g. the trebles are too strong), as it will give the engineer more control.

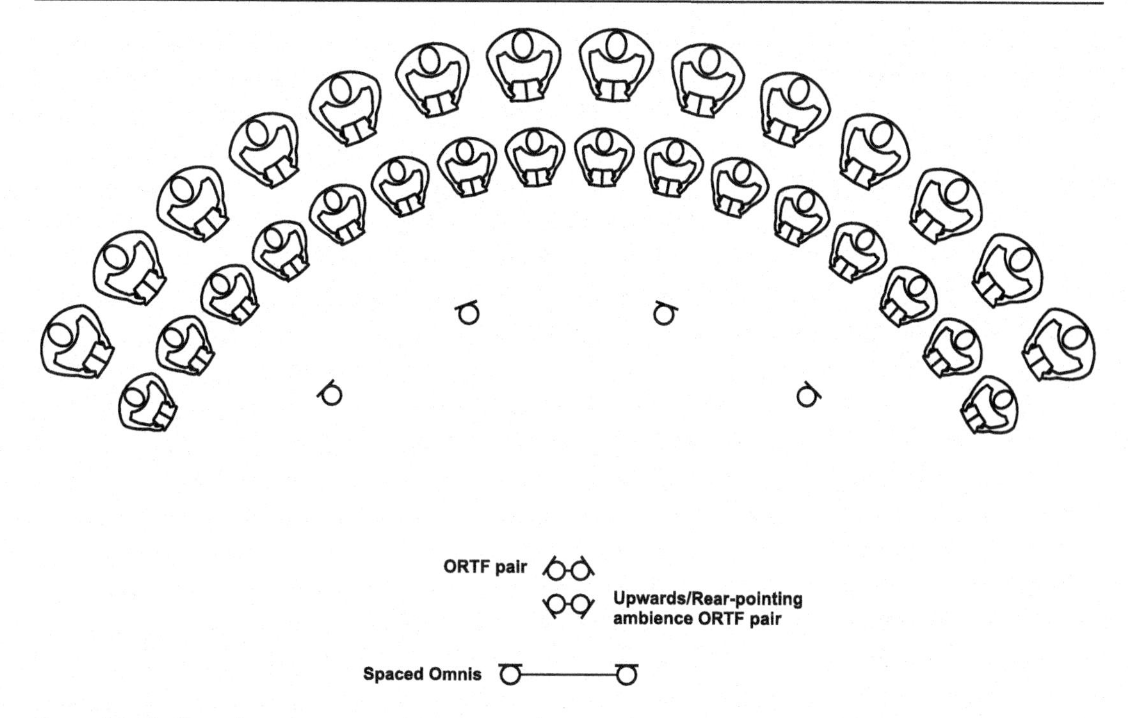

Figure 15.13 Chapel choir arranged in semicircle away from choir stalls; organ microphones can be added separately as in Figure 15.12

An overall ORTF pair or spaced omnis at 80–100 cm (2′8″ to 3′4″) apart is used, along with an upwards- or reverse-pointing additional ORTF pair for adding more indirect, reverberant sound, which is a very important and characteristic part of a church recording. Mounting the forward-facing and rear/upwards-facing pairs close together means there is a high degree of timing coherence between them, and the amount of reverberant sound can be adjusted without any significant occurrence of phase cancellations. The rear-facing pair is panned the same way as the front-facing pair (i.e. the left-hand microphone of each pair from the audience point of view is panned left).

Figure 15.13 also includes sectional spots in the same way as were indicated for the choral society choir, with wide cardioids being the preferred choice, but cardioids are an acceptable alternative (see sections 15.4.2 and 15.4.3 for discussion of section microphones and their panning and blending.)

Chapter 16

Solo voice, orchestra, and choir

The planning and logistics required to record a session or a capture a live performance of a choral symphony, oratorio, or opera are complex, and the aim should be to produce a convincing recorded image that conveys all the musical detail and audible words whilst retaining a believable sense of perspective. It is a demanding task, but we can bring together principles from previous chapters to construct a successful approach.

This chapter will start with choir and orchestra and move on to the addition of stationary soloists for oratorio and operatic repertoire. Lastly, given that recording studio opera is now outside the budget of most record companies, we will look at the recording of live performance as is done at the Royal Opera House, Covent Garden. For those interested in other approaches, Appendix 1 gives details on how opera was recorded at Decca from the 1950s to the 1990s – that is as a semi-staged performance with soloists moving according to stage directions but under session conditions.

16.1 Orchestra and choir

To record a large choir and orchestra at the same time, we can draw on the techniques outlined so far in Chapters 8 and 15, with some adjustments for different layouts of the choir and the orchestra.

16.1.1 Orchestra and choir: concert layout

Figure 16.1 shows an idealised starting point for an overall set-up for a concert, or a session recording using this conventional layout.

In a concert setting, the usual arrangement is to place the choir behind the orchestra, either on a stage or on some sort of raised and tiered seating so that the singers can both see the conductor and project over the heads of the orchestra to the audience. From the engineer's point of view, having at least 3 m (9′10″) separation between the back of the orchestra and the front of the choir would be ideal, but in practice this is unlikely to be achievable in a concert venue of normal size. It might be achievable in a session scenario in a large studio or other venue. From the conductor's point of view, allowing the choir to be positioned too far beyond the orchestra makes for difficulties in communication and sight lines. This is likely to be especially important in amateur choirs, where there can be a tendency for singers to keep their eyes on their score and not on the conductor.

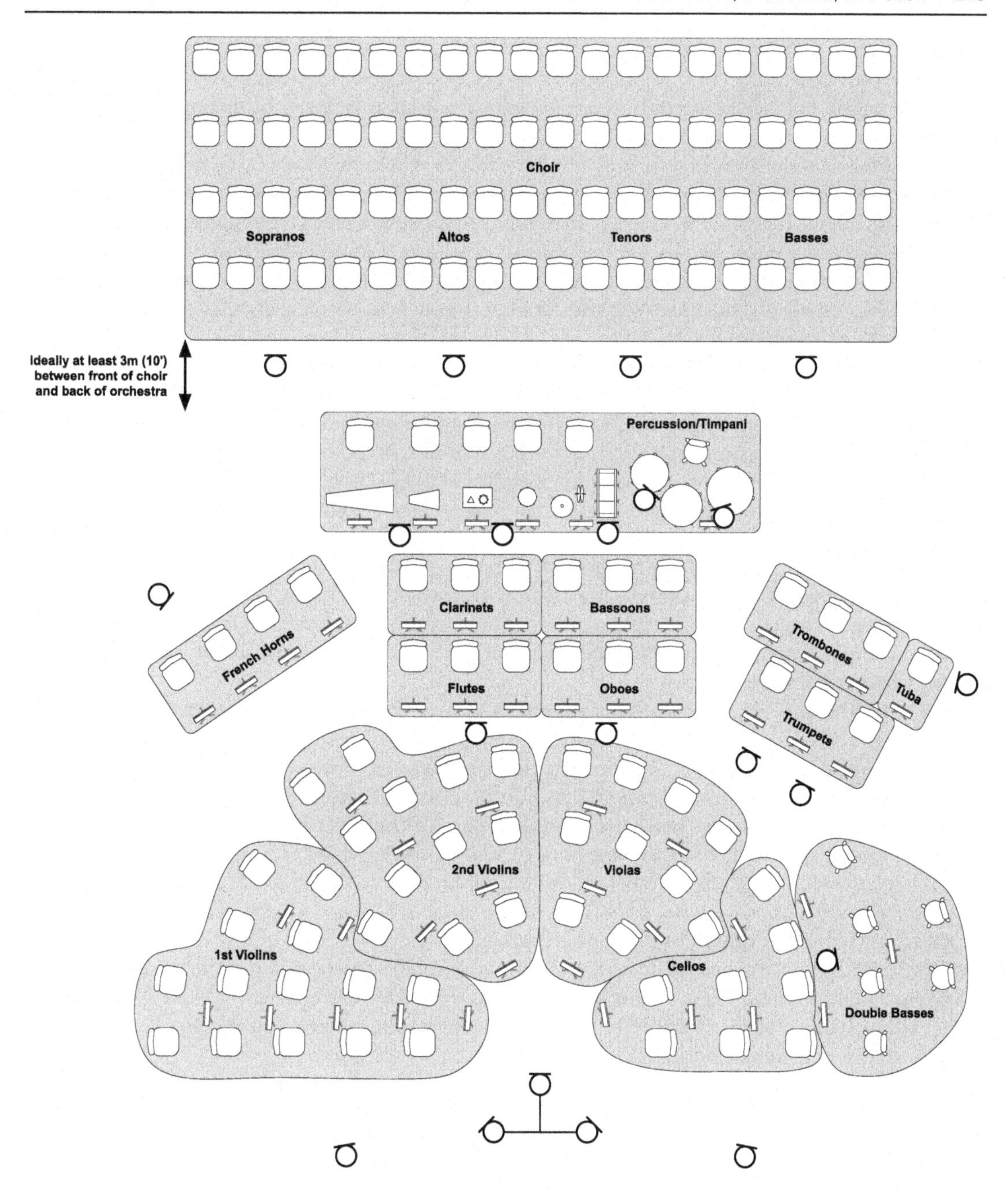

Figure 16.1 Orchestra with 80-member choir in concert layout with a sufficient gap between them

There are two main reasons for aiming for some physical separation between choir and orchestra; one is so that choir section microphones can be positioned far enough away from the choir to obtain a good balance between the blended choral sound and clarity of diction. This is harder to achieve with stands placed close to the front of the choir.

The second is so that percussion spill can be reduced; where there is a large percussion section that stretches across the back of a large orchestra, the choir microphones will pick up a great deal of this, especially if they are too close. The louder the orchestra and percussion section, the more of the choir microphones' signals will be needed to achieve a good balance in the recording, but there will be more percussion on those choir microphones. This is a big problem and not always easy to solve, especially in repertoire that is choral and percussion heavy, such as Prokofiev's 1939 cantata, *Alexander Nevsky*.

In a concert situation, solutions can only be partial, and the first choice of the engineer would always to be to negotiate moving the percussion section to either side of the rear of the orchestra. In a recording session, the second-best option after moving the section would be to use screens (if available) to reduce the percussion spill. The only other solution, if you must have choir and percussion right next to each other, is to place microphones closer in to the choir, but this will need more microphones to obtain even coverage and still retain a blended choir sound. Using fig of 8 microphones to discriminate against the percussion is also a possibility illustrated and discussed in Figure 16.2c.

Figure 16.2a shows reverse booming microphones so that the backs of the booms are above the orchestra. This would be useful on a session and where there is no loud percussion section directly in front of the choir. The outer microphones could be boomed in from the sides of the choir in a similar way. Slinging all these microphones would be a less visually intrusive way of getting them into a good position.

Figure 16.2b shows using more microphones to cover the choir in the event that the microphones have to be placed closer than is ideal. This might be because of physical restrictions on stand placement and height or deliberately closer placing to reduce the amount of percussion spill. Going back to our lighting analogy, in order to cover the choir evenly, the closer the microphones, the more of them are needed to avoid patches of the choir being more 'brightly lit' (i.e. sounding louder) than others. If the option is available, using wide cardioids when closer in will help reduce the localising effect. Where close microphones have to be used, adding a high ambience pair pointing upwards above the rear of the choir will pick up very useful blended choir sound to help bind the other microphones together.

Figure 16.2c shows the use of fig of 8 microphones to help ameliorate the effects of percussion along the back for the orchestra, but because their front lobes are narrower than a cardioid or wide cardioid, more of them would still be needed to obtain even choir coverage (analogous to using a narrower beamed spotlight).

Where there is a good distance between the back of the orchestra and the front of the choir, it might be necessary to add in some delays when mixing. This would mean delaying the choir microphone signals to be more in time with the choir sound as it arrives at the main orchestral tree pickup, which might be 10 m (35′) away (a delay of around 35 ms). (See Chapter 17 for more on using delays.)

16.1.2 Orchestra and choir: alternative layout for recording sessions

For a studio recording of choir and orchestra, the concert layout outlined in section 16.1.1 can be used, but if there is a lot of percussion or the venue lends itself better, an alternative layout can

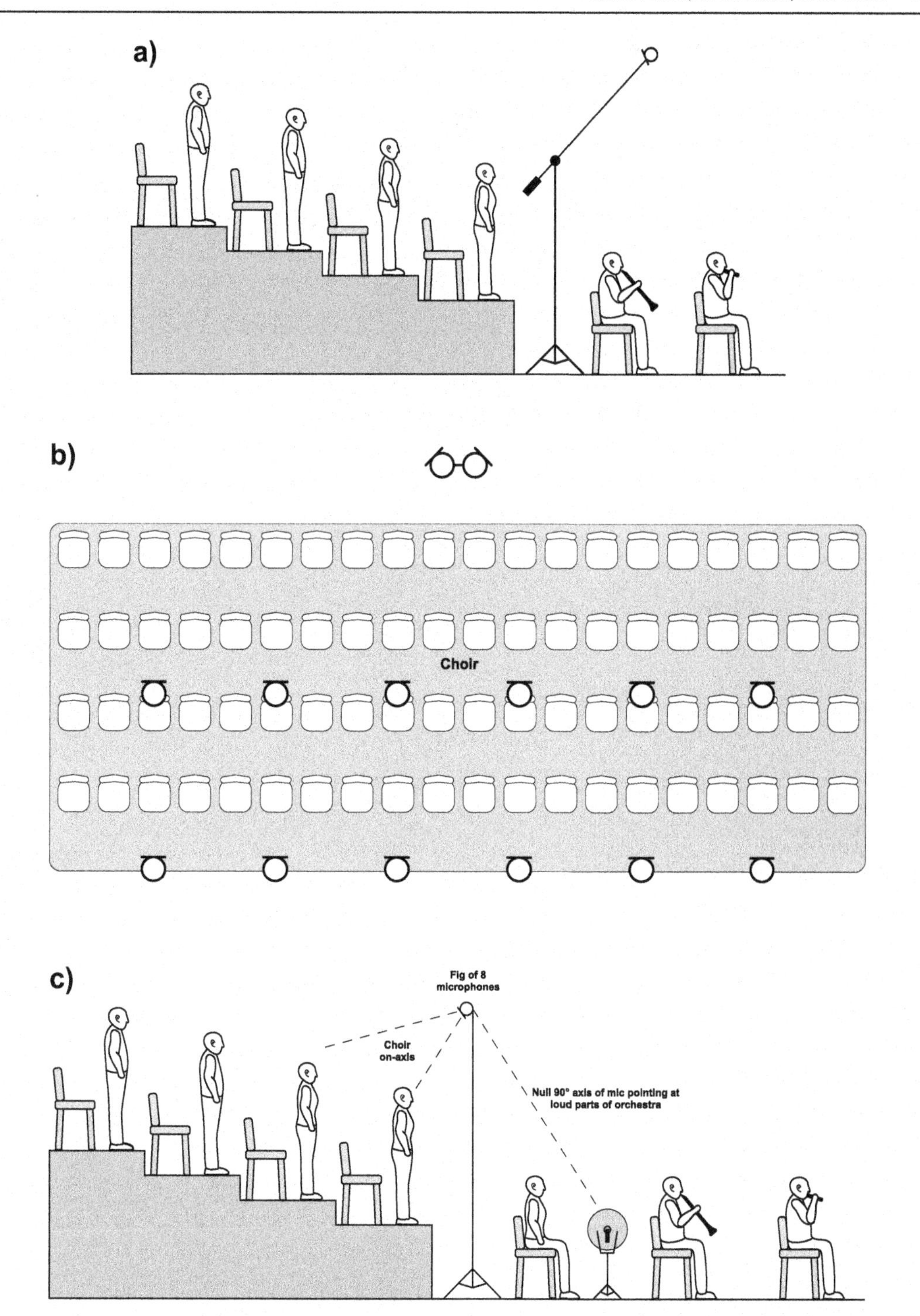

Figure 16.2a–c Alternative solutions to getting satisfactory coverage of the choir where there is no gap between choir and orchestra

be used with the choir on the rear side of the main orchestral pickup (i.e. behind the conductor). This has implications for communication, in that the conductor has to turn around to address the choir and split their attention between choir and orchestra. Therefore, it is probably not a good set-up for an inexperienced ensemble or conductor who might be thrown by a very different seating arrangement. In a large recording set-up, an assistant conductor can be used for the chorus if the choir has poor sight lines to the main conductor. This will also be the method used for live opera, where the chorus are offstage and cannot see into the pit. See Figure 16.3 for layout with choir behind the conductor, and note that the usual section order (SATB from left to right) is reversed so that they appear the right way round on the tree microphones.

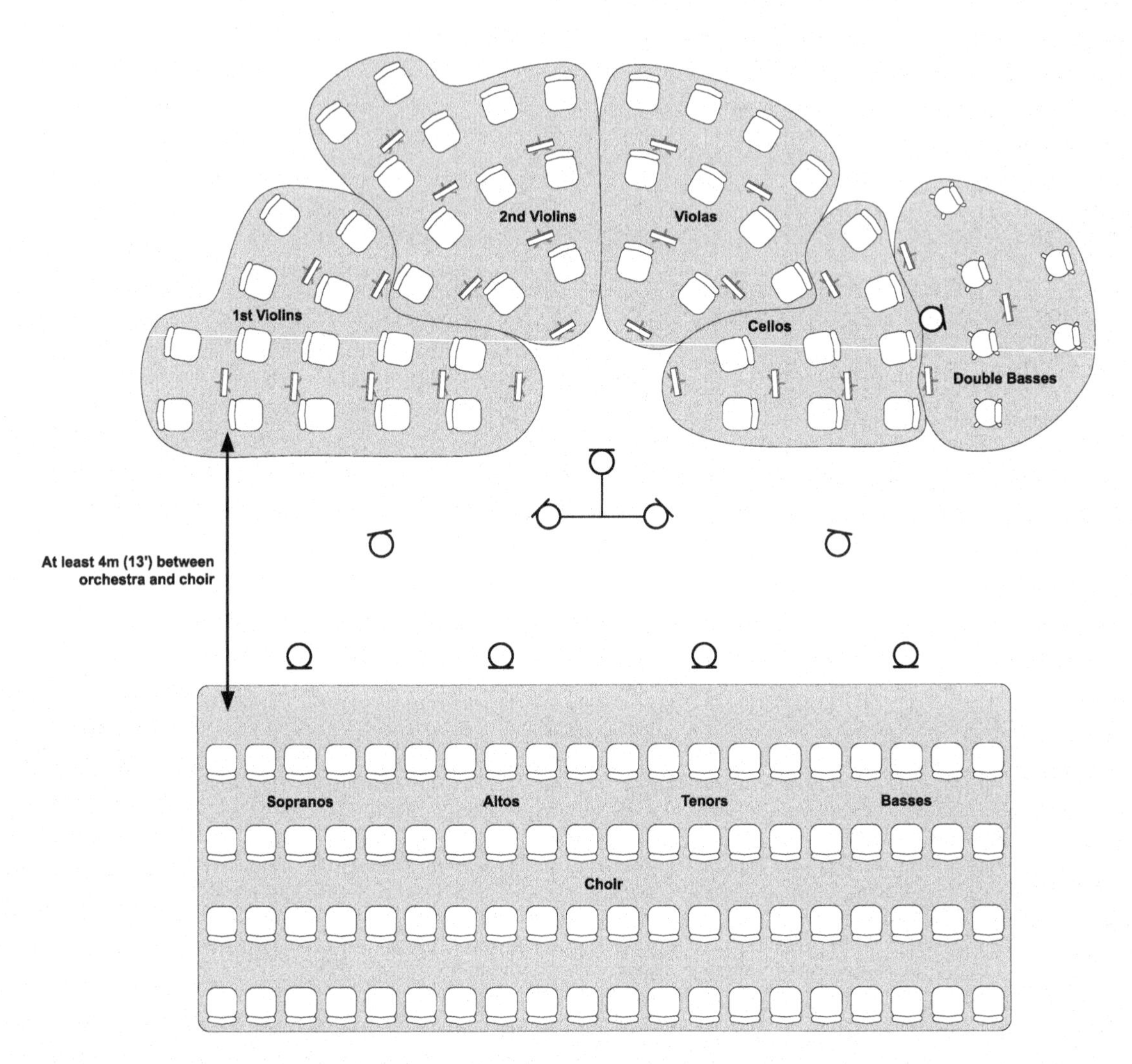

Figure 16.3 Choir on the far side of the tree for recording session

Using the position shown in Figure 16.3, it is important to make sure that the choir is far enough away from the main orchestral pickup; this approach should not be used unless the choir can be set back by at least 4 m (13′). This is to avoid close orchestral spill onto the choir microphones and also to create an appropriate sense of choir perspective by ensuring they sound sufficiently distant on the orchestral tree. The orchestra and choir images can be laid on top of one another in the final recording, although the choir microphones should be panned appropriately left/right reversed; the end of the choir nearest the first violins must be panned towards the left, and the end nearest the double basses panned towards the right.

If there is a suitable balcony in the venue, this can be used for the choir as it will introduce some distance between the choir and the orchestra without having to place them horizontally further back. Recordings made at Walthamstow Assembly Hall or Watford Colosseum have made use of the balcony for the choir, with the orchestra positioned just in front of the balcony so that the conductor can face everyone at the same time.

16.2 Orchestra and choir with stationary soloists

The main concert repertoire categories that include soloists who will be standing in one position are oratorios and some choral symphonic works. Opera can also be recorded with static soloists, but this will usually be done as a recording session, with the inclusion of offstage effects, chorus, or soloists where they occur in the score. Appendix 1 contains details of Decca's opera recording sessions of the 1950s to 1990s, where the singers were positioned on a stage in order to capture stage movement and perspective rather than positioning them in one place. Sections 16.3 and 16.4 consider how to record live, fully staged opera.

16.2.1 Orchestra and choir with stationary soloists in concert

In a concert, the soloists are most commonly presented in front of the orchestra, usually on the violin side but sometimes split on either side of the conductor. The immediate problem with this position is that the soloists are likely to be picked up strongly on either the left or right microphone of the overall orchestral pickup. Without additional spot microphones, the singers' images can be quite unfocussed, hard to locate, and biased towards one side. The singers will need additional spot microphones – preferably one each – in order to stabilise and focus their images. It will be sufficient to add a single spot microphone each rather than using a pair of spot microphones (as we looked at in Chapter 6), because the additional microphones in the room will add width and bloom to each individual.

Occasionally, soloists will stand within the choir, and either sing solo from their usual choral position, or step forwards a little for the solo. In this case, the layout is closer to that used for session recording, and the methods in section 16.2.2 can be adapted with more discreet soloist microphones. If you rig a soloist spot for someone who remains in situ within the choir during solos, remember to only fade it up when there is a solo or you will spotlight the singer throughout the work. You will need to bring the microphone up very gradually, in advance of the solo, to 4–5 dB below the target level because any fast fader movements will be audible. To do this successfully, you are likely to need a score and/or a lot of rehearsal time; if you can persuade them to come to the front where you have rigged some soloists' microphones, so much the better!

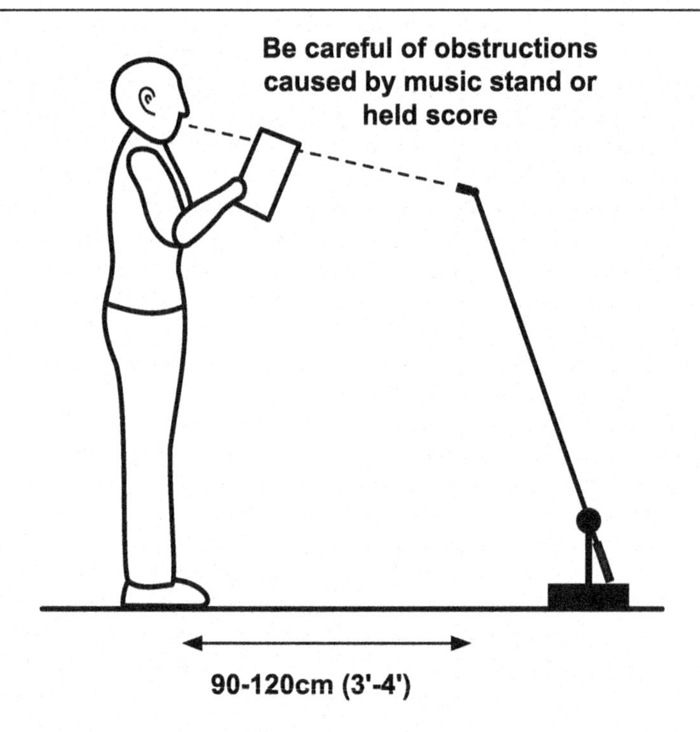
Be careful of obstructions caused by music stand or held score
90-120cm (3'-4')

Figure 16.4 Low stand spot microphones for soloists in oratorio

For stand-mounted microphones to be reasonably discreet, they will need to be low down, with the microphone below the level of the mouth and pointing upwards about 1 m (3′4″) away from the mouth. These could be normal pencil cardioids of some kind, or 'Pavarotti' microphones (as discussed in Chapter 6) if they are available. This position is quite good for capturing some diction and avoids large changes in tone as the singer moves their head. If there are music stands in use, or the performers are holding a score that causes an obstruction between the microphone and the singer's mouth, its position will have to be a compromise. It will need to be set up looking in from the side of the score/stand to minimise loss of diction and higher frequencies. Care should be taken not to ruin the singer's tone if you try to add some presence back into the voice with EQ.

When using low stands on the stage, use an anti-vibration mount if there is one available to avoid any seismically transmitted energy, such as footsteps producing LF and subsonic signals from the microphone. These will be unnoticeable in a compromised monitoring situation, and they will only show up later and have to be removed. Given that monitoring on location is almost always lacking in LF, it is sensible to avoid applying an HPF to the soloists' microphones until you can really hear what you are doing.

Alternatives to mounting spot microphones on stands include slinging microphones or using radio microphones. Slinging is very time-consuming; it means setting up before the performers arrive, and all your plans will come to nothing if they decide to stand somewhere else during rehearsals, or they accidentally stand away from their microphones during the performance. There is nothing you can do to quickly move slung microphones, whereas you can potentially move microphones on stands to where the performers are standing. If you were to sling, the microphones will have to be above the singers and pointing down towards them from a height of about 2.1–2.4 m (7′ to 8′). If they are much lower, they will probably be causing obstruction to audience sight lines.

The difficulties inherent in using radio microphones for classical singers will be discussed in the context of the Royal Opera House in sections 16.4.3 and 16.5.2. Because they produce a close, dry sound, radio microphones will need a generous amount of artificial reverb or actual ambient sound and some delays to blend them into the overall mix. You also need someone who is experienced at fitting radio microphones and with dealing carefully with artists, as this involves running cables through the clothes of performers who might be nervous (and very particular).

Whether the singers' close sound comes from radio microphones or normal microphones, they will need additional reverb to help them blend in. This could be artificial reverb or an upwards-facing ambient pair of angled cardioids positioned about 3 m (10′) above the singers' heads. This pair should be positioned with the aim of picking up more indirect sound from the singers than from the string section.

Any microphones aimed at singers standing in front of the orchestra will pick up something of the strings, particularly violins or cellos depending on the singer's location. When these signals are added to the mix, they will potentially affect the orchestral sound and image. The soloists' microphones need to be faded up enough to solidify the singers' image and draw the ear to them, without pulling them artificially close to the listener or affecting the orchestral sound too much. If you can sense the presence of a spot microphone on the singer as a patch of dry or close sound, it is faded up too high. If enough of the signal is used, it can be panned away from the singer's image as presented on the main pickup (especially if this is a tree and the singers' images are imprecise). The usual practice would be to pan them towards the centre as far as is possible without this confusing the orchestral image. How much artificial panning is possible will depend on where the singer was standing in the first place, and how much of the spot signal is necessary to focus their image.

16.2.2 Choir and orchestra with stationary soloists for recording session

Given the problems of orchestral and singer imaging that arise from having soloists on either side of the conductor in front of the orchestra, when a recording session is undertaken, alternative soloist positions can be explored.

The most common position is to place the soloists behind the orchestra and in front of the choir, as shown in Figure 16.5. Staging is included in the illustration, but less frequently, singers might also be standing on the floor.

The layout shown in Figure 16.5 enables the singers to see the conductor easily and to feel support from both the orchestra and chorus. Anything that supports the singers in their performance should result in a better recording. From the engineer's point of view, this position avoids

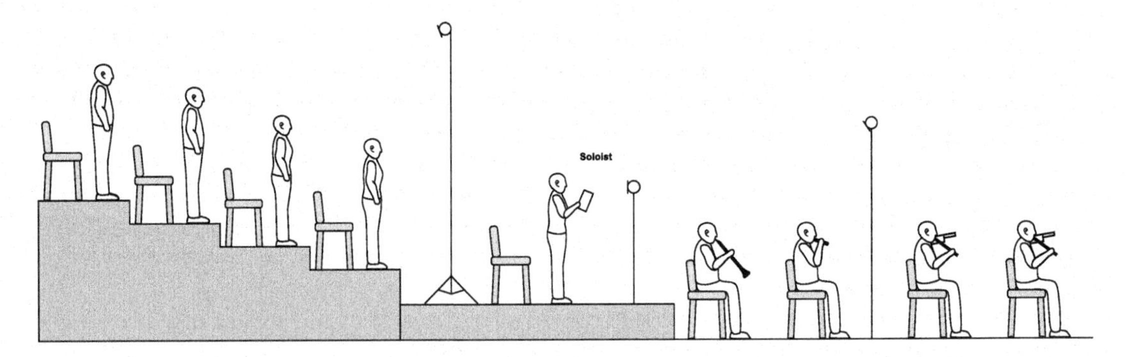

Figure 16.5 Orchestra with soloists behind, and chorus

close voice spill onto the orchestral microphones and it allows the singers to be placed centrally if required. It also produces a nice diffuse reverberant voice sound on the orchestral microphones which will enhance the setting of the voices into the room acoustic while having little influence on their lateral image position. The singers' exact position in the image will be controlled by the panning of the spot microphones, but the singers should also be standing more or less where they are required to be heard in the L-R plane to make imaging more secure.

For the recording of Handel's *Messiah* with Christopher Hogwood and the Academy of Ancient Music (July 1979)[1] recorded at St. Jude's, Hampstead, the chorus was raised on wooden staging and the soloists positioned one at a time behind the woodwinds and standing on the floor just in front of the choir microphone stands. This gave them some space to move in, but with closeness to the orchestra. See Figure 16.6 for the original session orchestral set-up sketch.

Although the usual place for the singers will be behind the orchestra, alternative positions for soloists can be used if this will work for the repertoire and suits the conductor and singers better. For the EMI studio recording of *Tristan und Isolde*[2] (2005 in Abbey Road Studio 1), the soloists were positioned within the orchestra as shown in Figure 16.7. This was done to facilitate good connection between the soloists and conductor, and it enabled the singers to hear everything, thus feeling enveloped and supported by the orchestra. Because the chorus in *Tristan* is only ever offstage, a large chorus at the rear of the orchestra was not needed. With the soloists in this position, their individual microphones had to be a little closer than usual to reduce orchestral spill, but a vertically facing pair was added about 3 m (10′) above the singers' heads to pick up some more reverberant sound from them whilst avoiding too much woodwind and brass.

The other most common place for soloists on a recording session is behind the conductor by about 5 m (16′). This is usually done because the conductor wants to be able to hear them properly and feel in contact with their singing. There are two obvious potential drawbacks to this layout: one is the loss of eye contact between conductor and soloists, and the other is a reduction in the feeling of orchestral support experienced by the singers.

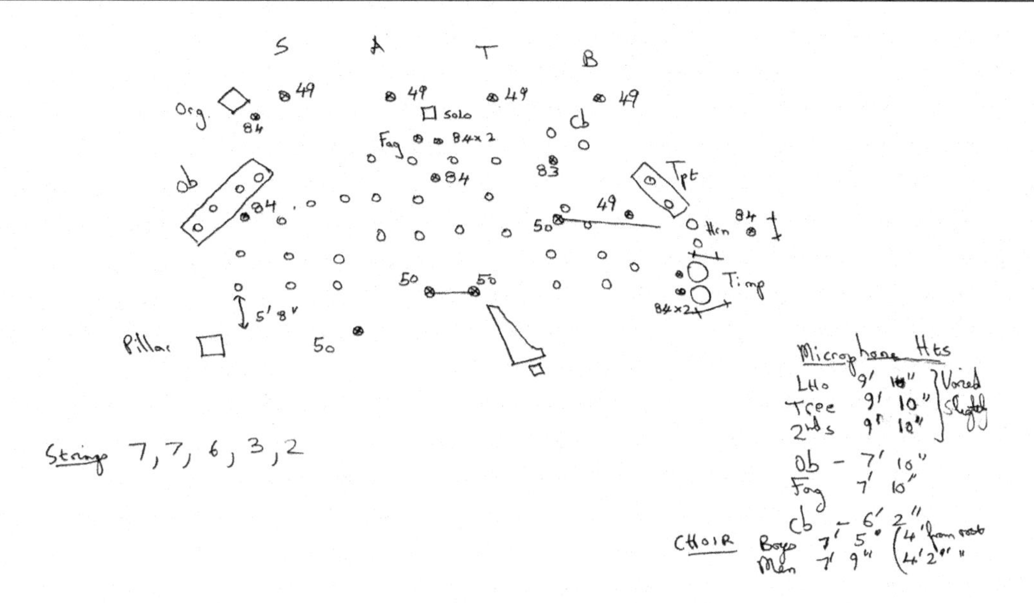

Figure 16.6 Session and soloists layout for Handel's *Messiah*, 1979. See Appendix 3 for the electrical set-up.

Source: John Dunkerley.

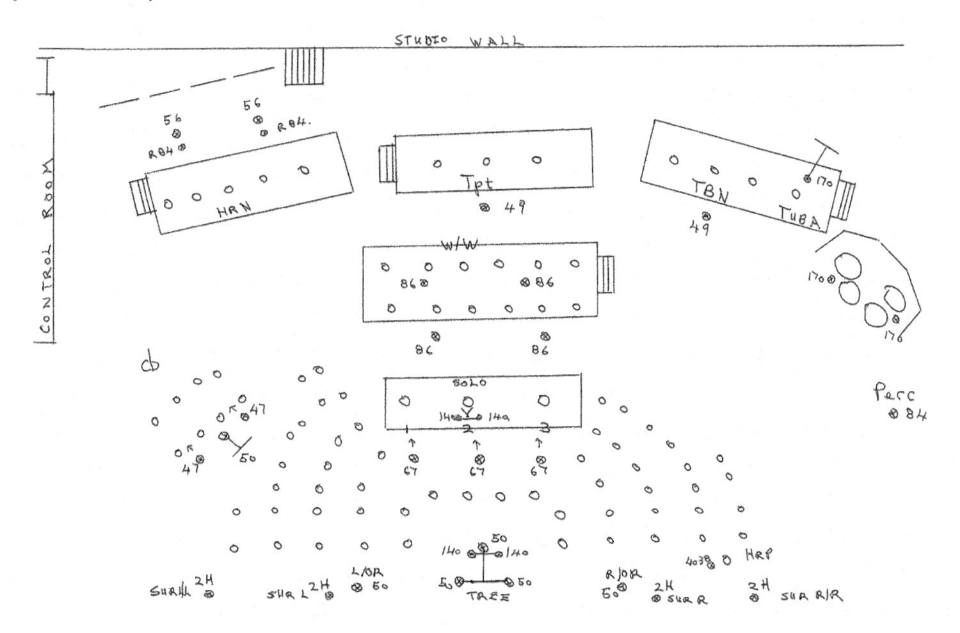

Figure 16.7 Session layout showing position of soloists for *Tristan und Isolde* (2005) in front of the woodwinds. This necessitated closer soloists' microphones, and an additional ambience pair was used to ameliorate this.

Source: John Dunkerley.

The first can be addressed by the conductor using a swivel high chair so they are able to turn around to face the orchestra or soloists as required. The second is less easy, and if the singers are at all uncomfortable with the reduction in orchestral support, it would be better to allow them to stand behind the orchestra and for the engineer to deal with any difficulties that arise from spill. Figure 16.8 shows this alternative position for the soloists.

Examples of recording made in this way include Beethoven's Ninth Symphony with the Academy of Ancient Music (AAM) and Christopher Hogwood (1988),[3] which was recorded in Walthamstow Assembly Hall with the chorus in the balcony and the soloists behind the conductor. Also recorded with soloists behind the conductor was *La Bohème* with the Berlin Philharmonic and Karajan (October 1972) recorded in Jesus Christus Kirche (Berlin).[4] In this case, the soloists were positioned on a stage erected over the altar steps with the typical grid layout found in Decca's opera recordings (see Appendix 1 for discussion of Decca's semi-staged style of studio opera recording.)

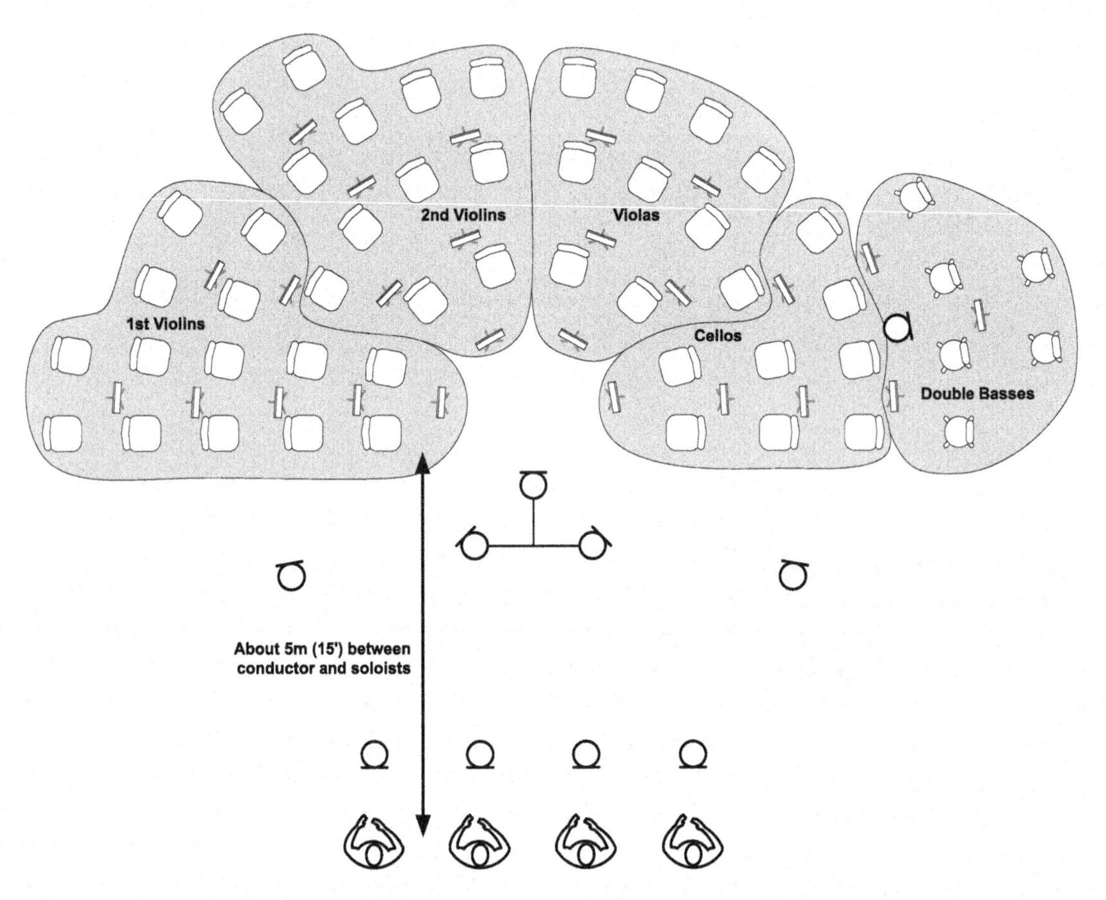

Figure 16.8 Orchestra with soloists behind the conductor. The choir position could be behind the orchestra or behind the soloist.

When the soloists are placed behind the conductor, the chorus is most commonly placed behind the orchestra in its natural position, but given an unusual venue or other restricted layout, it would be possible to work with both the chorus and the soloists behind the conductor. Figure 16.9 shows Christopher Hogwood and the AAM recording Haydn's *Creation* with the split chorus and soloists laid out behind the conductor.

Although the conductor can now hear the soloists well and verbal communication between them is straightforward, there will be higher-level and closer-sounding soloist spill onto the main microphones than would be the case with the singers placed at the back of the orchestra. This will not be as hard to manage as the spill that arises in a concert situation (see section 16.2.1) as it will not be as close or high in level, and therefore it should not cause imaging problems. The relative closeness of the voice spill will mean that some additional reverb might be needed on the voices to blend them in, although the further back the singers are placed, the more distant the spill will become. However, with greater distance comes poorer communication, so the improvement in spill quality might not be worth it.

Figure 16.9 Recording Haydn's *Creation* with Christopher Hogwood and the AAM at Walthamstow Town Hall.[5] Split chorus and soloists behind the conductor with orchestra in the foreground.

Photo: Mary Robert; courtesy Decca Music Group Ltd.

It might be thought that screens could be used to help to reduce spill between soloists and orchestra, but adding screens in a classical recording context can have negative as well as positive effects. (See Appendix 1 for experimental use of screens in older Decca recordings.)

One of the important messages of this book is that nice-sounding spill can be a very useful component of the overall sound provided it is considered as part of the microphone's signal during the set-up planning. When screens are added, there will inevitably be some sort of boundary effects and colouration in the form of additional very early reflections for those players near the screens. If the players can be moved to a position where the spill becomes a positive attribute, the negative aspects of screen use can be avoided. Of course, if the performers cannot be moved, and the quality of the spill is detrimental to the overall sound, using screens might be a better solution. To reduce boundary effects from screens, directional microphones should be positioned with their null axis towards the screen to reduce pickup of reflections.

16.2.3 Offstage chorus, soloists, and sound effects

Where an offstage chorus is specified, the best approach will be to place them at some distance away from the orchestral microphones to produce a natural offstage effect; a distant balcony would be one option. The sense of distance can be enhanced with artificial reverb if necessary.

Positioning a singer 'offstage' can be achieved by reducing the level of their individual microphone so that they are placed further back in the perspective without also moving them off to one side. In the era of Decca's studio opera recordings, a singer could also be asked to turn around to face the back of the stage to produce a distancing effect.

Any effects specifically required by the score (wind machines, car horns, boat sounds) can be easily added after the main recording, and nowadays there is also the option of overdubbing soloists as a last resort if illness requires them to miss a recording session where the orchestra, chorus, and other singers are already booked.

16.2.4 Balancing orchestra, choir, and soloists when mixing

There are three main elements to the opera balance: the orchestra, the chorus, and the soloists. (Chapter 17 gives more advice on which microphones to bring in first.)

The foundation of the whole sound will be the orchestra, as it is playing for the majority of the time. The orchestral microphones can be panned as for an orchestral recording (see Chapters 3, 8, and 9 for orchestral microphone techniques) to ensure the orchestra fills the full image width between the loudspeakers.

The next part of the puzzle to slot into place will be the soloists. Given the complexity of the medium, it is not surprising that there is a lot of disagreement among engineers and producers about exactly how the solo voices should sit in the mix, with stars of the opera world having their own views as well. Opera is a musical dialogue, and no one element is more important than the other. The sung words must be understood, but the voice must be supported by the orchestra and not appear to be a separate entity. The orchestral writing must be heard, but it must not get in the way of the voice. A useful contrast can be drawn with musical theatre or crooner-style recordings (such as Frank Sinatra) where the voice sits above and in front of everything. The operatic singer should be more embedded in the orchestra and in the natural acoustic, better reflecting the live performance in a medium where singers are not amplified.

The soloists should feel just in front of the orchestra in terms of perspective but must not be too close or dry. There will be soloist spill onto all the orchestral microphones; when the soloists are placed behind the orchestra in particular, this spill will be quite distant and useful for setting the singers into the same acoustic space as the orchestra. The individual soloists' microphones are used to capture diction and to give the image some stable localisation, but they must do this without dominating the sound and making each soloist into a narrow point source. It is really important that the soloists retain some bloom around their sound; *the listener should not be able to discern that they have individual microphones*. Therefore, care should be taken not to fade them up too far; levels will have to be frequently altered to keep the voice supported but not sticking out.

Another reason for avoiding pushing the singer levels too much is that this will start to ruin the orchestral sound because of orchestral spill onto the soloists' microphones. Opera recordings using the techniques that have been discussed have their basis in the Decca orchestral sound, which is a careful balancing act. The microphone balance that works is completely tied up in the microphone placement, and to achieve a vocal balance that is less naturalistic and more upfront would require the use of closer microphones. The material being recorded might also inform how the soloists and orchestra are balanced, especially if the engineer is sensitive to the musical intentions of the work; in Wagner's writing, for instance, the voice is very much only one element, and the orchestral parts can stand on their own in terms of musical interest. In Verdi's work, to take another example, the orchestral parts can sometimes be more accompanying in nature.

Finally, the chorus requires a good balance between diction and a solid body of sound, and it should be set into the acoustic space just behind the orchestra. This creation of appropriate depth is going to be an illusion; even in a concert, you might find that the natural depth is too deep and you need to foreshorten it. As we have seen, a session-based recording might have the singers and chorus standing in a number of places relative to the orchestra, but the end goal of mixing is to create a good orchestral foundation, a layer of chorus, and the soloists supported on top. This can be visualised as a pyramid (see Figure 16.10).

If the chorus is covered by five microphones, the outer ones can be panned fully left and right, with the central three panned half left, centre, and half right. Where the chorus is covered by four microphones across its width, panning the outer microphones inwards to about 80% left and 80% right will prevent the outer parts of the choir appearing to be located exclusively in either loudspeaker. Fully left or right panning of spot microphones and sectional microphones can have this localising effect, which detracts from the illusion of the orchestra filling the space between the speakers while not being obviously located in either of them. To avoid too much thickness in the LF, the choir microphones can be high-pass filtered, but care must be taken to avoid placing the HPF at too high a frequency and starting to remove the body and weight from the choir. It is best done when mixing in a good monitoring environment – probably after the recording.

The main danger in trying to make sure that both the orchestral detail and the choir diction is clear is that the perspective can become quite flat. There is likely to be some foreshortening of the natural distance to get enough clarity from the choir, but additional reverb can help provide cues that still place the choir slightly further back than the orchestra. The requirement for clear choir diction is also repertoire dependent; where the choir is being used as an orchestral colour (Ravel's *Daphnis et Chloé;* Holst's 'Neptune'), it can be placed behind the orchestra at a more natural distance. For something like Orff's *Carmina Burana*, the diction needs to be much more accessible.

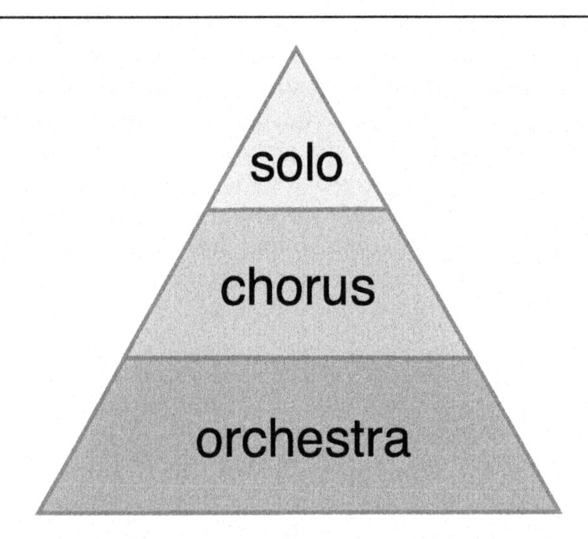

Figure 16.10 Mixing pyramid for opera balance

16.3 Recording live opera for cinema, DVD, TV, radio, or CD

Live opera is most commonly recorded and filmed for cinema or DVD production, but this is often done simultaneously with a live radio transmission or audio-only recording. Live recordings and transmissions of opera, plays, and ballet for local cinema distribution have become a popular way of disseminating the activities of our cultural institutions to those who cannot get to the theatre in person.

The sound quality requirements for each delivery method differ very little; radio microphones are more common where there is picture as well, but their use is increasingly accepted in audio-only recordings as a way of capturing the detail and contact with the singers that makes for more viscerally engaging recording. Radio microphones should never be obviously perceptible, and so they are very carefully blended to bring just enough detail without making their presence felt (see section 16.5.2). If a recorded sound is a good one, it should work adequately for both radio and TV; the aim should be for a good static balance that caters for 80% of the transmission and does not have to be drastically altered for a close-up shot, as large fader moves can distract from the listener's experience of the music. Therefore, to cater for close-up images without sudden audio perspective changes, the recorded sound for picture might need to be slightly more detailed.

The most serious limitation when recording live opera is in the inability to place microphones anywhere that is obtrusive or disruptive of audience sight lines; any differences between audio-only and audio-for-picture live recordings are small compared with the compromises in microphone placement that are required.

Sections 16.4 and 16.5 deal in some depth with how live recording and mixing to surround for cinema is approached at the Royal Opera House, Covent Garden.

16.4 Recording live opera at the Royal Opera House: microphone set-ups

Any microphone set-up for live opera can be divided into four main areas:

1 Orchestral microphones.
2 Stage coverage for soloists and chorus.
3 Individual singer radio microphones (if they are used).
4 Auditorium microphones.

16.4.1 Orchestral microphones

The orchestra in an opera house is physically constrained in the pit, so space is tight for accommodating stands, and no microphones can be rigged higher than the pit wall. This means that the use of an overall Decca Tree or ORTF-type technique is not possible, and other solutions have had to be found. Stands with heavy round bases are designed for use where a tripod-legged stand would cause some obstruction.

Figure 16.11 shows the orchestral layout in the Royal Opera House (ROH) pit, and it can be seen that the brass section is placed on the right, immediately behind the violas. For this reason, the overall string section microphone set-up is not symmetrical, and the distances between string microphones on the right-hand side are much smaller to avoid getting too close to the brass. The numbers in the following text refer to the microphones as numbered in this diagram.

The 'main pair' [1, 2] is usually a pair of DPA 4006 omnis placed on either side of the conductor in line with the music stands belonging to the front desks. They are as high as possible, which means about 2.4 m (8′) above the pit floor, and about 2.5 cm (1″) below the auditorium rail around the pit edge.

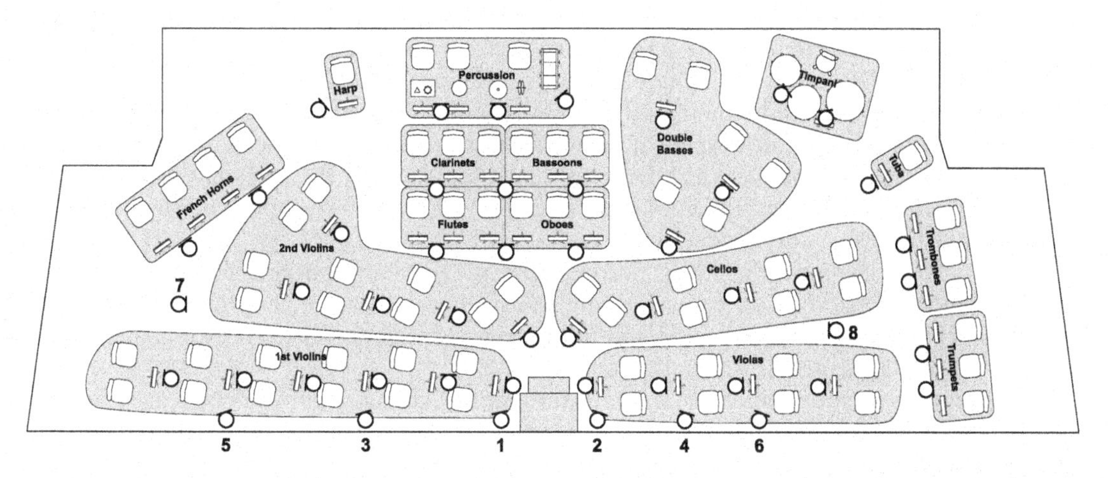

Figure 16.11 The ROH orchestral pit layout with overall pickup microphones numbered 1–8

Moving outwards from the main pair, there is another set of omnis (Neumann KM83) [3, 4] positioned at the same height as the main pair. On the left-hand side, this is in line with the third desk of violins, and on the right-hand side, it is in line with the second desk of violas.

Moving outwards again is a pair of 'outriggers' [5, 6], which are wide cardioids (Schoeps MK21) located in line with the fifth desk of violins and the third desk of violas.

Finally, there is another pair of DPA 4006s [7, 8] positioned within the orchestra in the gap between the string desks and horns on the left and a mirrored position on the right. The microphone on the right will sometimes pass through a little compression because it will pick up some brass, and this will automatically reduce the level when it does.

These eight microphones form the basis of the string section coverage, and the 'main pair' [1, 2] are panned to about 70% left and right, with the other microphones being panned fully left [3, 5, 7] and right [4, 6, 8], and also partly into the rear loudspeakers (see section 16.5.1).

In addition, the strings have one spot microphone per desk, the trumpets have two microphones between three players, and the horns have two microphones between four players. There is often a camera at the back of the pit to get shots of the conductor, and so any woodwind microphones have to avoid being in shot. They are therefore lower down than usual, and so one microphone is used for every two players to keep the coverage even. Because the pit layout is very shallow front to back, the woodwinds are much closer to the overall pickup microphones [1–8] than usual, and their spot microphones are used only a little.

16.4.2 Stage microphones

Capturing the stage in live opera is something of a challenge, as the options for microphone placement are either high above the stage or very low, near to the stage floor. A method for capturing the stage has evolved which uses low-mounted 'float' microphones at the front and 'bridge' microphones high up at the rear. This is illustrated in Figure 16.12.

The front 'float' array consists of three narrowly angled (about 25°–30°) co-incident stereo pairs of cardioids and two single cardioids at each end of the row. This is a similar method of stage coverage to that used by Decca for their studio recordings (Appendix 1), but because this is a live context, the microphones have to be placed as low down as possible to remain unobtrusive. If modular condenser microphones such as the Neumann KM140 (KM100 bodies, AK40 capsules) are used, they can be mounted very low at a height of about 5 cm (2″). The capsule can be separated from the output stage of the microphone to enable the use of a variety of accessories to adapt the microphone for different physical locations. If the stage has a front lip, or there is a false floor to the set, the microphones can be arranged to just peek up over the edge. An alternative float array of five microphones across the front of the stage can be made from a set of PZMs. The Schoeps BLM 03C is one example; it has a hemispherical pickup pattern and takes the form of a flat disk, so it sits very unobtrusively on the stage. This microphone is not as suitable if there is a lot of stamping and movement in the choreography, as it can easily get knocked out of place.

This basic array creates a series of 'stations' through which the singers will pass as they move laterally across the stage. The narrow stereo pairs give some coherency of image without creating excessive lateral image movement as the singers move. If the three inner stereo pairs are replaced with single microphones (e.g. if using PZMs), good coverage of the stage will still result, although

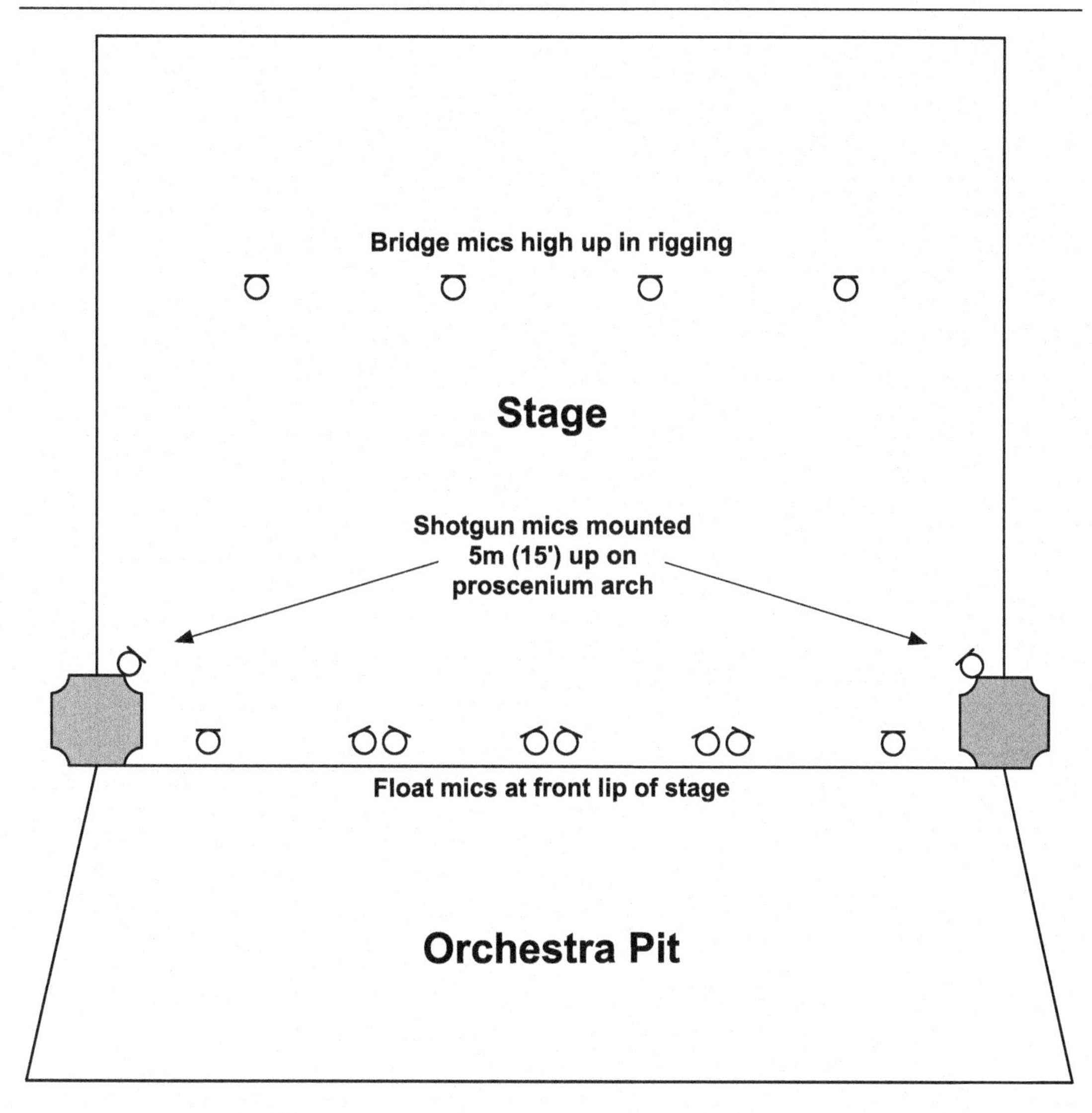

Figure 16.12 Stage coverage showing front float microphones, rear bridge microphones, and proscenium arch shotgun microphones

the image will be slightly less coherent. The float microphones are panned between the front channels of a surround sound mix, much as they appear across the stage. The outer microphones placed at around 80% left and 80% right, and the inner three pairs are panned so that they are centred around half left, centre, and half right, respectively, with a small amount of panning separation between the left and right microphone of each pair. This keeps the stage action contained in front of the listener, which will be particularly important when mixing for picture and in surround.

An alternative method of covering the stage is to use hyper-cardioids or shotgun microphones instead of cardioids which will produce a closer sound on the singers. This approach is often favoured by radio engineers if there are no additional radio microphones on individual singers (see section 16.4.3). The disadvantages of shotgun microphones are that they are physically larger, the singer's sound is rather brighter, the microphones are noisier, and the off-axis pickup is very coloured, which can detrimentally affect the orchestral sound.

To support the chorus and any other singers located towards the back of the stage, four 'bridge' microphones are placed high up in the theatre rigging aiming towards the back of the stage. These are usually shotgun microphones, as they are some distance above the singers. At the ROH, there are two additional shotgun microphones placed on either side of the proscenium arch at about 5 m (16′) high; these are used as fill-in microphones to give a fuller feel in the chorus scenes.

16.4.3 Soloist microphones

The 'float' microphones will be used as the primary source for the stage, providing most of the tone colour and body of the sound. To provide some more contact and detail, each soloist is fitted with an omnidirectional radio microphone which is very small and discreet and exhibits very little handling noise. Because the radio microphones are very close to their sources, there is very little spill from other stage performers or the orchestra, but the closeness of the sound itself presents problems in mixing.

Unlike for pop or musical theatre work, the personal microphone signals should never be conceived of as the main vocal source and will never be used separately from the overall 'float' stage microphone sound. Their role is to act as a small part of the mix in order to add some detail and image stability to the individual singers; all the radio microphones are panned to the centre channel when mixing to surround, as is the convention with cinema dialogue. Time delays have to be used to avoid the radio microphones being perceived as the dominant source; the singers' voice will arrive at the radio microphone tens of milliseconds before arriving at the 'float' microphones and will dominate due to the precedence effect. (See section 16.5.2.)

Fitting radio microphones to performers so that they are hidden in clothing and wigs is a specialist job that requires excellent interpersonal skills. The microphone and pack must be hidden, and also allow for costume changes where necessary. Any singers that sweat a lot will be fitted with two microphones to allow for microphone failure if moisture enters the capsule and starts to cause pops and bumps.

As mentioned in 16.4.2, in a situation where no radio microphones are to be used, it is usual to use shotgun or hyper-cardioid microphones across the front of the stage to obtain more focus on the singers. This will produce good results, but it will not give such a nuanced degree of control over the singers' sound as using radio microphones in conjunction with cardioid float microphones. The more the hyper-cardioid/shotgun float microphones have to be raised in level, the more the orchestral sound will be adversely affected by them, and so there is a limit to the amount of contact with the voices that can be achieved with this approach.

16.4.4 Auditorium microphones

The number of auditorium microphones used will depend on whether the project is to be mixed in stereo only or surround sound (see section 16.5), but auditorium microphones tend to form part of any permanent installation. As a minimum requirement for picking up reverberant sound and audience applause, there are four 'drop' microphones suspended from the auditorium ceiling very high above the audience, more or less above the orchestra pit wall. The outer ones are omnis, and the middle two form a near-coincident pair of cardioids.

There are two additional wide cardioid microphones rigged on the side balcony above the lowest tier of boxes and roughly in line with the front of the orchestra pit. On the rear balcony there are two more ambience microphones, and in the dome of the auditorium ceiling is a Hamasaki square of fig of 8 microphones, which is an array designed to produce four discrete channels of reverberant sound to use in surround mixing (see section 10.3). See Figure 16.13 for the location of auditorium microphones.

In general, a surround mix can support more reverb without losing clarity, as some of it can be sent to the surround loudspeakers. This gives the listener the experience of more reverb without muddying the image coming from the front loudspeakers. Conversely, if a surround mix has to be folded down into stereo, some reverb will have to be removed in preference to sending it all to the front loudspeakers.

16.4.5 Offstage effects

In a live performance, any offstage chorus or other sound source is placed behind the set and picked up by a pair of microphones placed there for the purpose. They will also be picked up as a distant sound on the main orchestral and stage pickups, and a balance between these and the backstage microphones can be used to give an appropriate sense of perspective. When mixing in stereo, the offstage chorus can only be made to sound as if it is coming from somewhere other than the main stage by use of appropriate reverb, reduced HF detail, and lowered level. When an offstage chorus is mixed in surround, the sense of location away from the main stage can be reinforced by placing the chorus into the rear speakers.

16.5 Mixing opera to 5.1 surround for cinema

When mixing audio for picture, the soundtrack should not pull the audience's attention away from the images being shown in front of them. For this reason, almost all the direct sound is presented as coming from the front channels, with the rear channels reserved for reverb and offstage sources. The difference between listening to classical music in stereo and in surround is a subtle one (see Chapter 10), but the reverb and the sense of space envelop the listener and can really add to the excitement of the performance. If the technology distracts the listener from being immersed in the moment, it is not being used to best effect; the surround speakers need to be as involving as possible but not to the point of being annoying or distracting. The engineer has to remember that the goal is enhancement of the musical experience rather than showcasing the technology. During the mixing process for the ROH productions, time is allowed for refining mixes in a cinema setting ahead of production.

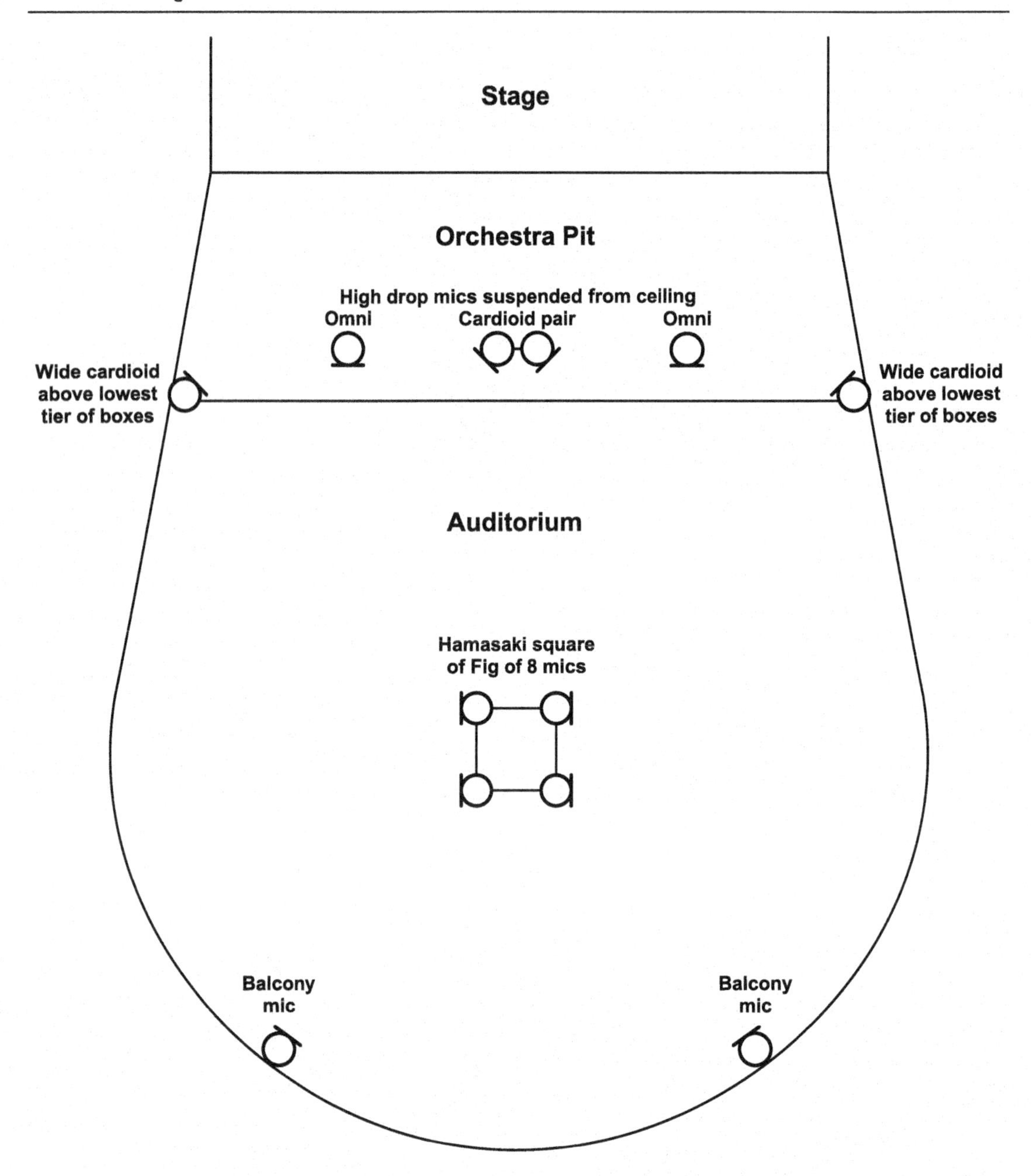

Figure 16.13 Auditorium microphones – drop microphones, side balcony microphones, rear balcony microphones, and Hamasaki square

16.5.1 Orchestra

The centre channel of the surround set-up, C, is usually mainly for the soloists' radio microphones and the central stage float microphones; this helps with the perception of vocal clarity, particularly in a large auditorium like a cinema. Therefore, the orchestra is mixed primarily between the front LF and RF channels, with some divergence into the centre, C, and a small amount sent to the rear channels, LS and RS. (See section 10.2 for wider discussion of the centre channel and divergence control in orchestral mixes.)

The 'main pair' (microphones 1 and 2 in Figure 16.11) are panned about 70% LF and RF to avoid a hole in the middle, but the overall orchestral string section is spread a little wider than in stereo by panning the outer string microphones (microphones 3–8 in Figure 16.11) partly into the rear LS and RS speakers. They are panned between 30% and 50% into the rears, beyond which the effect becomes distracting. The effect is one of greater envelopment in the orchestral sound by bringing the outer edges of the orchestra partially around the sides of the listener. (See section 10.2 for similar technique when panning a Decca Tree in surround.) All the spot microphones in the orchestra are sent to the front channels only, keeping the woodwinds, brass, harp, and percussion firmly in front of the listener.

16.5.2 Solo voices

The main body of the soloists' voices will come from the 'float' microphones at the front of the stage which are panned evenly spaced across LF, C, and RF (see section 16.4.2), keeping the stage image in front of the listeners. The float and bridge microphones are not all left fully open at all times, as this is detrimental to the orchestral sound, so the engineer will mix in more of the microphones nearest to the soloist (as outlined in the old Decca stage method in Appendix 1). In addition to overall stage microphones, small amounts of the radio microphones are added to provide some singer detail *without ever allowing the presence of the radio microphones to become discernible*. To help the very close and dry radio microphone signals to blend in, they are delayed so they are never ahead of the signals from the float microphones. When working live, one engineer is assigned to changing the delay in real time on each radio microphone, depending on where the singer is standing at the time. For singers towards the front of the stage, the delay will be in the region of 10 ms (because in this position they are approximately 3 m (10′) away from the float microphones, and for those standing at the back, the delay will be up to about 28 ms). It is better to play safe and add too much delay to be certain that the radio microphone signal is never ahead of the float microphones; if it arrives earlier, it will be perceived as the dominant signal and will be very audible, even at a fairly low level in the mix. The delays enable an appropriate level of the radio microphones to be used to bring in some detail and stability of image.

The radio microphones are all panned into the centre channel and then diverged (see section 10.2) by 10% to 20% into the LF and RF channels. Their presence will anchor the singers' images a little more to the centre, but enough of the float microphones are used to retain some degree of lateral movement. In theory, it would be possible to follow each singer's position with panning as well as delays on the radio microphone, but to do this live demands a lot of engineers and it is generally not practical. In a cinema setting, perceived lateral image positioning will vary across the audience according to seating position, and so keeping the voices somewhat anchored around the centre has the effect of reducing this variation. This follows the usual practice in film dubbing, where the dialogue is all placed in the centre channel unless there is a very good reason not to do so.

The next aspect of the voice mixing to consider is how much to take into account different camera perspectives (e.g. whether a singer's close-up shot should sound less reverberant than a wide angle shot). Film and TV sound usually avoids large changes in dialogue audio perspective with a shot change, because sudden audio changes can be particularly distracting from the story. For many TV dramas and soap operas, the dialogue is almost uniformly processed regardless of the environment or shot angle. A recorded live play is a slightly different genre in that the viewer is aware that the actors are in a theatre performing a live event. In these circumstances, it is common practice to make some gesture towards realism in the mixing by increasing the ambient sound for wider angle shots, although this can only be performed live with sufficient rehearsal time.

Live opera could be considered as an extension of the live play scenario, but dealing with music makes it different again. A noticeable change of audio perspective in the middle of a sung phrase would really detract from the beauty of the musical line, and anything that takes the viewer/listener out of that moment would be less than good practice. Therefore, consistency of sound is prioritised over perspective realism, and changes in sound are mainly reserved for when a singer leaves the stage whilst singing (reducing the radio microphone level) and for very obvious wide shots. A close-up of a soloist might be enhanced with the slightest increase in radio microphone level, but all the audio perspective changes used are subtle; the engineer is aiming for a balance that feels live, with the singing clear but not close. The radio microphones are there to provide diction clarity without being audibly present.

16.5.3 Chorus

When the chorus are on stage, they will be picked up on the float and bridge microphones, which will be faded up as required. Float microphone panning has been discussed, and the four bridge microphones will be panned in a similar way across the width of the front image, avoiding fully left and right. The microphones on either side of the proscenium arch (see section 16.4.2 and Figure 16.12) are used to add greater fullness to large chorus scenes; they are panned fully LF and RF, respectively. In this way, the direct chorus sound will come from in front of the listener, even though the chorus will be present to some extent in the reverb signals from the auditorium microphones that are set to front and rear (see section 16.5.4).

When the chorus is offstage, however, it can work very well to place the direct sound and some additional reverb in the rear speakers (LS and RS). This places the chorus spatially away from the stage, and it feels further away because of the increased use of reverb.

16.5.4 Ambient microphones and artificial reverb

The audience microphones and the rear balcony microphones are placed completely in the rears and sent only to the rear artificial reverb unit, so that any audience applause and noise is located predominantly behind the listener (although there will be some audience pickup on the orchestral microphones). The Hamasaki square is designed to pick up four channels of reverb, and these are sent to LF, RF, LS, and RS as they are laid out. (See section 10.3.)

As with most classical recording, artificial reverb is used to enhance the sound, but when working in surround, four discrete channels of reverb are needed. This can be provided by a surround reverb unit, but depending on what is available, it is common to use a stereo unit for the front reverb and another one for the rear reverb. They do not need to be using the same algorithm, although the overall reverb time and general characteristics should match. (See section 10.4 for fuller discussion on use of reverb in surround.)

The singer's radio microphones are usually kept dry, unless there is a weaker singer who needs more of the radio microphone to match in with more experienced, stronger singers. In this case, some reverb will be added to the radio microphone to allow it to be mixed in at a slightly higher level. This will usually be some sort of general ambient reverb to soften the impact of the radio microphone signal, not the same hall type reverb as is being used on the orchestra. Where special effects are needed, such as a scene set in a church with organ playing, the front reverb programme could be altered to a slightly more church-type reverb to help with the theatricality of the staging, as long as it serves the artistic purpose well. Creative decisions of this type will need to be taken in conjunction with the producer.

Notes

1 *HANDEL/The Messiah/AAM/Choir of Christ Church Cathedral, Oxford/Hogwood/L'OISEAU LYRE (1980 LP 1984 CD) UPC 0 28941 18582 2*
2 *WAGNER/Tristan und Isolde ROH orchestra and Chorus/Pappano/Domingo/Stemme/Fujimara/Bär/ Pape/Bostridge/Holt/Rose/Villazón EMI Classics (2005) CD 7243 5 58006 2 6*
3 *BEETHOVEN/Symphony No 9/AAM/Hogwood/L'OISEAU LYRE (1989) CD 425 517–2*
4 *PUCCINI/La Bohème/Berlin Philharmonic Orchestra/Karajan/Pavarotti/Freni/Ghiaurov/Harwood/ Panerai/Maffeo/DECCA (1973) CD (2008) UPC 0 28947 80254 9*
5 *HAYDN/The Creation/AAM/New College Choir, Oxford/Hogwood/Kirkby/Rolfe Johnson/George/ L'OISEAU LYRE (1990) CD 430 397–2*

Part III

After the recording session

Chapter 17

Mixing

Mixing or balancing classical music recording is the art of blending microphone sources that contain varying amounts of spill and reverb as well as direct sound from each source. EQ is generally minimal and subtle, and the use of appropriate artificial reverb is an essential skill. At all times, the aim is a balance of clarity with a sense of real space, perspective, and depth, and the listener should never be aware of the presence of individual microphone sources that are too dry, too mono, or both.

With the advent of inexpensive multi-track recording and editing, classical workflow has become more fluid with regard to whether editing comes before or after mixing. The skill of mixing straight to stereo on session is still valuable, if only to test your microphone placement and give you something to use for playbacks, but because it is now easy to remix after editing, a remix stage has become common where the budget will allow.

The usual workflow is as follows:

1. Recording session, capturing a good stereo mix alongside a multi-track backup of microphone feeds.
2. Editing process (see Chapter 18), where the editor listens to the stereo mix but edits the multi-track alongside.
3. Optional remixing stage using the final edited multi-track master, if it is decided that the session mix can be improved.
4. Mastering.

Sections 2.6.4 and 2.6.5 offer further advice on efficient session workflow, reasons for mixing live to stereo on session, and the advantages of using faders rather than a mouse.

For use of software such as Cedar Retouch and iZotope RX for the removal of clicks, thumps, coughs, bangs, hiss, and rumble after mixing, see Chapter 19. For discussion about mixing in 5.1 surround, see Chapter 10.

17.1 Choosing and blending microphone sources into a static balance

The core of classical recording technique is getting the microphones in the right place when on the session, and the main pickup will form the majority of the mix. Ancillary or spot microphones can be blended into this sound where an orchestral section or instrument needs highlighting, focussing or bringing forward in perspective. Any main pickup needs to collect a suitable blend of direct and reverberant sound and to produce an image width that is appropriate to the source. Chapter 8 explains what to listen for when positioning a Decca Tree, and Chapter 3 discusses placement of other main pickups.

Even if you have rigged a lot of additional microphones for coverage, it is important to remember that *you do not have to use them all* if they are not doing anything useful or are making things worse. A good example might be the rigging of a brass section microphone in an orchestra; its presence alone will discourage excessively loud playing, and this might be its main purpose. There is no need to use it in the mix if it is not needed, that is if the brass are clear and well punctuated and do not feel excessively distant. If at any time you are aware of a spot microphone in a mix, it is too high in level; being able to hear that a source has a microphone destroys the illusion of 'naturalness' that we are striving for.

When ancillary or spot microphones are blended into the sound from the main pickup, they will need to be panned to a position that matches the source location in the main image. This can be done most quickly with one finger on the ancillary or spot microphone fader and the other hand on the associated panpot. Alter the pan position incrementally while testing the position by fading the source up and down again. If the panning does not match, the image will either move or become smeared when the source is faded up; once the image is stable whilst the spot microphone is altered in level, the panning is correct. An exception to this might be percussion microphones that are often panned a little wider than their actual position for clarity of lines; see Chapter 9 and section 17.1.1 for more on percussion microphones.

The 'solo' function can be misleading when mixing because it takes the individual microphone out of context, and so the engineer should avoid getting too focussed on the individual sound of one microphone at a time. In this recording style, where spill is to be used and embraced, the main use for the solo button is for listening to the quality and amount of spill on a microphone rather than the direct sound of the instrument it is pointing at. If we are using good microphones at an appropriate distance from a nice-sounding instrument, the direct sound will generally be quite acceptable. What might not sound good is the spill of the other instruments, so this is where the engineer's attention needs to be directed. The usual caution should be observed with 'solo', that is not to use destructive solo (whereby all the other channels are cut) whilst in record. Stereo AFL should be used instead. Using PFL or mono AFL is usually quite unhelpful in this context as they present the solo signals in mono without panning information.

Each of the following sections deals with basic balancing of microphones to arrive at a good static mix in several scenarios. It is usual to ride the fader levels of ancillary and soloist microphones depending on musical content; a classical orchestral balance is not static. See section 17.2.

17.1.1 Orchestral

Whether the recording is based on a tree with outriggers and a bass section microphone or another overall pickup technique, proceed in this order:

1 Fade up the main pickup (with a tree this includes the outriggers and bass section microphone) and check that overall image width and amount of reverb/direct sound is acceptable. See Chapter 8 for notes on setting up the tree and Chapter 3 for alternative overall pickups.

2 Bring in the woodwind microphones, taking care of panning and avoiding pulling the section too far forwards.
3 Bring in the horns in the same way.
4 Bring in the brass if you decide you are going to use their section microphones.
5 Bring in the timpani. If there are more than two timpani and you therefore have more than one microphone, pan them so they are separated a little.
6 Bring in the percussion microphones. Note here that because the percussion section do not usually play continuously, it is usual to leave their microphones at a low level and fade them up when they are needed. This requires score following and great attention to detail to keep on top of this. However, if there are a lot of microphones left open at the back of the orchestra, their contribution is not helpful to the overall image. Additionally, you need to reduce the sound of the percussion players moving things around to get ready for the next section.
7 Bring in any ancillary microphones for instruments placed in the front (at least for a recording session), such as harp and celeste.
8 Bring in any microphones put there for ambience/reverb collection, panning all pairs fully.
9 Finally, bring in any artificial reverb (see section 17.5).

Take care if you are mixing live on session that the conductor's talk microphone is not left open and not sent to the main mix bus.

17.1.2 Opera: orchestra, choir, soloists

For opera, proceed in a similar way, but with the addition of the solo and chorus voices early on into the overall scheme. (See also section 16.2.3.) The soloists will usually be added to the balance first, as there is usually more solo work than chorus in an opera.

1 Main pickup.
2 Woodwinds.
3 Soloists – these will not be in use all of the time; see section 17.2 for discussion on riding the level of ancillary and soloist microphones throughout a performance. (See Chapter 16.)
4 Chorus (see Chapter 15 for discussion of obtaining even blended coverage of a choir).

Then follow steps 3 through 9 (from section 17.1.1) for the rest of the orchestra.

17.1.3 Piano and soloist (e.g. violin)

The quality of spill on both the piano and soloist microphones is very important because both pairs will be contributing a significant amount to the overall sound; neither should dominate the sound to the extent that a main orchestral pickup does. As seen in Chapter 7, the basis of the sound will be the piano microphones, which are placed to obtain the best piano sound and give a good stereo width to the overall space. They will also contain a great deal of ambient soloist spill.

1 Start with the piano microphones, panned fully.
2 Bring in the soloist's microphones, also panned fully, or nearly so; check these in 'solo-in-place' mode to assess the quality of the piano spill. If this is muddy and bassy, a small amount

can be gently filtered off at no higher than 100 Hz if the solo instrument's range is not affected (this should be fine for violin, upper woodwinds, and brass, or female and tenor singers). If the problem is more significant, do not try to cure it with more severe filtering, but move the microphones to be at least 1.2 m (4′) back from the piano. Changing the microphone to pick up less piano is also an option.
3 Any ambient pick up.
4 Artificial reverb.

The soloist and pianist should feel as if they are occupying the same room, but the soloist should feel as if he or she is placed very slightly in front of the piano.

17.1.4 Piano trio

This applies to the methods looked at in Chapter 12, using a small tree for sense of space on the strings, individual string spots, a vertically mounted rear ambience pair, and a main piano pair (which will provide the starting basis for the sound.) As with the piano and soloist, the quality of spill is important to the success of the mix.

1. Start with the piano microphones, panned fully.
2. Bring in the individual violin and cello spots (single microphones) panned to the location that you would like them to be, either side of the centre, in front of the piano (the small tree image should also align with this).
3. Bring in the small tree that is centred on the strings (panned left, centre, right) for some 'space' for the string sound. This is needed because the spot microphones can be too localised and the string spill onto the piano microphones too distant to produce the right feeling of perspective for the strings.
4. Bring in the rear vertical pair for additional 'space' for the piano.
5. Artificial reverb.

17.2 Riding levels on ancillary microphones

Riding levels means to alter the static balance when it is needed, and in an orchestral context, it will be the ancillary microphones more than the main pickup that will be altered in level. This is usually done to ensure that music lines are supported to come across clearly but without sticking out inappropriately. For singers in particular, it is usually necessary to ride the soloists' levels, keeping the voice audible and supporting it against the orchestra or piano as required.

The key to successfully riding levels on any ancillary or soloist's microphone is to anticipate the necessary fader moves so that sudden large gain changes are avoided. Because of the amount of spill on most microphones, fast level changes will almost always be audible as they will affect the sound of other instruments. When a singer's microphone is changed in level on a piano and voice recording, the piano sound will be affected, even if the effect is subtle, so care has to be taken. Fader movements will usually be in the 2–3 dB range and not more than about 5 dB. Younger opera singers will usually have less control over their voice, and a greater degree of fader riding will be needed to help them smooth out exuberant plosives and other unevenness of line.

As mentioned earlier in the context of orchestral percussion microphones, where you have a microphone that will not be used often during the course of a piece, its level should be kept low but not faded out entirely. This will avoid a large change when that instrument needs to be faded up for a solo line.

For woodwind and horn sections that will be playing through most of the piece, the ancillary microphones can remain faded up to provide a satisfactory static balance, but small fader movements of up to about 3 dB will be useful in bringing out any part that has a solo (usually indicated in the score as 'solo' to distinguish it from general part writing). If larger fader movements are needed, they can be accomplished more slowly so that the changes remain imperceptible. The extent to which this is necessary will depend on the repertoire.

17.3 Use of EQ and high-pass filters

17.3.1 EQ

For classical work, the aim is a natural-sounding instrument, and the initial approach always involves trying to choose the right microphone and put it in the right place. Because classical recording technique involves avoiding very close placement wherever possible, capturing a good overall representation of an instrument's sound is easier than in pop recording, where close microphone placement inevitably means that the sound could be quite localised. This then requires some EQ either to restore a good tone or to make sure that a part can sit well in a mix. Even so, there are times when some EQ will be helpful in classical recording, but it is used gently, usually in amounts of around 2–4 dB.

If it is remembered that the aim is to retain a natural-sounding instrument, but with a small adjustment, possibly to brighten a dull tone or to calm down some over-bright key noise in the absence of a ribbon microphone, then EQ should not go too far wrong. For example, to brighten timpani, a combination of small reduction in the mid-bass range (–2 dB at 150 Hz) and a boost in the presence range (+1–2 dB at 2.8 kHz, and a +2 dB shelf at 10 kHz) can be used.

An interesting technique used to add more sense of space and ambience to an overall sound is the addition of a gentle shelf boost of about +2–3 dB at around 10–12 kHz. A more subtle version of the same could be placed as high as 16 kHz. Removing the same amount will give the effect of drying a recording out a little bit.

When adding EQ, it is surprising how quickly an altered sound is one that you become used to, so take frequent breaks in order to give your ears a rest and keep some perspective. Then play the passage with and without EQ in quick succession to make a final judgement. (See also Chapter 19.)

17.3.2 High-pass filters

The HPF is used quite a lot in classical recording when trying to clean up the lower frequencies, particularly in a recording context where there are a lot of microphones, such as an orchestra. The troublesome LF comes from picking up both room tone (rumble) and the lower orchestral frequencies on many microphones. Another source of unwanted LF in a room full of people is from seismically transmitted vibrations finding their way through microphone stands, so it is a good idea to use a cradle support for a microphone rather than a rigid clip where possible.

The overall pickup should usually be left unfiltered (especially if you have been able to decouple the microphones from the stand), but if a hall is particularly boomy, applying a gentle 6 dB/octave roll-off no higher than about 100 Hz to the centre microphone of a three- or five-microphone tree can help. (See also section 8.2 for centre microphone filtering to help LF imaging.) If this isn't sufficient, add a similar filter to the other tree microphones rather than increasing the steepness of the filter on the centre microphone.

All the ancillary microphones can have some contributory effect to an overabundance of LF, although where they are cardioids, they will not be supplying a great deal of really low frequencies (50 Hz and below). It can be helpful in cleaning up the lower end of the recording to roll-off the ancillary microphones at no higher than around 150 Hz at 6–12 dB/octave. The 24 dB/octave filters are a little brutal in their effect, so they will usually be avoided except in very difficult circumstances. Caution should be applied with microphones that are being used on bass instruments to avoid removing important content. On a DAW, there is usually a spectrum analyser available which can help with the diagnosis of where any LF problem lies. However, you should beware of over-reliance on a visual representation of EQ; your final decisions about HPF and EQ should be taken aurally.

17.4 Use of delays

Use of digital delays is much talked about in classical recording, perhaps because of the relatively large distances between microphones, and because instruments spill onto almost all the other microphones in the room to a greater or lesser extent. Given 30 microphones on an orchestra, the order of complexity in calculating which player is delayed relative to which, and on which microphone, begins to render the whole question unanswerable and certainly not simple. Following are some principles that you can use, as well as some examples of where delays are helpful and perhaps even necessary. However, given the number of great recordings made before the arrival of digital delays on every channel, it could be concluded that they are a potentially useful but not essential tool in most circumstances.

They key thing to remember if you are going to use delays is that there has to be one set of microphones that is dominant in the mix, and these are then a reference point to which other microphone signals can be delayed. Any other approach will result in confusion, as signals are all delayed with respect to one another depending on each microphone's location in the room.

As a straightforward example, let's consider the woodwind ancillary microphones in an orchestra. The sound of the woodwinds will arrive at the ancillary microphone before it arrives at the main tree. It can be argued that this means that less of the ancillary microphone might be needed in the mix (due to the precedence effect), but it is equally valid to argue that these microphone signals should be delayed so that the wavefronts arrive at the same time as the woodwind signals on the main microphones. Excellent recordings have been made with and without use of delays, so there is no objectively right and wrong way to proceed. If you want to apply delays during mixing, you should have a good idea of the actual distances involved – this is another good reason to measure your orchestral microphone set-up accurately. Each metre will require 3.4 ms delay; this means 1 ms delay per foot. These delays are not large, and provided that the distance does not push the delay above about 35 ms (10 metres' distance), leaving them uncorrected should not result in any apparent double attacks at the start of notes. Uncorrected delays of more than this do have the potential to cause smearing of transients, especially if they are percussive in nature.

Low frequencies can also be adversely affected by large delays as there is the potential for some partial phase cancellations. For example, 35 ms represents a whole wavelength for about 28 Hz, 1.5 wavelengths for about 43 Hz, and so on. If choir or opera chorus microphones have to be placed more than 10 m away from the main pickup, using some delay on the chorus microphones to correct for this will usually have a beneficial effect on cleaning up the bass end.

In Chapter 14, we looked at the scenario of having several sets of organ pipes distributed around the church. This might seem a situation where delays might be useful, but again, unless there is an obvious overall set of microphones that are dominating the mix, it can be hard to decide which microphones should be delayed with respect to which. Leave playing with delays until mixing after the event; it is easy to get into quite a mess, and it is best not to record delays on the microphone feeds at the time.

The situation where use of signal delays during mixing is essential is when recording live opera using radio microphones as part of the set-up. The personal radio microphones are very close and dry, and their signal will be significantly ahead of that on the main stage microphones at the front of the stage. This means they will stick out of the mix very prominently even if faded up only a little. Their purpose is to capture some detail and provide some focus to the voices, so to enable them to be used at an appropriate level in the mix, they are delayed by variable amounts depending on where the singer is standing at the time. Once the delay is adjusted appropriately (and any error is made in the direction of having too much delay rather than too little) the individual voices blend right back into the mix. See the end of Chapter 16 for more discussion of how this is done.

17.5 Reverb: natural and artificial

The most obvious characteristic of classical recording as opposed to pop recording is the use of greater recording distances, and when engineers are learning to record classical music, a common error is to add a large amount of reverb that obscures the detail like a thick layer of varnish on a painting. Artificial digital reverb is an essential tool for most classical recording, but it has to be used with skill in order to blend it into the natural reverb of the space used for the recording.

Before digital reverb existed, additional reverb was added to classical recordings by playing them back through a good-sounding room or studio and recording the results through a pair of omnidirectional microphones. This new signal could then be used as a reverb return and blended with the original. Kingsway Hall was frequently used for this purpose by Decca, although there was a downside in the audibility of tube trains that would be recorded in the reverb.

This live playback method has also been used in more modern times as an effective way of adding a natural sense of early reflections to a recording from a dry space. Using a single artificial reverb on a very dry recording will leave an obvious artificial signature, but recording some real early reflections that are then augmented by artificial reverb is much less obtrusive. (See also section 17.5.6 for discussion of cascading reverbs for the same purpose.) Live reverb was created in 2001 for a recording of Angela Gheorghiu from ROH Covent Garden.[1] Three B&W 801 loudspeakers were placed at the back of the great hall at Air Lyndhurst Studios to play back the original recording, and the results were picked up on two Neumann M50s spaced about 4.5 m

(15′) apart and placed a long way further back in the room. Abbey Road's Studio 1 has also been used in this way to enhance recordings made elsewhere.

For discussion of using artificial reverb in surround sound, see Chapter 10.

17.5.1 Implementing artificial reverb

It is worth taking a moment here to discuss how reverb should best be implemented both in a traditional mixing desk and within a DAW. Using a mixing desk means using a mono or stereo aux bus to take a feed from each microphone source to send to the external reverb unit and bringing a 100% reverberant signal back into the mixer as stereo reverb returns. On a DAW, reverb might be available as a channel insert, but there are some very good reasons to set it up as an internal aux bus with a send from each channel. A single stereo plug-in is then used on the aux bus and the 100% wet reverb signal is added to the mix.

Reasons for doing this are:

1. You will only have a single set of reverb parameters to alter.
2. You will not be wasting processing power by running a separate reverb plug-in on every channel.
3. Adding reverb returns into the mix is much easier than altering a 'wet/dry' control on each individual channel. The disadvantage of a wet/dry control is that as you increase the 'wetness', the dry signal is turned down in level. This changes the fundamental balance between your microphones every time you alter the amount of reverb and makes mixing harder.
4. Unless your DAW does clever things on your behalf, reverb inserted into a mono channel will be mono reverb.
5. Even if your DAW makes a channel into stereo when you insert a reverb, you cannot alter the reverb return panning or EQ it separately from the original microphone source.

17.5.2 What is the purpose of additional artificial reverb?

Reverb is very appealing, and we use it to make things sound 'nice', but we do not necessarily aim to make them sound further away. When too much reverb is added, it starts to have a smearing effect on the details in the music and eventually stops adding to a greater sense of distance and becomes muddy and confusing. Artificial reverb can be used to augment the natural sound of a room and to extend reverb tails when necessary; it is generally less successful when used alone to salvage a recording made in a very dry room.

Because reverb sounds good, it is particularly easy to overdo things. If you can obviously hear the reverb, you have too much of it either in terms of reverb length or amount in the mix. A suggested rule of thumb is to set the reverb algorithm to how you like it when you can hear it well, then adjust the level to what you think sounds good, and *then* reduce the level of the reverb returns by about 2–3 dB. Reverb amounts are particularly difficult to judge if you are monitoring on headphones; one suggestion to help with this if you have to use headphones is to pan the headphone feeds in slightly (you will have to pass it down some channels to do so), so panning the channels to about 75% L and 75% R. This will introduce a little crosstalk between channels and mimic loudspeaker listening to a small extent.

The early reflections of natural reverb are very characteristic of a particular space, and these are usually the least natural sounding part of digital reverb algorithms. Therefore, the most successful approach is to rely on the microphones in the room to pick up the genuine early reflections but to use the artificial reverb to augment the tail part of the reverb if necessary.

17.5.3 Choosing a reverb programme

Artificial reverb can be provided by plug-ins or by stand-alone units such as the Lexicon range (480L, 960) and the Bricasti M7. In terms of cost, the outboard units used by professional recording engineers are at the highest end of the range, and it is advisable to pay as much as you can afford for your outboard reverb or plug-in as there is a huge variation in quality and what you can achieve with them. The most expensive reverb units will give you control over many more parameters and reverb characteristics than a basic free plug-in will allow.

Programmes such as plates, halls, spaces, and rooms are all available as pre-sets, but these are best thought of as a good place to start; given control over all parameters (and you might not be able to access many parameters in a basic plug-in), you can learn to adapt any of these to work as you wish. The primary parameters that you will have control over are discussed in the following sections.

17.5.3.1 Tone colour

This is often expressed by means of HF and LF multipliers, which are then applied to the headline reverb time. A pleasant-sounding room for classical recording generally has slightly longer LF reverb, so using an LF multiplier of about x1.1 and an HF multiplier of x0.9 is a good place to start. A more sophisticated programme will allow you to choose where the crossover from mid-range to HF or LF reverb times occurs. Be careful about using over-bright reverb as this is where artificial reverb tends to show itself by sounding a little 'tizzy'.

Church reverb programmes tend to be very bright and long, and you shouldn't feel you have to use a church reverb on a recording that was made in a church. Layering reverbs can work very well, provided you are really listening to what you are doing and avoid making an audible double decay due to a mismatch between the real reverb and what you have added. See section 17.5.4 about blending reverb with signal from ambience microphones.

17.5.3.2 Reverb time

It is quite easy to overdo the length of artificial reverb, and this can give its presence away. The headline number given as the reverb time in seconds can be misleading, so avoid getting too stuck on the number of seconds given, and listen to what is happening instead. Listening to the ends of phrases is an essential part of judging how your reverb settings are working, and you should err on the shorter side in the first instance. A useful thing to listen for is whether there is a drop in level between the end of the direct sound and the reverb. Something to avoid is a long, low-level reverb tail; the results are more satisfactory if the end of the direct sound transitions smoothly into slightly higher level reverb. Therefore, using a slightly higher level of a shorter reverb is often a good thing. If you still feel that the recording needs more reverb, try a higher level of reverb returns first, and if that doesn't work, try a slightly longer reverb time.

17.5.3.3 Pre-delay

This can be quite a crude parameter in the reverb controls, as it usually simply delays the early reflections that give us the characteristic sense of a space. The longer the pre-delay, the further away the walls should feel. However, even in a very large real room, there will be some early reflections of a few milliseconds from the nearby floor, making the sound from a real room much more complex. The natural early reflections captured on your recording by overall pickups and ambience microphones in particular will give a far better sense of the character of the room, and when you are trying to enhance a recording that already contains some natural reverb, it will often work best if the pre-delay is removed from the artificial reverb. With additional pre-delay added, the artificial reverb can tend to separate out from the real thing. The Hall programs on the Lexicon 480L can work very well for piano, provided the pre-delay is removed by setting it to zero. Section 17.5.4 looks at how to use natural early reflections in combination with artificial reverb for the complex build-up of later reflections.

17.5.4 Blending with ambience microphones

Reverberation can be divided into the early part of the room response, where there are discrete reflections that provide us with information about the size and nature of the room, and the later part of the reverb, where the reflections have multiplied rapidly to form a rich mix of complex reflections that make up the main body of the reverb. As noted earlier, artificial reverb tends to model the latter part of the reverb better than the early part, and a high-end unit such as the Bricasti M7 allows the user to control the amount of early reflections against reverb on a scale of ±20.

The vertically or backwards-facing pairs of cardioids that have been frequently used in the techniques discussed in earlier chapters (e.g. Chapters 12 and 10) are designed to pick up the early reflections that are characteristic of the room, and the artificial reverb can be used to enhance the reverb tails with its pre-delay set to zero. These microphones can be sent to the artificial reverb unit so that the early reflections are then used to generate reverb tails. If you are using more than one ambient pair in a stereo mix (such as an upwards-facing pair at the front of the tree and a rearward-facing pair behind it), you will find that you have to choose one of the pairs to dominate by at least 10 dB or the result is confusing. Each pair will have a different sounding reverb contribution, and either can be used as the main one.

17.5.5 Which microphone sources are sent to the reverb?

It is possible to send only the final mix to the reverb, but this does not allow you to customise the amount of signal sent from individual microphones, which can be a very useful mixing tool. The starting point when using an aux bus is to send the tree microphones, outriggers, and woodwind ancillaries to the aux bus (post-fade). All the channel aux send level controls should initially be set to the same value so that the amount sent to the reverb from each is in proportion to that channel's level in the mix. This can then be altered if necessary.

Common reasons for deviating from the initial settings include:

1 *A very reverberant hall* – reduce or remove the send from the main microphones and leave the ancillaries only.

2 *A live set up with a closer-than-preferred soloist's microphone* – more will need to be sent to the reverb, although using a different or additional reverb programme for the soloist is another option. Sometimes simply sending more just makes the artificial reverb more audible. (See section 17.5.7 for notes on treating a dry source with more than one reverb.)

As noted in section 17.5.4, microphones designed to pick up the natural early reflections in the room can also be sent to the reverb unit.

17.5.6 Stereo versus mono aux send?

There will usually be an option to send a stereo aux signal to the reverb unit rather than a mono one. Assuming that the channel aux send occurs after the channel panpot (if it doesn't, you will have to manually set up the stereo aux bus panning to match the channel panning), this will then generate reverb that gives a sense of lateral source location in the return reverb signal. If any significant amount of reverb is used, this imaging information will simply reinforce the location of the various sources. Where a mono send is used, this spatial information is not present in the reverb returns, and if a significant amount of reverb return is added to the mix, lateral image smearing can occur.

17.5.7 Using more than one reverb

Using more than one reverb is a useful technique in some situations, but this doesn't mean putting a different reverb on every microphone, given that the aim is to create the impression that everyone is located in the same room. Occasionally, a soloist might benefit from a different reverb to the rest or from sending a little more to the main reverb. An additional reverb is useful because it opens up more options in that the RT length and send and return levels can all be altered. With a single reverb, the only option available is to send more of it.

Cascading reverbs – that is, taking the output of one reverb unit and putting it straight into the input of another (in dual machine reverb units such as the Lexicon 480L, this is called 'cascade mode') – is an unusual technique, but it can be a very elegant way of adding reverb to a very dead hall. Rather than adding a single long reverb at a high level (which can sound obviously artificial), the desired length is built up in stages. The first programme should be a shorter reverb with no pre-delay, and the second reverb can then be a longer programme. This makes a much denser reverb tail and avoids an obviously artificial reverb signature.

17.6 Riding overall levels

Classical music often has a very large dynamic range, and while it can be contained within the dynamic range of a 24-bit recording system, the listener cannot generally reproduce the loudest sections at their true acoustic level in a domestic setting. Therefore, the quietest parts of a large orchestral piece might fall too close to the acoustic noise floor for the listening environment. A recording such as the Concertgebouw/Chailly Stravinsky's *Firebird* suite[2] has a dynamic range of about 45 dB, and while it is a very exciting and dynamic recording, this does not equal the real dynamic range that would be experienced in the room. Automatic compression might seem like a solution to this, but compressors are designed for use over the time frame of the envelope of individual notes rather than to gradually alter the dynamics of a phrase to reduce the overall dynamic range. Therefore, it is usual to alter the overall level of orchestral music by careful riding of the main mix bus faders. Because the engineer can use the foresight that comes from knowledge of the piece or the ability to follow the score to decide when and how to effect any necessary gain changes, this process can be referred to as 'intelligent compression'. This is a process that demands some experience and confidence to execute at the time of recording; to make editing workable, resulting levels need to be fairly consistent between takes. For recording rather than live broadcast, there is a good argument to make for altering levels in this way as part of the final mastering process (see section 19.4.2), although for a very dynamic piece some level manipulation might occur at both the recording/mixing and mastering stages.

Notes

1 *Live From Covent Garden/Orchestra of the Royal Opera House/Ion Marin/Angela Gheorghiu EMI Classics (2002) CD 7243 5 57264 2 1*
2 *STRAVINSKY/The Firebird/Concertgebouw Orchestra/Chailly/DECCA (2008) CD 473 731–2*

Chapter 18

Editing and post-production

18.1 Aims and philosophy of editing

Once the recording session is completed, a complete version of the repertoire has to be edited together. It usually comes as a surprise to those new to the industry just how much editing takes place, and given that the musicians are usually very much on top of the repertoire, the next questions are often 'Why are you editing this at all?' and then 'Why are you editing this so much?'

To answer these questions, we can look at what the purpose of editing is: to produce a complete and best representation of the player's interpretation of the piece. There are many reasons for editing – some technical, some musical, some negative, and some positive.

Technical reasons include:

- To cover accidental noises from the instrument, the player or the environment (including creaks, squeaks, page turns, touched strings, and knocking of music stands).
- To cover sections where there is a problem with the recording (distortion, reverb malfunction, or drop out).

Musical reasons include:

- To cover mistakes (wrong notes, wrong rhythm, or missing notes).
- To cover poor intonation, timing, or playing together (ensemble).
- To choose the best or most beautiful interpretation, intonation, and phrasing.

Given that this is a recording and will be listened to hundreds of times, it makes sense to at least replace any obvious flaws with a better take. Some flaws are straightforward (e.g. a wrong note) and others are a matter of degree (e.g. intonation), and hence the limit to editing is determined by how particular the editor, producer, and musician want to be in correcting more minor imperfections. This is on a sliding scale, with classical musicians of international standing exhibiting the highest degree of perfectionism about their own performance (this is, after all, why they are at the top of their profession). When dealing with professional musicians playing repertoire they know well, there is usually a great deal of excellent recorded material to choose from, and correcting small flaws is usually a case of finding another take from several potential alternatives. The software that is available to us nowadays enables editing to be as detailed as desired, provided the editor is skilled in retaining musicality and is sensitive to performance nuances while piecing work together. When you are dealing with players at the top of their game, correcting errors becomes only one part of the craft; maintaining and enhancing musicality of performance is an essential skill.

If you are working with classical players who do not perform for a living, the judgements you might make about what intonation is acceptable, for example, will be different. The playing or singing techniques will be less refined, and it is likely that flaws that would be corrected for a commercial recording will be left alone for lack of any better takes. Inconsistency of interpretation between takes is usually a feature of less skilled players, which makes heavy editing harder to achieve while retaining smooth flow. Nevertheless, a lot can be done in the way of correcting mistakes while maintaining musicality, provided you don't try to edit too heavily.

18.2 Requirements of a classical editing system

Classical editing is much more exposed than the editing on a pop recording because all the tracks are edited simultaneously, the sound is more reverberant, and there is little in the way of subsequent processing that might mask crude edits. It is not possible to edit a single instrument within an ensemble because of the microphone spill that is part of the classical recording approach. Therefore, the crossfade tool needs to very flexible, and fine adjustments to both timing and level changes must be easy to implement in order to make edits inaudible.

Other requirements for fast classical workflow are:

- Multi-track editing and playback of heavily edited material across many tracks without any loss of playback responsiveness or repeated playback stalling.
- A system that can work at 24-bit and higher sample rates where clients require this.
- A good waveform display. Although editing should be done by ear, a good detailed waveform can be a help when pinpointing events on the timeline.
- A way of quickly putting together an edit sequence in some empty tracks whilst leaving the session takes untouched in their original tracks. This is usually managed by a semi-automated copy and paste process known as 'source-destination' editing (see section 18.3). Retaining the session takes as complete clips is invaluable when needing to compare the flow of an unedited take to the edited version. It is also essential when working alongside a producer or musician who will frequently want to listen to extended, unedited takes during the process.
- A way to place edits accurately by ear; this usually involves instant playing up to and from any edit point or sync point.
- A way to quickly change the level of the outgoing or incoming audio clip at the edit point without having to create a new audio file (audio clip gain or automated gain control).
- Edit rehearsal that happens immediately (without having to wait for any computer rendering).
- Edit rehearsal time that is easily adjustable to any length (so that you can listen to the effect of edits with a large run-up to check the impact on the flow of the music).
- Crossfade tool that enables takes to be overlapped for any duration, and the fades independently adjusted for length (to deal with problems in overhanging reverb and sustain). (See Figure 18.5.)

- A way to store different versions of an edit for instant comparison.
- Mastering facilities to enable production of delivery formats (e.g. CD images) directly from an edited project.

Facilities that are nice to have:

- Being able to have many projects open at once for copying sections between them or for compiling a final master. (A classical CD is usually 60–74 minutes long, and it is sometimes easier to work on individual pieces in separate projects if the editing is intensive. They will need to be joined up at the end.)
- Offline, non-real-time generation of mixes and CD images for speed in running off listening copies.

There is some emphasis in the preceding list on speed of response of the system, which is for two reasons. One is that post-production takes a long time and is labour intensive, and anything that can be done to speed up the process will help to meet tight financial constraints. The second reason – more important from the creative point of view – is that when you are making tiny changes to an edit to see if it improves, it is very valuable to be able to hear the 'before' and 'after' immediately after each other. Where changes are subtle (and working at the higher-skilled end of the job, small details are all-important to the overall sense of musicality), a time lag in between making a change and being able to hear it can make it harder to judge whether there is an improvement.

Systems that are currently in common use in the classical industry are Pyramix (from which screenshots have been taken for illustrations in this chapter) and SADiE, as both of these can be operated in a 'source-destination' mode and are also mastering platforms. (See Chapter 19 to note the rather more fluid boundaries between mastering, mixing, and editing in classical workflow.)

Pro Tools is not commonly used for commercial classical work because the editing tools and process are not optimised for classical workflow, and mastering cannot be done from within the same project.

18.3 Source-destination editing

Source destination editing is a method of editing that allows the user to retain the original session takes intact in one area of the project whilst quickly compiling an edited version elsewhere. It can be implemented in a number of ways, using different sets of tracks, or track groups, but the goal is to make the task of choosing sections of material and joining them together a quick and efficient process that can be performed mainly with keystrokes. It is usual practice to separate out the fine-tuning of an edit (often on a separate page) from the initial creation of a series of joins in more or less the right place.

The work area is divided into tracks for source material (session takes) and tracks for edited material where the edited master will be built up. Using the keyboard or edit controller, the user marks out a section in the edited material that needs to be replaced with a different take using 'in' and 'out' points (see Figure 18.1, top two tracks). Similarly, they mark 'in' and 'out' points in the appropriate source material (an alternative session take) (see Figure 18.1, bottom two tracks).

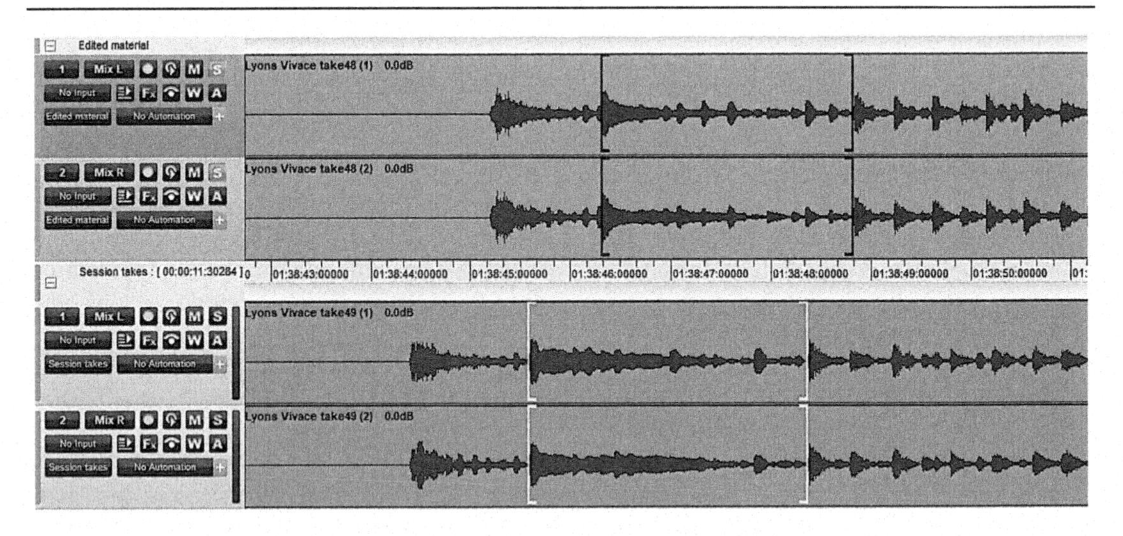

Figure 18.1 Showing a stereo source destination layout with 'in' and 'out' points marked in the source material (bottom two tracks) and destination material (top two tracks)

Image: Courtesy of Merging Technologies.

This marking of edit points can be done 'on the fly' as the audio is playing, and it need not be terribly accurate as detail can subsequently be addressed in the fade editor window.

This is known as a 'four-point edit' as all four points have been defined by the user. Using a single command, the audio between the 'in' and 'out' points in the destination tracks will be replaced with the audio between the 'in' and 'out' points in the source tracks. (See Figure 18.2.)

This is extremely quick to use and makes building up a complicated edit list very time-efficient. In the case that the inserted audio is shorter than the audio that has been removed, all the audio later in the timeline ('downstream') will be moved earlier by the correct amount – a process known as 'rippling'. Similarly, if the inserted audio is longer than the audio that has been removed, the downstream audio will be shifted a little later to accommodate the difference. (See Figure 18.2.) For classical editing, where there is no click track and the tempo can vary, there is no need for the replacement audio to exactly match the original in length.

It is also possible to perform a three-point edit in this sort of system. This means that three out of the four possible edit points have been user defined, and the programme will work out the fourth point by assuming that the replacement section of audio is exactly the same length as the section that is to be removed. Any one of the four points can be omitted for the DAW to calculate the fourth. (See Figure 18.3.)

This feature is useful for two main reasons. Firstly, it can be very useful when roughly compiling an edited version to only have to bother locating three out of the four edit points. This saves a bit of time at the initial compilation stage of the editing process. As long as the tempo is roughly similar, and the section of audio not overlong, the edit point that has been generated by the programme will be within a note or two of the right place, and any error can be corrected in the fade editor window. Secondly, if you are working on an overdub or sound-to-picture editing where it

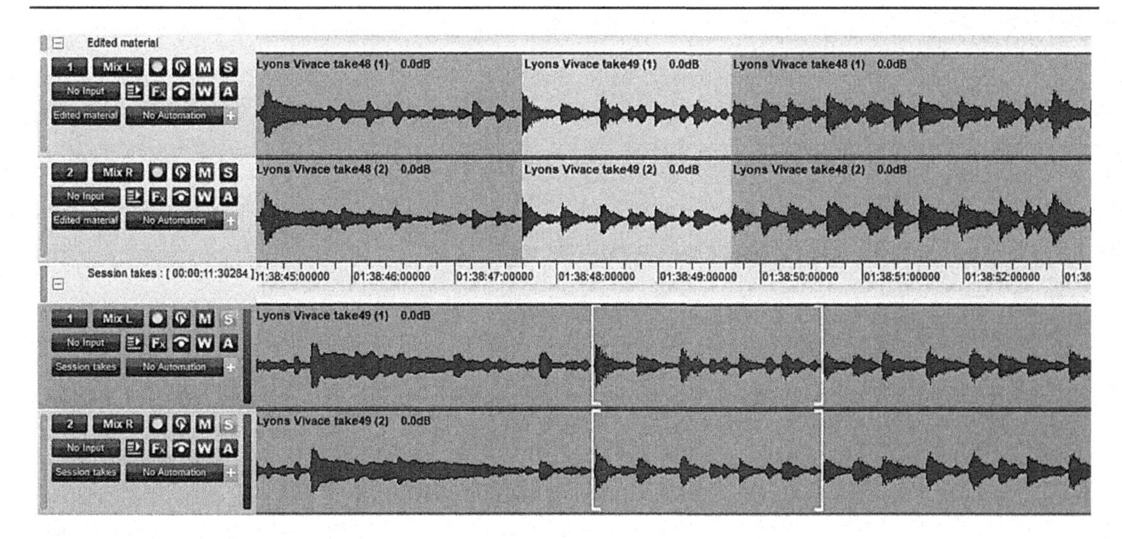

Figure 18.2 Shows the edited timeline in the top two tracks with a new take (lighter colour) inserted from the selection in the bottom two tracks.

Image: Courtesy of Merging Technologies.

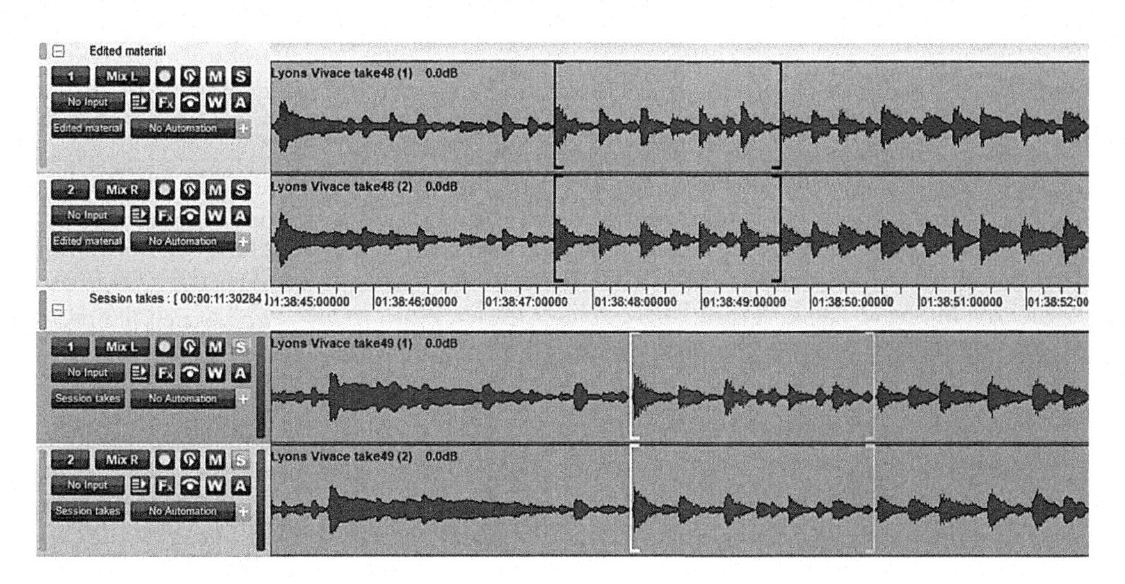

Figure 18.3 Showing a 3-point edit. The 'out' point in the session takes (bottom two tracks) is greyed out and has been calculated by the DAW instead of being defined by the user.

Image: Courtesy of Merging Technologies.

is important that the audio that is removed is replaced by something of exactly the same length to maintain synchronisation downstream, using a three-point edit will ensure that this is the case.

Once an edit sequence is put together using the 'source-destination' commands, all the edits can be refined in the fade editor, and it is generally a waste of time to spend too long in carefully placing the 'in' and 'out' points beyond hitting them 'on the fly'.

18.4 Classical post-production workflow

Normal commercial recordings spend longer in post-production than on the recording session. The usual pattern of workflow is as follows:

1. During and after the session, the score is marked by the producer to indicate which takes are to be used.
2. The recording is edited as per the marked score, usually over several days (first edits).
3. A listening copy is sent to the producer and artist.
4. Comments are sent from the artist back to the producer concerning musical points they are not happy with.
5. The producer/editor finds alternative takes where possible.
6. The recording is edited again as per the new markings.
7. A new listening copy is sent to the producer and artist (second edits).

Repeat from 4 to 7 as required! Depending on the nature of the project, and the recording contract if there is one, second and third editing stages are fairly usual. If an artist is newly signed, they might be limited to second edits only, and if they are paying for the project themselves or are of established international stature, it is likely that the editing will extend to as many times round steps 4 to 7 as are needed to produce something that everyone is happy with. Something that can be difficult to avoid is losing sight of the overall project and becoming bogged down in tiny details so that the musician or producer is never happy with the result. There will always be some tiny flaws that have to be disregarded; after all, it is doubtful that something as subjective as a musical performance can ever be considered to be perfect. However, tiny flaws can assume enormous proportions in the minds of the performers involved, and part of the producer's job is to gain the musicians' trust and convince them that it is safe to let go of the project. If you are able to prevent a musician acting as their own producer during post-production, then you should do so; they are usually too emotionally involved to take the necessary pragmatic decisions.

18.4.1 Monitor mixes and remixing

In the olden days (up to around the mid-1990s), all classical recording was mixed live to stereo on the session, with multi-tracks taken as a backup for large projects where the budget allowed, beginning with 8-track backups at Philips, Deutsche Grammophon and EMI during the 1970s. The stereo recording was edited straight from the session, and there was not usually a remixing stage except for partial remixing of an opera.

Nowadays, with the advent of DAWs, it is very easy to capture both a stereo mix and a large multi-track backup of a session as part of a large project. It is normal practice to edit both the mix and the multi-track at the same time by performing each edit across all tracks; it makes the workflow outlined in section 18.4 simple, as it is a quick job to use the session mix to make listening copies at each stage without having to waste time on recreating a mix. (See also sections 2.6.4 and 2.6.5 and Chapter 17, introduction.)

However, the presence of the multi-track throws up the question of what the editor should listen to whilst editing; for all but the simplest edits, the microphone balance has a significant impact on whether edits work or not. The editor needs to monitor a good session mix that is as close as possible to the final mix in order to judge how well the edits are working. You should be aware that if you remix an edited multi-track, you will need to listen out for edits that worked for the session mix but have subsequently been made more audible by the change in microphone balance.

18.4.2 Putting an initial edit sequence together

If you are working with a producer who is going to provide you with a marked-up score, then part of the job is already done for you. However, if you are faced with an unmarked score and a large collection of session takes, the task of putting together an edit sequence will fall to you. If you produced the session, you will already have some idea of where to start; if not, you will have a lot of listening to do, although there should be some session notes to guide you, including artists' requests (e.g. 'Use take 46 for the end').

It can be a lot more efficient to mark up the score and put the sequence together at the same time. In effect, you are going to build up the edit sequence in the DAW and retrospectively mark what you have done into the session score. The working score is a record of what has been done, and it is important to keep it up to date. At any stage, you might have to be in discussion with a musician or producer and you will need to be able to answer the question 'Which take are we in here?' with absolute certainty.

In order to keep the overall structure of the piece making sense, it can work well to build a sequence of extended takes in the first instance, roughly editing them together to make sure that the flow is good. Choose takes for their general musicality and sense of direction, but note where there are problem areas. Then go back and attend to more of the detailed problems of noises, mistakes, tuning and timing. See section 18.5 for notes on retaining and enhancing musicality in the edited performance. The advantage of being able to mark up and edit at the same time is that you can more quickly discard solutions that will not work due to performance incompatibilities.

18.5 Refining edits: how to solve problems and maintain musical flow

Once you have your edit sequence roughly in place, the process of refining the edits can begin. This will take place in the fade editor (in Pyramix terminology), which will be a page where fine adjustments can be made to the edit placement, the crossfade length, shape and overlap, and the level difference between the two takes. The aim of all these alterations is to produce an edit sequence that is technically inaudible (no clicks, double attacks, or other lumps and bumps) and musically imperceptible (no inexplicable changes in tempo, pitch, phrasing, or dynamics). There should also be no effects that are impossible on a given instrument, such as piano overhang increasing in volume. The tools at your disposal to achieve this smooth and musically coherent result are edit placement, crossfade shapes, and timing and level adjustments.

Many edits will be relatively straightforward, requiring little more than accurate placement of the cuts and application of a suitable crossfade, but others are made difficult by mismatches of dynamics, tempo, pitch, and ensemble between adjoining takes. These are all manifestations of difference in performance, mood, and phrasing and will need to be addressed in order to join takes together while creating a coherent musical flow. Sometimes they are a global problem (one take is at a consistently faster tempo than another) and at other times they are local to the bars on either side of the edit (there is a difference between the pace of accelerando or crescendo on alternative takes). Problems with overall pitch most commonly occur with unaccompanied choir during a long take, but local difficulties with intonation of a single note can occur anywhere. Poor ensemble (players not sounding notes together) can be one player coming in early or late for a single chord or a whole run of fast notes where the players are out of time with each other. It should be noted that there are times that, no matter how skilled the editor, two takes are incompatible at that point in the piece and the edit will not work. In this case, the best solution is usually to try and find a take that can act as a bridge between the two takes in terms of mood, timing, and dynamics and use it for a bar or two, or even a few lines.

18.5.1 Where to place edits within the note, bar, or phrase

In the days of analogue tape editing, edits were placed on loud chords, stronger beats, and always at the start of the transient, as these positions were easiest to locate and known for having a masking effect to cover any changes; any editing during a sustained note would cause an audible bump. Editing at the start of a note will still be the approach for most edits; they will be placed where there is some sort of transient, however soft, as this will mask a lot of reverb or overhang changes. However, in a minority of cases, this simply doesn't work, and given the flexibility of the tools available, editing during the sustained part of a note with a long crossfade is now a feasible approach for many ensembles. Crossfades can be of very extended length and of variable shape, and although these more unorthodox options will not work on all instruments, this flexibility can enable musical phrasing to be maintained more easily.

The simplest form of editing involves joining together long sections or complete phrases, and it is fairly easy to retain musicality taking this approach. However, for any detailed work, it is guaranteed that you will have to frequently replace a few notes during a phrase, and this places greater demands on your ability to keep the result musical. It is important to place edits as accurately as you can so that the pulse of the music is not disturbed by the presence of a join that brings in an incoming take slightly early or late. A nudge resolution of about 5 ms is usually sufficient for timing corrections, although smaller nudges of 1–3 ms can be useful for reducing phasing effects through longer crossfades.

Because classical editing involves working from a score, it is easy to be drawn to bar-lines when placing edits, but the results are often more inaudible and musically flowing if at least some edits are placed on weaker beats. To escape from the lure of the downbeat, forget the printed score and listen to the ebb and flow of the music. Each take will have its own slightly different shape – small fluctuations in tempo and dynamics that create an individual phrase. Listen to both takes and try to work out where to place the edit so that the flow of one take moves naturally into another. Making small adjustments in the timing and/or loudness of the outgoing or incoming take will help to shape the musical sense of a phrase, and this attention to detail is essential to retaining the musicality of the artist. The more editing there is, the more important the editor's attention to tiny details. To give one small example, a dramatic downbeat is often most musically coherent when taken with its associated upbeat, and using both together from a single take can help retain real musicality in the playing.

Edit placement on the transient of a single monophonic instrument is straightforward. What causes more problems in classical editing is where there are two or more instruments not quite playing together; which start of the note do we use as a reference point in order to keep the timing feeling natural? Consistently poor ensemble is not usually a feature of professional players, but the performance practice of singers in particular is to lag or lead the accompaniment, and how to deal with this is discussed in section 18.5.6.

18.5.2 Crossfades

Crossfade durations can be anything from 5 ms to 500 ms or more, depending on circumstances. It is never particularly helpful to make prescriptive statements about crossfade length, and different editors develop their own style, but as a rough guide, a very straightforward 'average' edit at the start of a note will have a crossfade in the order of 20–50 ms. Shorter crossfades of 5–20 ms sometimes work very well where there is something tricky like a piano pedal change, but at other times the short crossfade makes its presence felt in a subtle 'hardness' to the transient of a note, and if it is used during the sustained part of a note or there is a lot of background reverb, it will produce a definite 'blip'. Increasing the crossfade length in these circumstances usually results in the 'blip' becoming more of a 'bump', but audible nevertheless. If you place an edit during a note, you will usually have to use a longer crossfade of more than 250 ms or so to avoid bumps.

For most edits, an 'equal power' curve (see Figure 18.4) will work to avoid any surges or dips in level across the crossfade duration, although these level problems are only easily identifiable as such during a longer crossfade; during shorter fades the edit might feel bumpy, but it is not easy to discern that the crossfade shape is causing the problem. Other bespoke fade shapes might be useful during longer fades if you are trying to tame a change in the overhang during the crossfade; for example, the incoming take might need to be brought up in level more quickly if its overhang is quiet. Having a crossfade shape that can be customised by the user can be very useful in controlling what happens during a longer crossfade.

There are many occasions in classical editing where the ability to offset the crossfade to overlap the takes is invaluable, and using a DAW that offers this facility will make this quick and easy to implement. The most obvious is in the case of joining takes where one stops and the other is a cold start, but the music should have been continuous. The reverb from the outgoing take needs to be retained under the start of the incoming take or it will be audibly chopped off early at the edit. See Figure 18.5 for an illustration.

Other uses for overlapping takes include when a pause length needs to be shortened or when joining takes where a pause in the score has been interpreted as a different length in each take. (See section 18.5.5.)

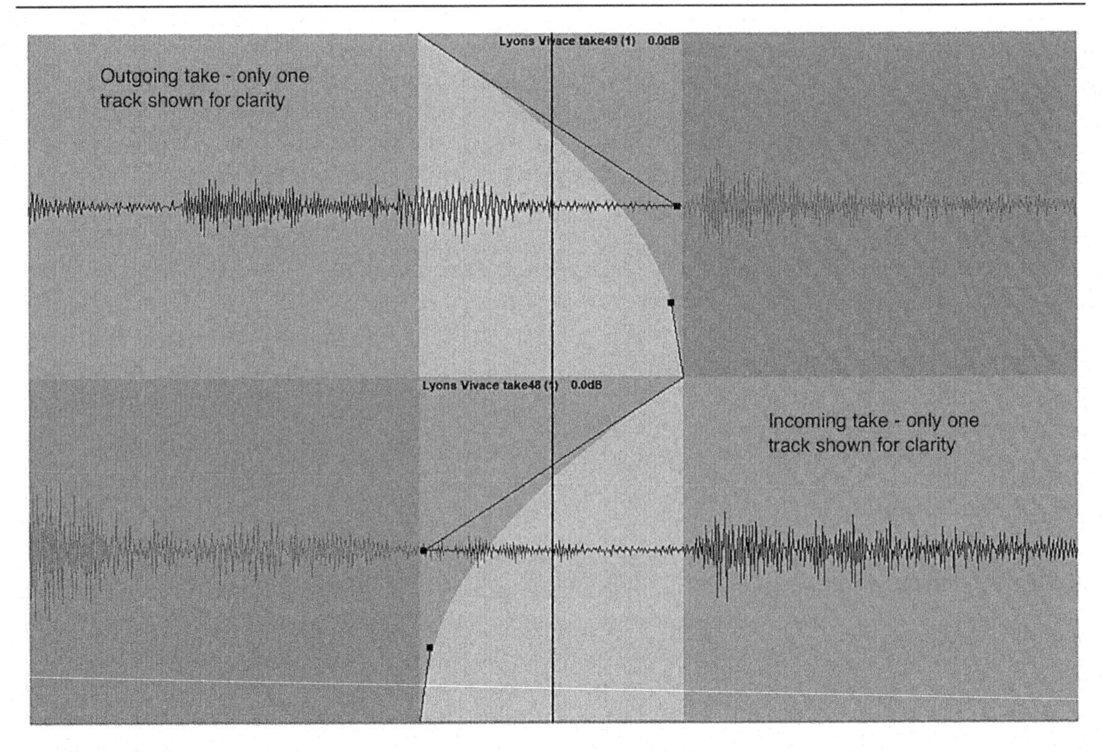

Figure 18.4 An equal-power crossfade curve – detail from the fade editing window. The outgoing take is shown at the top, the incoming take at the bottom.

Image: Courtesy of Merging Technologies.

18.5.3 Dynamics: making level changes

The ability to change the level of the outgoing or incoming take to match up different dynamics across an edit is invaluable in preserving musicality and inaudibility. The level change can be effected by changing the gain of the whole clip (in which case the edit at the other end of the altered clip will also be affected). The potential problem in only using clip gain is that the effects can be cumulative; if you have to alter the gain upwards for several edits in a row, you end up raising the level of the piece gradually.

Where possible, it is a better option to use some sort of gain automation, change the level at the edit point, and restore the change in level gradually during the course of a take. This can be applied to either or both of the outgoing and incoming takes. To restore a level change inaudibly, a rule of thumb is to allow a change of about 0.5 dB per second. Level changes at edit points of more than about 4–5 dB will usually start to be audible as the background noise level of the recording changes. When you are altering level to match subtle nuances in phrasing, you will be dealing with level changes down to about 0.5 dB, and even 0.3 dB will often make an edit ‘feel’ different.

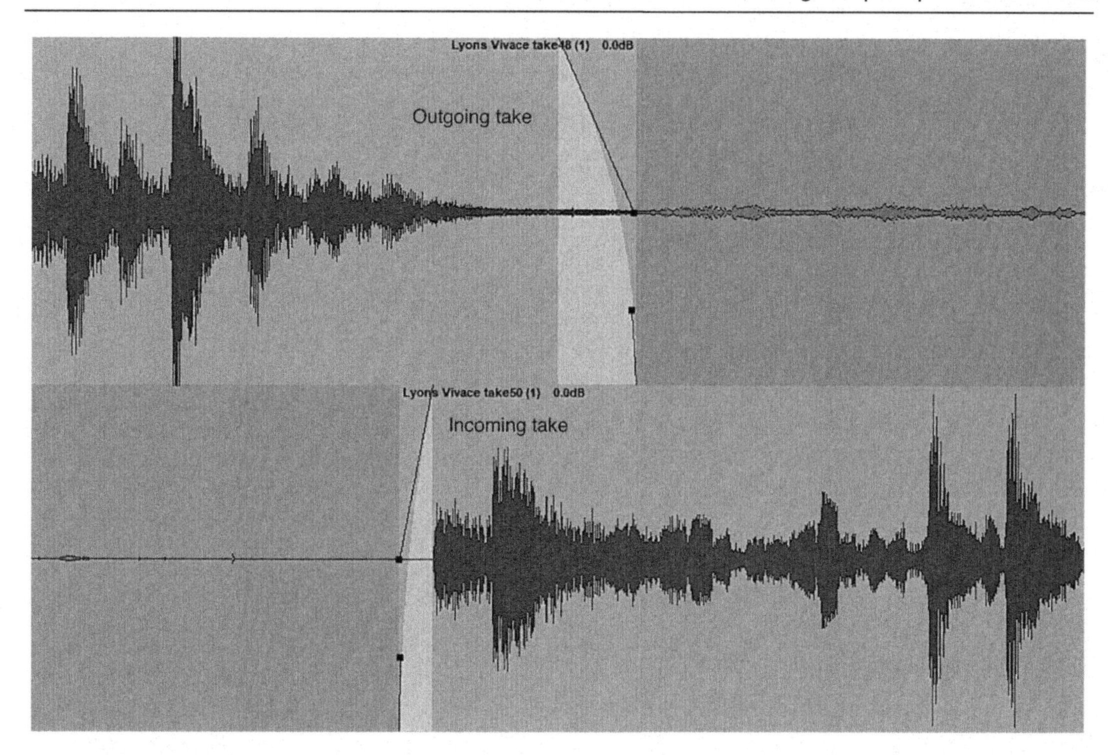

Figure 18.5 Offsetting edit points to overlap takes – in this case used to bring in a cold start under take that stopped where the music should have been continuous

Image: Courtesy of Merging Technologies.

Simply changing the gain of a clip, however, does not add in the change in timbre that comes with playing louder or quieter on a real instrument. Take the piano as an example: when the incoming take has a louder chord than the outgoing one, simply lowering the gain of the incoming take might not help; it will sound like a chord played loudly (with brighter timbre) but with the gain turned down. Where the incoming take is unavoidably louder, with more energetic playing, it can work much better to place an artificial, audible crescendo in the outgoing take (possibly with a slight reduction in level of the incoming take). This will not only match levels at the edit, but it will also give some musical direction to a new phrase as it builds towards the incoming louder playing. See Figure 18.6 for an example; the level alterations are not prescriptive.

In general, once you move beyond changes of 1–2 dB, turning up the gain of quiet playing is more successful than turning down the level of loud playing, especially with an instrument where timbre changes a lot with loudness. In order to shape phrasing through an edit point when the takes are not quite matched, always consider the option of altering the level of the outgoing take as well as the more obvious option of altering the incoming take.

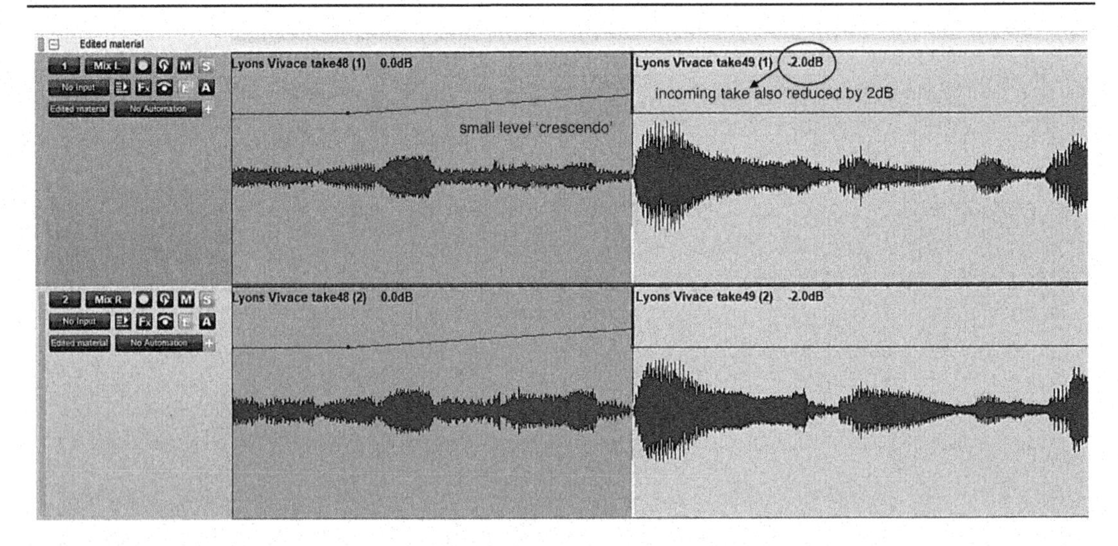

Figure 18.6 The incoming take is too loud to match the outgoing take's phrasing; a small 'crescendo' has been added to the outgoing take in conjunction with a slight level reduction of the incoming take

Image: Courtesy of Merging Technologies.

18.5.4 Pitch differences

Because all the players are in the room together, solving the problem of a single player playing out of tune is not easily addressed by using a pitch shifting plug-in, and it is best addressed by using another take. It is possible to use some subtle pitch shifting with unaccompanied instruments or singers (such as for a cadenza), but it is best done manually, note by note, and with an ear open for disturbance in the reverb overhang from other notes.

Where a whole body of singers has dropped in pitch throughout a single take of a piece, some pitch shifting can be used to bring them back up again in order to match in with a different take at an edit point. A fresh incoming take that starts a few bars before the edit point should be at the correct pitch (assuming the singers were given a pitch reference on the session before each take), and the outgoing take will have to be matched to it. This should be done in stages by pitch shifting sections of the piece up in small steps of about 10 cents (10% of a semitone) at a time. Find some good transition points and make edits between audio clips that have been pitch shifted by different amounts until you have brought the pitch up to the required level.

Pitch shifting can usually be done as a simple sample rate conversion which will change the duration of the audio clip as well as the pitch. This will be the most transparent processing, and given the very small increments in pitch being suggested, the tiny change in length will be insignificant. However, if it is essential to retain the original length exactly, it is possible to alter the pitch only. This relies on the processing to extend or shorten the audio clip by creating new interpolated samples, and it can sometimes present audible artefacts, especially on headphones. It is a subtle difference, but it is perceptible, and so the more transparent method would be the first choice where possible.

18.5.5 Tempo and timing differences

Differences in basic tempo between one take and another usually go hand in hand with a very different mood in the playing as well. If the piece is played quite flexibly (such as romantic piano works), it is very often possible to ensure that the edit points occur during a natural accelerando or decelerando so that the change in overall tempo is disguised. Where a piece is more strictly rhythmic in nature (such as baroque dance movements and fugues), this is much more difficult to disguise. It might be necessary to time-squeeze or stretch sections of the music, but as with pitch shifting while retaining the same duration, this involves the processing algorithm making decisions about where to remove or add in audio samples. It is simpler with pop music, where there are percussive beats that can disguise the process, but for classical music the processing can become subtly audible. In this case, the piece can be sped up or slowed down manually by inserting or removing tiny sections on each main beat. This is, of course, hugely time-consuming, and it is far better to keep an eye on tempi on the session or to make the decision only to use takes that all have a closely matched tempo.

In order to manually alter timings with great transparency, it is essential to have a crossfade tool that allows takes to be overlapped, bringing the new take in before fading the old take out. To shorten a beat is the simplest; see Figure 18.7. Place a cut in the clip at the start of the transient. Apply a short crossfade to the start of the incoming transient and then alter the timing of the outgoing clip so that some of the end of the previous note is removed. Then extend the crossfade on the outgoing clip only so that it extends as far as possible underneath the start of

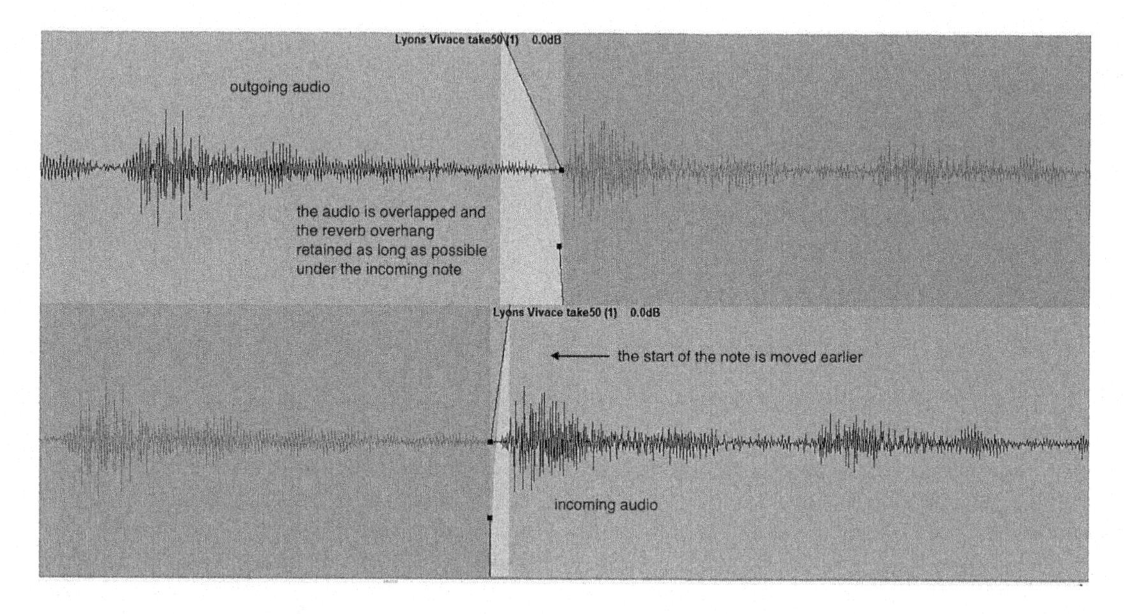

Figure 18.7 Shortening a beat or pause within a single take, including some overlap to preserve the outgoing reverb

Image: Courtesy of Merging Technologies.

the incoming note. This will prevent the effect of cutting off the natural reverb at the edit point, which will give away the edit's position. Where the music is fast moving, and the alterations are only for a short section, it is sometimes possible to give the impression of the piece moving faster by bringing forwards only the first beat in each bar, or every beat in the bar and not the half beats. If it becomes at all uneven sounding, each beat and half beat will need to be treated; it all depends on musical context.

Shortening a pause gap length is the same problem as shortening a beat, and it is tackled in the same way. Place a short crossfade at the transient of the incoming take, move it to the correct timing position, and continue the reverb from the longer pause of the outgoing take underneath the new transient so that it is not cut off at the edit.

To extend a beat (or a gap) is more difficult because it involves repeating a section from the end of the previous note, and it will usually mean there is a small jump up in level (the repeated section uses an earlier part of the reverb die-away that is slightly higher in level). Figure 18.8 shows the general approach, but it needs two joins close together. Make the first cut just before the start of the note, and slide the incoming audio to the right so that it is positioned later than it was originally. Extend a crossfade back into the previous note and reduce the level of the incoming take to ameliorate the effect of the slight increase in level or change of tone that will be present. The next step is to place another cut at the start of the incoming note with a small crossfade and use this to restore the level of the incoming take to its original level. Both edits will need adjusting as a whole; finding a successful combination of crossfade shape, length, and incoming clip level for the extension might be difficult (especially on piano).

18.5.6 Ensemble problems

Ensemble problems are usually of three types:

1 The chordal entry where the players don't all quite play at once, often, but not always at the start of the piece.
2 The fast passage where the players don't stay together.
3 The soloist that pulls ahead of or lags behind the accompaniment throughout a take, for expressive reasons; most commonly singer and piano.

The case of a chord that is not together can be approached in two ways. One is to use the attack of the chord from a take where it is together, and immediately crossfade into the required take. This approach can work very well particularly on string quartet or orchestra, but can be tried in most situations (usually with less success on piano, see section 18.6.1). The second method is to tidy up the take with the ragged attack, a process known as 'trimming'. In the case of a single player being ahead of the others, the idea is to remove as much of the timing difference between the first player to sound and the rest. Firstly, make a cut in the audio file in the attack of the note. In the outgoing take, you need to include only the transient from the first player to play. The start of the incoming take should include the transient from the rest of the players, cutting off the early player. These can be joined together and overlapped as much as possible. Figure 18.10 illustrates this. It is not possible to completely time-align the various transients, but the situation can be much improved. Removing some time from a take will usually upset the flow of the piece a little,

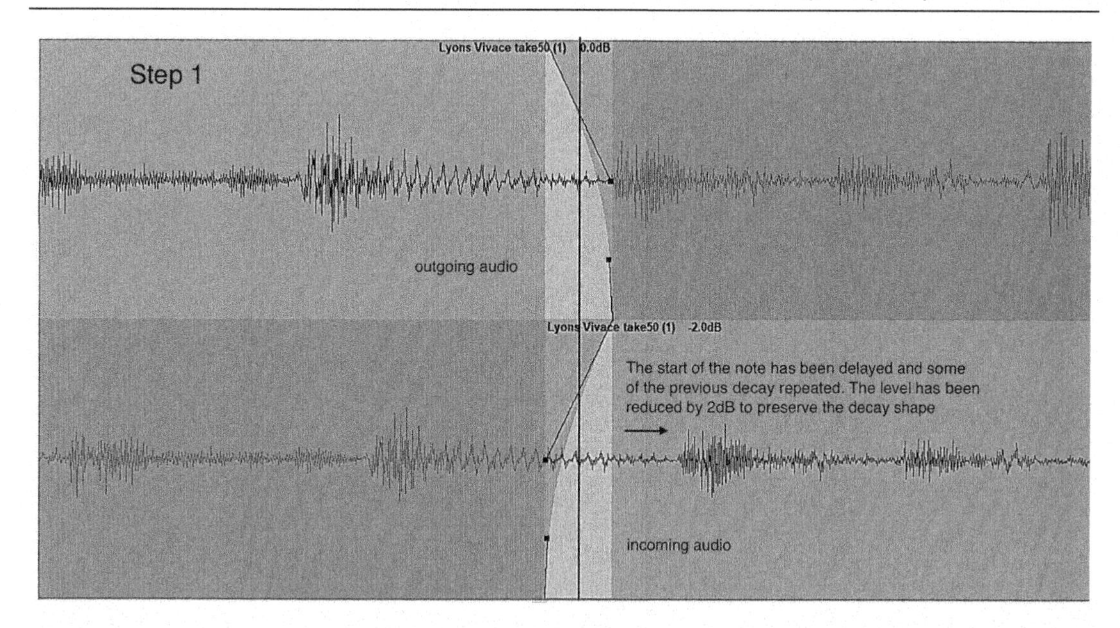

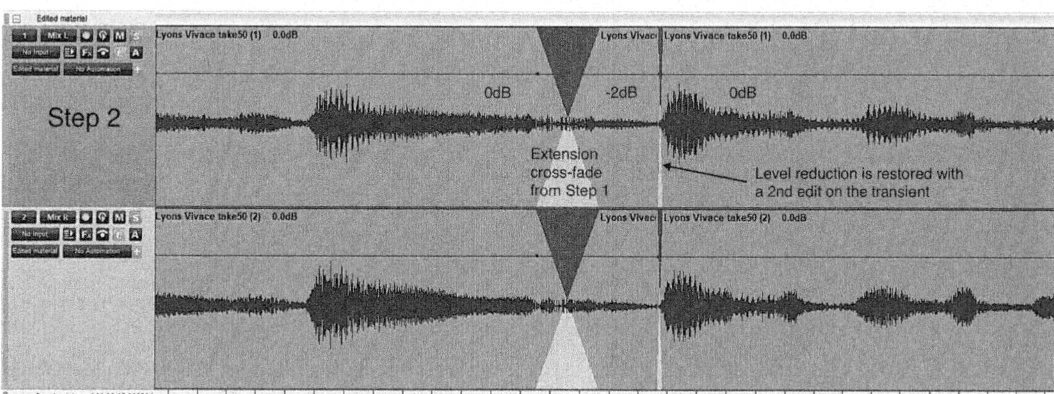

Figures 18.8 and *18.9* Extending a beat or pause within a single take. Part of the audio has to be repeated; the incoming take might need to be reduced in level to preserve any natural decay in the note's envelope (Step 1). This level reduction is restored in the timeline overview, shown in Step 2.

Image: Courtesy of Merging Technologies.

and a compensatory extension will need to be inserted either before or after the trimmed chord. (To extend a note, a section of audio will have to be repeated, as described in section 18.5.5, when extending a beat). If the first player to sound is actually coming in early, the act of trimming will bring all the other players in early too, and the extension will have to be inserted before the trimmed chord. However, if the first player to come in is in the correct timing and the others are late, any extension will need to come after the trimmed chord.

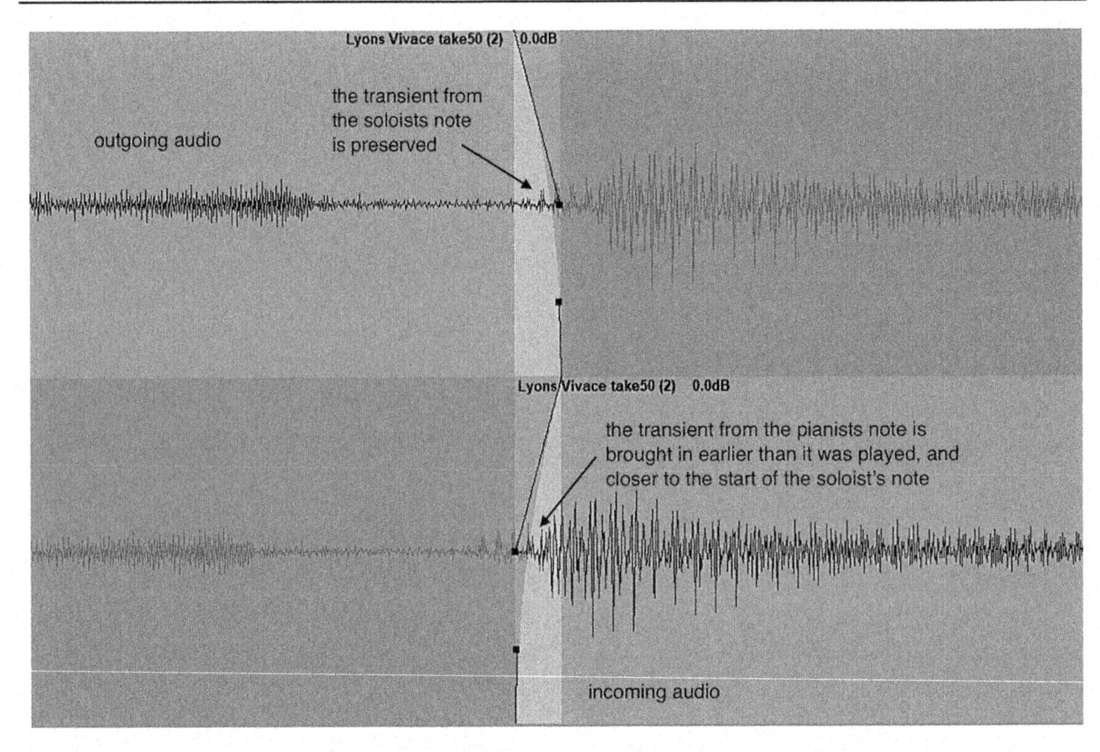

Figure 18.10 A simple trim on a chord where the soloist comes in before the piano. No timing has been compensated in this example.

Image: Courtesy of Merging Technologies.

Where there is a fast passage and players are not quite together, another take should be used. An extensive amount of trimming and extending could be undertaken if there is no other solution, but it will be extremely labour intensive and the results variable.

When editing a soloist where the timing against the accompaniment is expressively leading or lagging, it can be difficult to manage several changes of take without throwing the overall feeling. It is not unusual for the soloist to push ahead of the accompanist in one take and in another take to sit just behind, although the habit is particularly prevalent amongst singers. The key to keeping it feeling musical is to line up the edits using the timing of the accompaniment as a guide and then find a place where editing works for the soloist. This will mean accommodating the difference in timing between takes by losing a small part of a soloist's note or managing an increase in length of a note. (See section 18.6.2 for places in the sung word that can hide cross-fades, extensions, and trims.) Keeping the accompaniment rhythm going smoothly is essential to keeping edits hidden; if the timing and flow is disturbed, it really jumps out during extended listening.

18.6 Notes on working with different instruments

Editing an orchestra is usually more straightforward than editing a small ensemble, especially with piano included. This is because some individual player detail is lost within an orchestra, and there is usually more scope for editing during notes, placing shaping level changes and using longer crossfades, particularly over a large string section. Where very small details of an instrument can be heard, it is harder to disguise what you are doing. Not all instruments are included in the following sections, but the ones discussed have features to be aware of.

18.6.1 Piano

Piano is one of the trickiest instruments to edit, mainly because of the complexity of its sound. The sound changes timbre during the course of a sustained note; the sustain pedal means that there can be very complex mixtures of overhanging notes which can be very different on alternative takes; the transient of the piano is very easily distorted; both the transient and the tone change a great deal with the player's touch; and it can play more than one note at once, so there is immediately the possibility of some notes being sustained while others are sounded. Where you are trying to edit on moving notes in one hand while there are sustained notes underneath, the sustained notes are likely to bump unless you take great care with longer crossfades.

- Placing edits – avoid going into the piano transient with the crossfade (i.e. overlapping transients from different takes). A short crossfade just into the transient can be quite appealing initially, but you will eventually start to hear that there is additional 'hardness' and something slightly unnatural.
- Watch for bumps and discontinuities in the overhanging sound from reverb, pedalling, or notes that are held down. You will either have to use longer crossfades to solve these or move the edit. They can be hard to spot, as the tendency is to listen to the tune and not all the other parts, and transients in the tune will mask the effect a little. However, if you listen back to the editing in a 'relaxed listening' state, the discontinuities will be heard or felt.
- Watch for overhang getting chopped off or appearing from nowhere at an edit. Where piano takes are played differently, the overhang can build up quite differently and be very mismatched on different takes.
- Watch for wrong notes still being present in the overhang even a bar or so after a mistake. If you edit into a new take after a mistake in that take, the wrong notes might still be present in the pedalling or reverb.
- Pedal changes can sometimes sound like edits because they cause a stop to the overhang from undamped strings; if you edit at this point, check very carefully that any strange overhang effects are part of the take and not your doing! If it sounds peculiar, move the edit.
- Long crossfades can be used through a note's duration (avoiding the transient), but only with great care and careful positioning of the takes. Always compare with a real take to make sure that any changes in timbre during the note are natural and not due to your crossfade. Try moving one take 1–2 ms in either direction to help with phasing effects and/or zoom right in on the waveform and line up the peaks and troughs. Sometimes moving the incoming or outgoing take by just a few samples can make the required difference.

- Very short crossfades – these will sometimes work well, especially at a pedal change or where a longer crossfade (100 ms or so) is audible in the overhang change. Just be careful that there is no hint of 'blipping' in the overhang.

18.6.2 Singers

Placing edits in the sung word has some challenges, especially when it is accompanied. It is easier to edit a choir in the same way that an orchestra is easier; individuals are not as exposed.

- The sustained part of a solo singer's note will never take a long crossfade as it always sounds like there are two singers. (You might just get away with it if it is part of a huge operatic orchestral tutti and the singer is not exposed.)
- Where you need to use longer crossfades (if the singer is accompanied by an instrument that needs longer fades), the places that they can be hidden include through breaths, through rolled 'rrrrr', 'ssss', 'fff', and any other sound that is not pitched. These places are also good for hiding extensions and trims that are useful to accommodate the singer's variable timing (see section 18.6.4).
- The envelope of a singer's note can be split into a few stages: the initial hard consonant (if there is one); sometimes a short 'preparatory' sound that is not at pitch but can swoop up to it; the pitched part of the note; a final hard consonant (if there is one); and sometimes an 'end sound' where the pitch falls off. Any of the transition points between these moments can work for editing. Although the simplest is a hard consonant, this might not fit in with what is happening with the accompaniment.
- Edits can often also be placed at a diphthong or hiatus where there is a change in vowel sound but no change in pitch.

18.6.3 Strings

The main difficulty in editing strings is intonation, especially when working on chamber and solo music. However, as long as intonation and tone colour are very well matched, you will find that you can use longer crossfades during the notes on a string quartet or even a soloist without hearing two players during the fade.

A pizzicato note has a well-defined transient on which to locate an edit, but a bowed note usually starts with a small amount of woody scraping noise before the pitched note emerges. An edit can be placed before the bowing noise begins or just as the pitch becomes established, and the woody bowing noise will easily accommodate a generous crossfade. This means that when editing a bowed string quartet, crossfades can be successfully allowed to stray across into the start of the note or chord, provided everyone is playing together. If the ensemble is poor but tuning is well matched, it is usually quite straightforward to use the attack of one take joined to the sustain of another. (See section 18.5.6.)

18.6.4 Woodwinds

Woodwind instruments vary a little in how their transients work; the double reeds always have a sudden start to a note that quickly establishes a pitch, even if playing quietly, but a clarinet or saxophone can start very softly with a blowing sound before the pitch starts. As with the strings, it is possible to edit this sort of gradual onset of sound either at the start of the unpitched preparatory noises or at the start of the pitched note, and the blowing will often accommodate a long crossfade if necessary. The other important thing to remember with woodwinds is that breaths are needed to make the playing feel natural; they should not be too loud, but they do need to be there, especially if you have a wind octet about to start a piece. Breaths can also be another place to disguise long crossfades.

18.6.5 Harp and guitar

Both these instruments have easily identifiable transients, but because the sound rings on unless damped, the overhanging notes are what will cause potential problems in editing. A professional player is likely to be more skilled in controlling which notes are allowed to hang on and which will be manually damped down. The accumulation of overhanging notes must be kept smooth and free of bumps caused by edits, and it might easily be very different in tone between takes. Therefore, rather like editing piano, editing will be easiest when phrasing is consistent, and if there are wrong notes, they might still be hanging around a few seconds after they were played. Longer crossfades might have to be used to keep things smooth; avoid letting them extend into the transient and keep them in the sustained part of the note as much as possible.

18.7 Overdubbing scenarios

In classical recording, overdubbing is less common than in pop music, where it is an everyday occurrence. Because click tracks are not used, the approach to use if you have to overdub a part needs some careful consideration. When everyone plays to a click track (which imposes an external and regular time frame to the recording), it is fairly straightforward to edit any individual part at any stage as there is no time slippage of one part with respect to another. Without a click track, the timing is freely expressive, but it then becomes more difficult to alter parts once another layer of overdub has been added.

For this reason, the usual procedure for a singer or organ overdub is as follows:

1. Record the orchestral session in the usual way, with multiple takes, without the singer or organ.
2. Edit the orchestral session until everyone is happy with it; this will form the backing track.
3. Make sure that the project that contains the backing track is saved, and that the audio is locked to the timeline, as this will now be the time-stamped reference for any overdub files.
4. If it is necessary to create a new complete audio file of the orchestral backing, make sure it is a broadcast WAV file that has a timestamp.
5. Using the original project or a rendered BWAV file, play back the orchestral track to the singer or organist via headphones, and record the new takes *either* on a new group of tracks per take (as there will be multiple microphones even for an individual singer overdub) *or* using playlists (so that all new takes have the same time code as the original orchestral track.)

Next, the overdubbed takes can be edited against the backing track using the source-destination method and copying material into a set of empty tracks. Care has to be taken to retain the original synchronisation between the overdubbed takes and the backing while editing, so it will be essential to know how to move audio from track to track without slipping its timeline position in your DAW. Because singers in particular can be rather free with their timing (see section 18.6.4), it is often the case that the timing on one take is quite different to another, and edits do not line up well together. In this situation, you should note down how far any individual take has had to be moved from its synchronised position in order to make an edit work (move the audio clips using a 'nudge' of known value in milliseconds rather than dragging them). This allows you to correct this known offset as soon as is musically possible.

A word should be said about conductor videos. An opera singer is very used to watching a conductor, and if they are to record an overdub, it is usual practice to provide an edited conductor video. The conductor is filmed whilst conducting during the original recording session, making sure that the audio and video time codes are the same. Once the orchestral audio has been edited, the conductor video can be automatically edited to the same EDL by a process known as conforming. This produces a conductor video that contains edits in the same places as the audio edits in the backing track, and this can be used to guide the singer through tempo changes during the overdubbing session.

18.8 Emergency measures: sampling piano notes and note removal

Once the recording sessions are over, and you are back in the studio trying to make the editing work, it is not unusual to realise that a note is missing from a chord, or one note in a scale passage didn't sound very well, and you can't find another take that is quite as good as the one with the missing note. The conventional solution to this will be lots of complex (and possibly audible) edits, or using lots of EQ to try and boost a quiet note, which has an unpleasant audible impact on all the other notes.

A much better solution involves using your own individually sampled piano notes, recorded at the time of the session. It is easy to do this for piano because you can still use the instrument when the artist has gone home. Once the session is over, drop the DAW into record, head into the quiet studio, and play all 88 notes on the piano four different ways:

- Soft (piano) staccato – short, stabbed notes.
- Soft (piano) tenuto – held, so that after the initial attack, the note sustains – no need to use the sustain pedal. Don't let it fully die away, but lift your finger gently so that you hear the distinctive sound of the end of the note as the dampers drop into place.
- Loud (forte) staccato.
- Loud (forte) tenuto.

These recordings will have a sound that matches your recording, and will be invaluable in patching in missing and weak notes. Using extra tracks on your DAW editing system, you can drop in extra notes as and when necessary. On most DAWs, it is easy to adjust the gain of these individual clips to get the relative levels right, although sometimes a little EQ might be needed to match the tone to the session take given that piano tone changes a lot depending on how loud it is played. If the attack on the 'loud staccato' or 'loud tenuto' samples is too harsh, apply a low-pass filter and adjust the turnover frequency until it approximately matches the sound of the notes you are trying to augment – somewhere around 5–6 kHz is typically a useful starting point in taming an over-aggressive attack.

The duration of the note might also need adjustment, and you might also need the distinctive sound of the dampers going down at the end of the note. The tenuto notes can be used as a basis for notes of varying length. In most DAWs it is a straightforward matter to place edits during the sustain section of the tenuto note to shorten or lengthen it as appropriate; shortening is simple, as the damper sound can be brought in at any time to finish the note early. Lengthening the note needs a little more care, but when dealing with a single note, the repetition of the waveform should be visible, so just looping sections with a fairly long crossfade should be straightforward. Manually extending notes will usually give smoother results than any time-stretch plug-in (see also section 18.5.5 for discussion of lengthening notes).

It is worth building up a library of these note samples, particularly if you record in the same venue many times using the same recording technique. Notes from a previous session are not likely to be noticeable if used singly to cover the occasional weak or missing note from your current project.

Figure 18.11 shows an example of this technique in action. The main stereo edit (of a ragtime piece by Scott Joplin) is shown in stereo track 1. Just after the edit from take 104 to take 105, there is a chord that was missing an A above middle C (A_4).

A 'Soft Tenuto' A_4 was placed on stereo track 2 so that it lined up with the start of the other notes of the chord. The gain of this clip was reduced by 1 dB to match the main chords, and the sustain of the note was faded out gradually using the clip handles to match the die-away of the existing chord. The result was chord that contained all the notes, and unnecessary edits were avoided. Figure 18.12 shows a zoomed-in view to show how accurately the waveforms needed to be lined up.

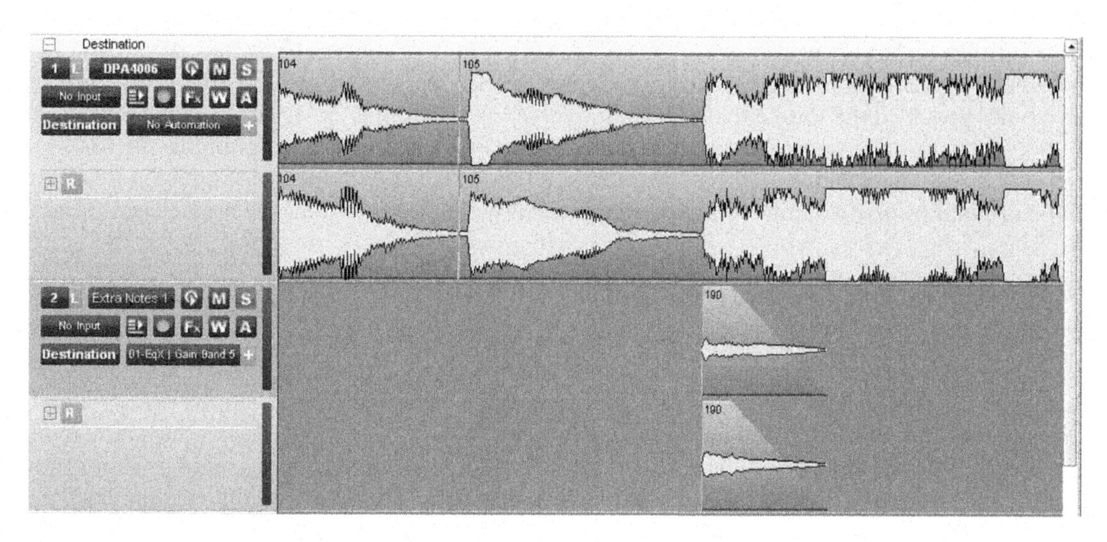

Figure 18.11 Excerpt from Joplin piece with additional A_4

Image: Courtesy of Merging Technologies.

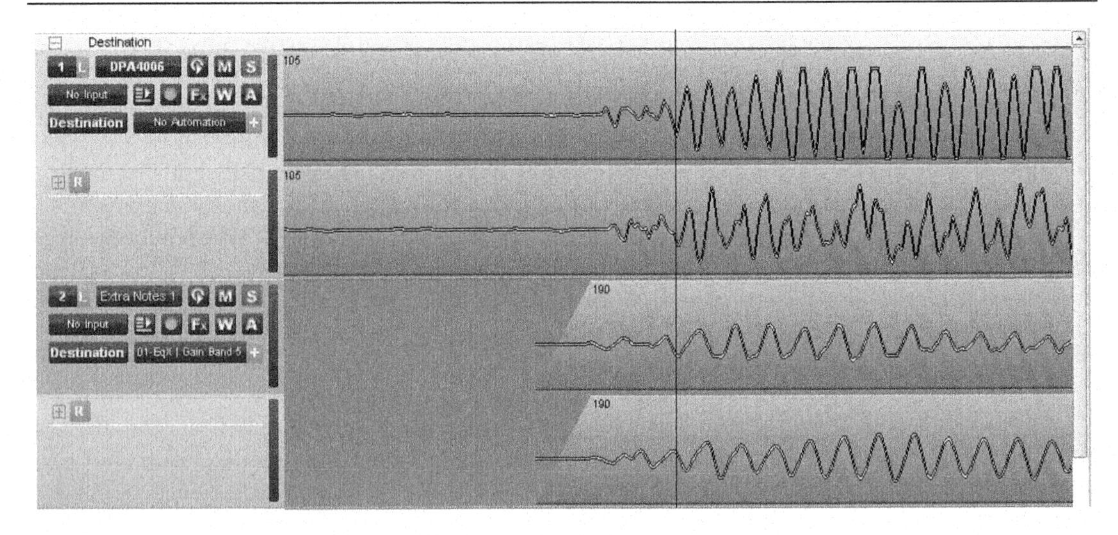

Figure 18.12 Zoomed-in waveform from Figure 18.11 at the start of the note

Image: Courtesy of Merging Technologies.

On occasion, it might be necessary to try to remove a wrong note and then replace it with the correct one, as outlined earlier. This will require the use of noise removal software, as discussed in section 19.1. Given the clarity of the display within such software, it is sometimes possible to locate a single note and remove it. Although it is far quicker and preferable to have enough correct takes in post-production, mistakes sometimes happen, even down to an artist having learned the wrong accidental in a piece. When attempting something as complex as whole note removal, the overtones must also be removed, as well as any associated reverb. Celemony Melodyne has a 'Direct Note Access (DNA)' facility which can work well at removing a single note and its harmonics provided the source material is quite dry. Unsurprisingly, it struggles with reverberant sound, and in this situation, treating each harmonic individually using a programme like CEDAR Retouch is recommended. If the removed note is to be replaced with a clean sampled note, some of the original transient can be left untouched which will help to 'bed in' the replacement note.

18.9 Professional finish: joining into room tone

When a CD of classical music is put together, room tone is used between the tracks to avoid fading into complete digital silence. The ending of each piece should have a clean reverb die-away that sinks seamlessly into a recording of the sound of the venue known as 'ambience' or 'room tone'. At the start and the end of the CD there will be a fade up of about 1.5–2 seconds into room tone just before the start of the first piece, and a fade-out of about 3–5 seconds in room tone just after the reverb finishes at the end of the last piece. A guide to gap lengths, dealing with the creation of clean endings where needed, matching different room tones and dealing with audience noise is included in Chapter 19.

Chapter 19

Mastering

Mastering for classical recording is different to that in the pop world, but in both cases, one of the main benefits of a 'mastering stage' in post-production is having another experienced person listen to the recording. Mastering for pop music as a very distinct stage of the post-production process has acquired a certain level of mystique, and many beginner engineers assume it can compensate for inadequate recording or mixing skills. This is really a misunderstanding of what it is all about, and while it is possible to improve a good recording with skilful mastering, significantly improving a poor one in the mastering process is a rare occurrence. Readers should refer to Bob Katz's book *Mastering Audio: The Art and the Science* for more on this topic.

Mastering classical music is part of a very different post-production workflow, and many of the necessary tasks will have already been dealt with as part of the editing process. These might include noise removal, addition of room tone between tracks, and alteration of gap lengths.

19.1 Noise removal

Noises off will always be a problem when recording a large number of people in a reverberant space; there is always someone who will accidentally knock a music stand or drop part of their instrumental hardware. Noise removal covers continuous background sounds such as rumble, hum, and hiss as well as discrete acoustic noises such as breaths, rustling, page turns, audience coughs, and thumps, bumps, and taps from audience or performers. The most well-known programmes commercially available at the time of writing include CEDAR Retouch, iZotope RX, and Algorithmix reNOVAtor. Celemony Melodyne can also be useful for occasional pitch correction and note removal.

Removal of continuous background noise is usually known as 'global de-noising', and at the higher end of the market, the programme can be given a sample of the noise which it will then analyse to work out what needs to be removed from a signal that contains music as well. CEDAR is the market leader in global de-noising, but it is also the most expensive. iZotope RX currently offers the best value for money, and it offers broadband hiss removal as well as a de-rustling mode which is very effective. If you have a poor-quality sample of room tone (see section 19.3.1) that contains some discrete noises, de-noising this can make it easier to loop repeatedly to create a montage that is usable.

Hum can be approached as a global de-noising problem, although it will usually be masked for a lot of the time during the recording and only becomes apparent in quieter sections or during the gaps between pieces. Its spectral components are usually quite visible, and something like iZotope will treat it with a series of narrow notch filters. Given the non-transparency of steep filtering and potential detrimental effect on the audio, a gentler approach would be to treat it only where it is audible. Of course, HF hum such as CRT monitor line frequency (15.625 kHz in the UK) might not be audible to you once you enter middle age (!), but it will be audible to younger listeners and should be addressed.

Removing individual noises can require more decisions from the user; the noise itself must be highlighted and then various types of interpolation and processing can be chosen to reduce its impact or remove it altogether. The display typically shows time along the horizontal axis and frequency up the vertical axis, with intensity being indicated by colour. The simplest noises to locate and remove are short transients with primarily HF content, such as clicks and taps. The duration of audio that needs to be replaced or interpolated is short, and the disruption to the original audio is minimised. Broadband LF thumps are more difficult, as are coughs and page turns. The longer the duration, and the wider the audio band covered by the noise, the more noticeable will be any artefacts created by the processing, and it should be remembered that all acoustic noises will carry some sort of reverb tail that will also need to be removed or reduced. Some complex noises are better approached with a few passes through the processing, and some trial and error will be needed. If you are considering the addition of a small amount of reverb as part of your mastering, it is likely to be best to leave that until after noise reduction. See section 18.8 for discussion of removal of individual wrong notes.

19.2 Changing the sound

Again, the differences between pop and classical workflow tend to reduce the amount of sound manipulation that is done in classical mastering. In pop, the initial concentration is on the individual sounds, and mixing and remixing is the main focus of the post-production period. In classical workflow, the focus is on the overall sound from the very beginning, and the post-production period is focussed on editing. This means that everyone involved has been listening to what is hoped to be the finished sound for a long time. Additional EQ and reverb are not always part of the mastering process, and they should only be used if they are improving the sound. Do not feel that you have to do something for the sake of it so you can feel that it has been 'mastered'.

19.2.1 EQ

Some improvements can often be made with small amounts of EQ, although this is a skilled job, and EQ amounts will usually be in the 1–3 dB range and occasionally a little more. Bearing in mind how quickly the engineer or listener gets used to a new sound, decisions about EQ will need a sense of perspective, and you will need to go away from the recording and come back to it. Your opinion when listening to it in the first few seconds is likely to be correct; make a small change, and then take another break before making a judgement between 'before and after' EQ. EQ judgements will become more assured the more experience you have of listening to recordings of a similar genre.

19.2.2 Reverb

This should be approached with caution, as reverb will have been dealt with at the time of mixing, and you do not want to add something on top that does not blend and feels like a layer of thick varnish on a good painting. However, if you have several items from multiple sources and locations, a little overall reverb can help the individual items gel together as part of the same album.

19.3 Tops and tails: room tone, breaths, noises, and fades

All tracks on an album have to start and finish cleanly with no noises before the start or during the final reverb die-away, and with appropriate gap lengths between movements and works. Gap lengths on a CD are part of the programme, and when the CD is played from start to finish, the gaps will be of the duration inserted by the mastering engineer (according to the wishes of the producer and artist). The running order will also have to be decided, particularly where there are many shorter pieces.

19.3.1 Room tone

Anyone who has listened closely to a classical album will have noticed that there is room tone (also called 'ambience') between the tracks. The presentation of the album is such that the initial fade-up brings the listener into the room where the performers are located, the performers play, they rest silently between movements, and when the 60–70 minutes are over, the listener leaves the room again by means of the room tone fade-out at the end.

For live albums where there are pieces from different concerts, the presentation follows the same pattern; where there is a change of venue, the fades in and out will go to digital silence to mark this change.

Room tone should be recorded at the time of the session; a couple of minutes of silent recording should be enough from which to select a quiet section that can be edited to itself if necessary. Most DAWs will perform an automatic looping function for you, but beware that if there is any noise at all in the short segment, automated looping will set up a subtle rhythmic repetition within the sound of the room tone. This is not acceptable for use between movements, and it is often better, and quicker in the long run, to loop some room tone manually, varying the length of the extract to avoid repetitiveness. When room tone hasn't been recorded, a lot of post-production time can be wasted on extending a fraction of a second's worth of silence between takes into the 20–30 seconds needed to comfortably manage the creation of quiet endings. As noted in section 18.1, when looping a poor sample of room tone, removing some of the noise from it before you start will help.

Room tone should be included in all gaps between movements of a work on an album. It should also be included between works, especially if the album consists of many short pieces (such as a Lieder recital). Where an album consists of a long symphony and a filler piece such as an overture, it is common practice to include room tone in the fairly long gap between the two works, although some mastering engineers/editors will choose to fade out to digital silence between the separate works. Generally, fading to silence and back into room tone is used to indicate a change of venue.

Joining room tone to the edited music can be a simple task if the background room tone of the takes is a good match to the room tone sound file that you are using. This should simply involve a straightforward edit at the start of the piece and probably a longer crossfade at the end to ensure that the transition from the reverb tail is smooth. However, if they are not well matched, you will need to use clip gain or EQ on the room tone file to help the matching process, and longer cross-fades to transition from room tone into the piece. If the room tone file is noisier than that under the music, the simplest way to transition is to offset the crossfades and bring the music in with a short fade at its opening transient, whilst retaining the noisier room tone underneath. This room tone can then gradually be faded out under the music. At the end of the piece, any noisier room tone can be gradually faded up under the ending of the piece (more gradually for quieter pieces) so that the reverb can die away into it and be faded out.

The ultimate goal is for there to be no audible change in room tone as the musicians begin performing and no audible change when they finish and the reverb dies away. All these transitions have to be smooth and seamless to maintain the illusion that the listener has entered the room, and that the performers have played the piece perfectly without any noise or chatter before or after.

19.3.2 Noisy endings

If there are any noises in the room during the final die-away, these will need to be dealt with, preferably by using a new take, even just for the reverb tail. However, if there is no suitable take that can be used, the noise in the die-away will need to be removed. This might be partly possible with noise-removal software, but it is likely that a false ending will have to be created as follows:

1. If you have noise removal software, try to remove as much of the noise as possible. This will probably result in removal of the transient and HF content but leaving some noise behind.
2. Fade the natural die-away early into some clean room tone, making sure you have faded out before the noise, or that it is at least a very low level in the fade. The ending is now missing its reverb tail, but the room tone should sound natural (see Figure 19.1a).
3. Send the tracks containing the new ending to an appropriate reverb unit or plug-in, and record the 100% wet reverb returns on another pair of tracks.
4. Fade this clean reverb up under the ending to replace the reverb tail that is missing (see Figure 19.1b).
5. Mix two sets of tracks together to create a new, clean ending with a full-length reverb tail.

19.3.3 Breaths and other preparatory noises

Immediately before the start of a take, there might be a variety of noises. An intake of breath before the downbeat can be left in the mastered compilation as it can feel very natural, especially in chamber music where there are wind/brass instruments or singers. Noises that are knocks and bumps should be removed; care should be taken to make sure that any reverb from the removed noise is no longer present under the start of the piece; if it is, either another take or noise-removal software will be needed.

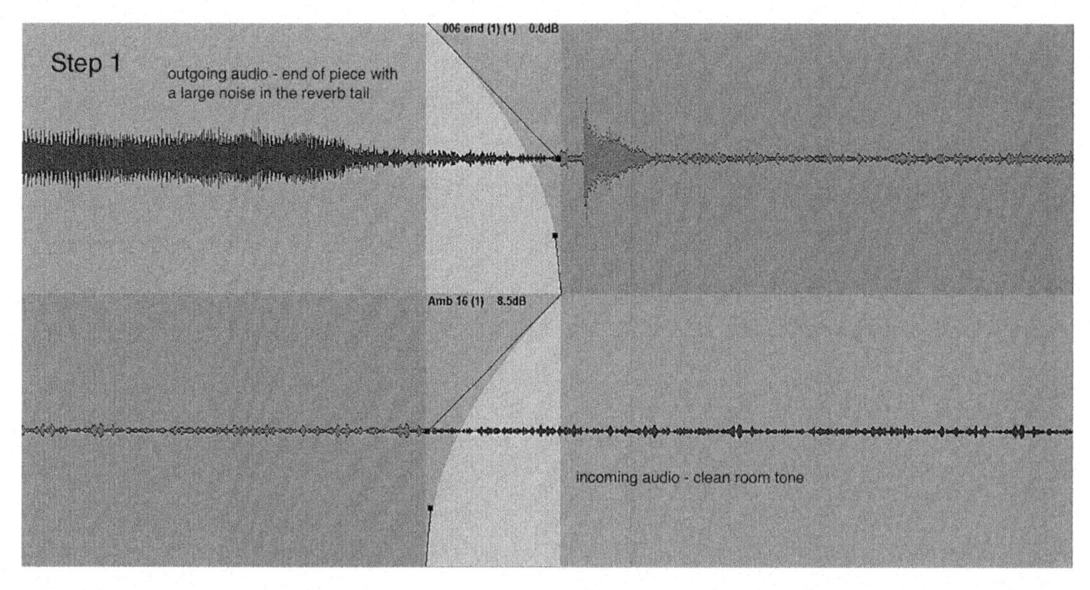

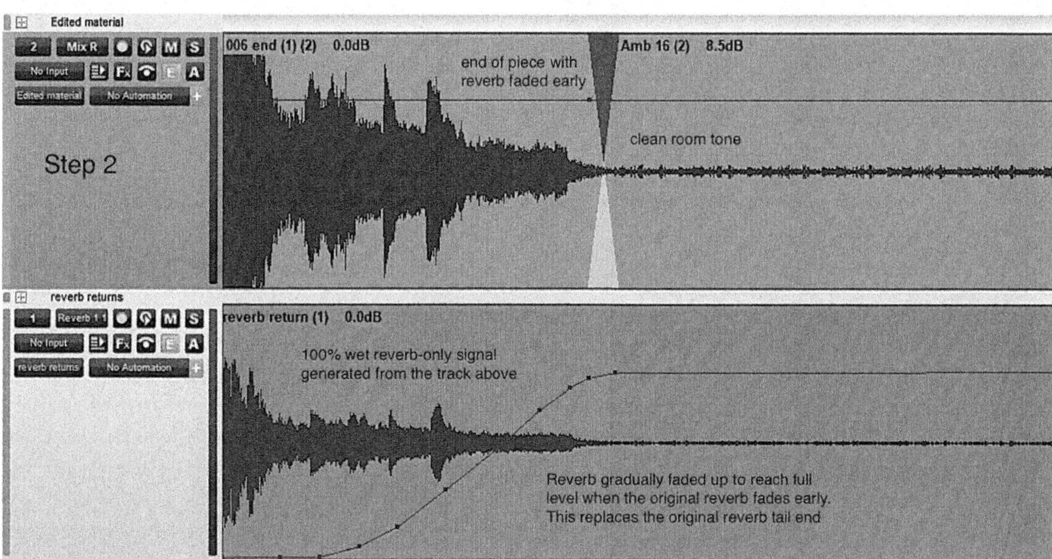

Figure 19.1a–b showing the process for creating a clean ending

Image: Courtesy of Merging Technologies.

19.3.4 Fades

Fading up or down can be performed on the DAW, but the shape of the fade should mimic a good manual fade. See Figure 19.2 for the S-shaped fade required; this will have different names in different DAWs, so be careful: what is termed a 'cosine' fade in one DAW can be a different shape to a 'cosine' fade in another DAW if it uses a different portion of the cosine curve.

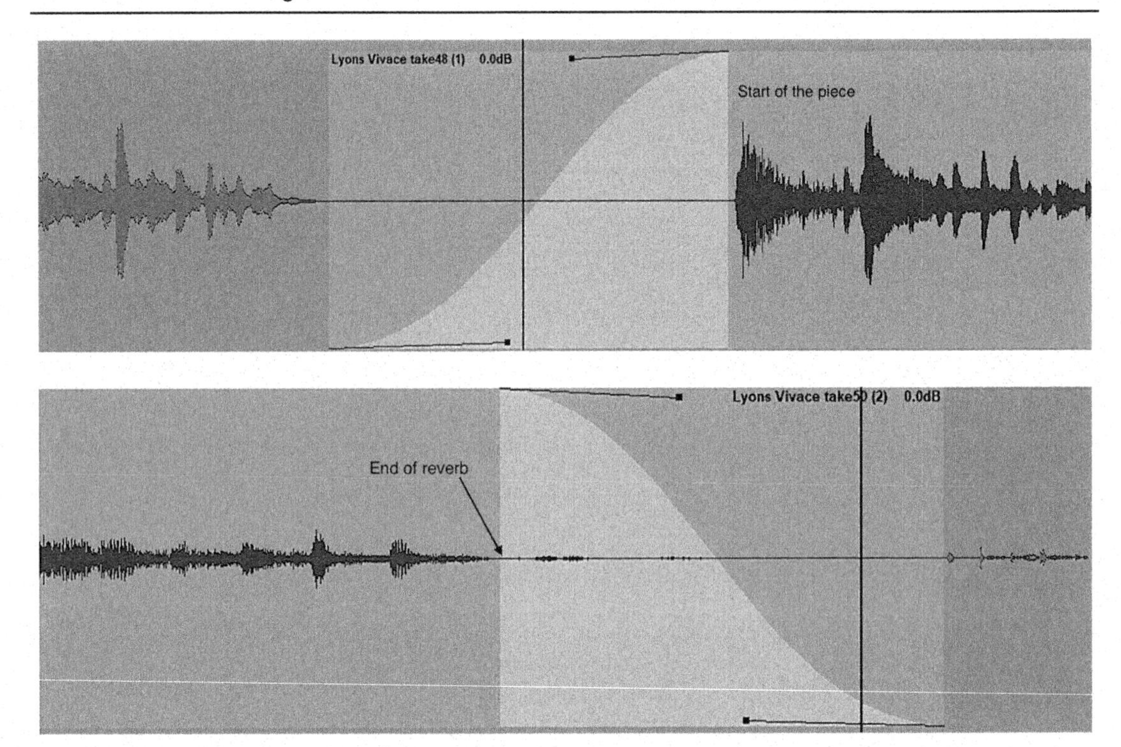

Figure 19.2a–b S-shaped curves used for fading in and fading out. The shape is important to avoid too fast rates of level change when approaching $-\infty$ dB

Image: Courtesy of Merging Technologies

At the start of the album, you will need to fade up during clean room tone over about 1–2.5 seconds, and the music should start as soon as the fade up reaches full level; no one wants to sit listening to room tone for a second or two while wondering what has happened to the music. The duration of the fade-up that works really depends on the noisiness of the ambience, and the scale of the musical forces that are to follow. The shorter durations tend to work for piano and chamber music, with longer fades used for large-scale works.

The noisier the room tone, the slower the fade needs to be so that the listener can become accustomed to the sound and identify what it is. A fade-up into noisy room tone that contains organ blower, for example, will sound quite alarming if it is done very quickly, as the listener experiences a sudden surge in hiss and rumble at quite a high level. If time is taken over the fade, the listener can work out that they are coming into a church where the organ blower is switched on and a choral album is about to start.

At the end of the album, you will need to fade out during clean room tone, starting to fade just as the natural reverb tail enters the noise floor of the room tone. Try to match the pace of the fade-out with the rate of die-away of the reverb, so that they feel as if they are one continuous thing, and never cut the reverb audibly short by starting the fade too soon. If the piece dies away at the end with little obvious reverb (e.g. strings diminuendo to *ppp*), then try to make the fade-out slow and reflective of the same mood as the end of the piece. Never rush a fade in order to get it completed before an intrusive noise. Remove the intrusive noise (see section 19.3.2 for dealing with noises in reverb tails) so that you have a clean reverb tail with plenty of quiet room tone following. Then use this audio as the basis for the end of album fade.

19.3.5 Managing applause

When recording a live concert (even if there is to be editing patches taken from rehearsal material), applause at the start and/or end will need to be managed convincingly. The first thing to note is that applause should sound enthusiastic and loud, so if it doesn't, you will need to try and help this by joining sections together, double tracking, and compressing a little if necessary. Good-quality applause will only be captured if audience microphones are considered at the recording stage.

The next thing to note is that applause at the end of a piece should never be allowed to peter out before the fade-out. You must make sure you have enough applause duration to leave it at full level for about 10 seconds, and then another 10 seconds or more in which to gradually fade out, without it losing enthusiasm. The listener must leave the concert while still under the impression that the applause went on for a long time, with great enthusiasm. Again, if you don't have enough of it, create a montage of applause, avoiding any obvious repetition of individual shouts and noises.

If you fade in at the start of a concert and choose to use applause rather than audience room tone, the fade should be extended to 5–10 seconds, and then the applause should be allowed to linger for a few seconds at full level before it settles down to silence and the piece begins. In real life, the gap between the applause petering out and the piece beginning can be quite long; it contains the conductor tapping the stand, and the orchestra settling down ready to play. These preamble noises should be removed, and if the applause petering out takes a while, this could also be foreshortened with some creative editing. The goal should be a fade-up during rapturous applause, the applause coming to a natural end, a couple of seconds of expectant room tone, an unobtrusive collective intake of breath if there is one, and then the start of the piece. The job of the mastering engineer or editor is to create this series of events from the material available, and although it is artifice, it must be believable. For this reason, it makes no sense to crossfade from clean ambience into applause; this is not something that would happen in real life, and when applause is faded in or out, the room tone will be naturally faded in and out with it; it is this that takes the listener into or out of the concert hall.

19.3.6 Gap lengths

It should be noted that when mastering for CD, the gap lengths that are included between pieces or movement on the final continuous WAV file will be reproduced exactly as they are when the CD is played from start to finish. The gap lengths on an album are an integral part to the feel of the whole; where they are short, the listener is moved on quickly from one piece to the next, and this can keep the energy levels high throughout. Shorter gaps are more suited to albums of fast movements, shorter pieces, and chamber music. There are no hard and fast rules about gap lengths, and they must be judged by ear, usually with the input of both the producer and artist.

As a rough guide, typical gap lengths between movements of a piano sonata or string quartet will be in the order of 3–5 seconds (assuming there are no composer instructions to play the next movement with barely any pause). This is measured from the perceived end of the reverb die away and the start of the next movement. Sometimes the artist will play across the gap between movements on session which will give you a guide to the gap length, although this is more common where two movements are to be played *attacca*. Gap lengths between classical symphonic movements will be a little longer at 5–6 seconds, and romantic symphonic movements longer still at 6–7 seconds. Gaps between works on an album will be a couple of seconds longer than gaps between movements. This might sound very long, but following the experience of listening to 50–60 minutes of a large romantic symphony, a gap of 9–10 seconds can feel completely appropriate. Gaps tend to need to be longer where there is a change of pace or mood.

To judge how long a gap should be is a matter of feel: listen to a good length (a few minutes) of the end of the previous piece followed by silence, and use the marker system on your DAW to mark where you think the next track should start. Repeat this a few times; if you are consistent, you will have a cluster of markers as a guide to where to start the next track. If you only listen to the final 10 seconds of the previous piece, your gap lengths will come out rather short and will feel wrong when you listen through to the whole album. All gaps should finally be judged by a play through the whole album at some stage.

19.4 Levels between tracks, compression, and loudness meters

Classical music is generally given as much dynamic range as is practical, although reproducing the natural dynamic range of a large romantic orchestral work can prove a challenge in a domestic listening environment, and concessions will be made to reduce the range using manual compression; see the following sections.

19.4.1 Levels

For something like an album of piano pieces, there is no need to alter the dynamic range, but the overall level of each track should match up well with the others. The easiest method for checking levels between tracks is to spot-check through the album once it is compiled onto a single DAW timeline. This should show up any pieces where the recording level feels different, although if everything was recorded in a single set of sessions with an experienced engineer, there is not likely to be any significant problem. For a compilation of pieces from different sources, levels might need some adjustment, and attention should be paid down to a perceived difference of about 0.5 to 1 dB. Any EQ changes that you make to compensate for different recording locations will also affect level perception.

Typically, a classical mastering engineer will normalise the whole album. This means that each track should have a natural level relationship to the others but that the highest peak level on the album should be just below 0 dBFS.

19.4.2 Compression: manual and automatic

Automatic compression is rarely used on classical recording as it affects the envelope of sounds as well as reducing dynamic range, and so it is not considered transparent enough. Part of the joy and excitement of classical music comes from its dynamic range and the fact that it has long passages of quiet and loud playing; this should be assumed to be normal and completely acceptable, with no comparisons drawn with pop or crossover music recording. If automatic compression is required for some reason, parallel compression is more transparent for classical music; the most important thing to pay attention to are the reverb tails, which will rise up in level if the release time of the compressor is not set long enough. See section 17.6 for discussion of dynamic range reduction on orchestral recordings as part of the mixing process at the time of recording.

When the dynamic range of a classical piece needs to be reduced following the session, manual compression by means of fader riding will generally be used. This will usually be either for broadcast purposes or to make a CD more listenable in a domestic living space where the loudest orchestral tuttis cannot be reproduced at their full original SPLs. Using manual compression, a dynamic range reduction of up to 10 dB can usually be achieved over the course of a piece. The process will involve gradual gain changes to lift up the level of the quietest passages by the required amount while preserving as much as possible of the contrasts in loudness that are part of the score – moving from *ppp* to *fff*, for example, needs to feel dramatic. An example would be the sudden jump in level at the start of the 'Infernal Dance' (in Stravinsky's *Firebird*) that follows on from the very quiet passage that precedes it. This extended quiet passage would have been raised in level by 6–8 dB earlier in the piece but gradually reduced in the approach to the sudden tutti, so that the tutti has enough headroom and the contrast between the quiet and loud sections is preserved. The opera recordings from the Royal Opera House, London, usually have a loudness range of about 23–24 LU (loudness units) after mixing, but once these have been manually compressed by fader riding, this is reduced to about 20 LU. (See 19.4.3 for further discussion of loudness measurement.)

Another common situation in mastering that calls for a reduction in dynamic range is when the album as a whole is peaking 4–6 dB below 0 dBFS, and the only thing to take the level to 0 dBFS is a single orchestral bass drum hit. In this situation, rather than a gradual level change that might affect adversely the dynamics of the climax of the piece, the level of the drum transient can be manually reduced momentarily to allow the overall album level to be increased. This involves locating the very short section of the stereo master file that causes the level to peak, and taking its level down by up to about 4 dB using clip gain for the shortest possible time. The duration of the drop in level will be in the order of a couple of milliseconds and should be inaudible.

19.4.3 Levels and loudness measurements

Normalising the level of a complete album does not guarantee that two recordings of the same instrumentation will sound equivalently loud. Our perception of loudness is dependent on average as well as peak levels, and when level-normalised, material with a high peak-to-average level ratio will sound quieter than material where the peaks do not greatly exceed the average level. Albums of different 'loudness' are not generally a problem for the home listener playing back CDs, provided the loudness perception throughout a single album is properly managed by the mastering engineer.

Outside the world of classical recording for release on CD, 'loudness' or 'volume' calibrations are now part of the audio signal chain in cinema, TV broadcast, and music streaming services. Loudness analysis tools record the total sound energy over the whole length of a TV programme or an individual audio track, and a figure for its 'loudness' is arrived at by integration. Loudness

AES	*Spotify*	*Apple Music*	*YouTube*	*Tidal*	*Amazon Music*
−16 LUFS	−14 LUFS	−16 LUFS	−13 LUFS	−14 LUFS	−14 LUFS

Figure 19.3 Table showing the standard loudness in LUFS for various streaming platforms

measurements have been standardised in both the US (ATSC A/85) and in the EU (EBU-R128). For UK TV broadcast, for example, the integrated loudness across the whole programme must read –23 LUFS ±0.5 and the true peak level must be no more than –1 dBFS. The loudness range (LRA) measurement gives an excellent indication of perceived dynamic range.

Spotify, Apple Music, and other music streaming services use loudness measurement to calculate how much relative gain to apply to each audio file at the point of streaming. This carries implications for both pop and classical mastering engineers, as the application of different amounts of gain is used to present a consistent 'perceived loudness' of tracks to the listener. The 'standard' loudness varies across the streaming platforms, although the AES Technical Council have produced a recommended standard of –16 LUFS.[1,2] Figure 19.3 provides a comparative table.

Assuming the loudness standardisation is enabled in the user settings, tracks will be altered in level to match the standard. For example, if a track arrives at Spotify with a loudness of –20 LUFS, it will be increased in level by 6 dB at the point of playback. *This carries important implications for tracks with high peak-to-average levels, such as classical music, as any original transients peaking above –6 dBFS will then be passed through a limiter and removed, possibly audibly.* Conversely, a heavily compressed track arriving at Spotify with a loudness level of –6 LUFS will be reduced in level by 8 dB at the point of streaming. This means that it will no longer sound 'louder' than its uncompressed neighbours, which carries implications for the use of heavy compression in mastering to make whole albums sound 'louder'.

It is usually possible to turn off the loudness/volume normalisation feature using an 'album mode'. Listening to an entire album in this way will preserve the relative levels of individual tracks as they were mastered. This is particularly important for classical music, as individual tracks are likely to have very variable loudness readings due to the musical content of each.

19.5 Placing track markers for CD mastering

Although much music is streamed nowadays, it is still necessary to place track markers (known as PQ markers from the PQ Subcode embedded in a CD) in the timeline for CD production in order to define where the tracks stop and start. A Red Book CD consists of a single stereo WAV file that includes the gaps between pieces as they are arranged on the DAW timeline when the CD master is made. The track and index start and stop markers on a CD are aids to navigation within this long file. There are a few conventions to note when marking the tracks on a classical album:

1 The first track start is placed just before the fade-up at the start of the album, and PQ software will usually add a further two seconds to this initial start time to comply with the CD Red Book standard.

2 Track starts that are the beginning of a piece should be placed about 0.25–0.5 seconds back into the room tone before the first note and should include any breath that you have left in as part of the editing/mastering. When the user selects an individual track to play, it should not start suddenly, but neither should there be a long section of room tone before the music starts.
3 Tracks starts that are placed during continuous music (e.g. for through-composed opera) are usually placed closer to the start of the note in question, and they will always sound rather sudden if the track is played in isolation. The overhanging notes from the preceding music will jump in at the track start, and there is no way of avoiding this. It does mean that the user can jump to their favourite aria.
4 Where there is room tone between tracks, the gap is assumed to belong to the end of the previous track, and the end of one track and the start of the next occur at the same time. With some software, this means that 'end-of-track' markers are not used at all if there is room tone, except the one at the end of the album.
5 Where there is no room tone between tracks, the track start and end markers will be placed so as to incorporate the full extent of the room tone or applause fade-in and fade-out.
6 The final track end marker of the CD is placed at the end of the fade-out.

Disc Description Protocol (DDP) image files are the usual way to transfer a PQ encoded master file to a factory for CD production.

Notes

1 *Katz, Bob – Mastering Audio: The Art and the Science (Focal Press, 2014) ISBN-13: 978–0240818962*
2 *AES Technical Document TD1004.1.15–10 'Recommendation for Loudness of Audio Streaming and Network File Playback'*
www.aes.org/technical/documents/AESTD1004_1_15_10.pdf

Appendices

Appendix 1

Opera recording: practices at Decca from the 1950s to the 1990s

Although this is not a history book, there are a few occasions where describing how things were done during a previous era can be useful in understanding how those iconic recordings were achieved and how older practices compare with where we are today. Recording opera in the studio, especially in the way it was done at Decca during this period, is an extremely expensive undertaking. For financial reasons, a great deal of opera recording today is captured live from a theatre, even if there is subsequent editing between performances (see Chapter 16).

Philosophically, it is possible to approach recording opera much as one would an oratorio, without taking into account the fact that an opera is a staged genre where the singers move around during the action. However, under the initial influence of John Culshaw, one of the additional goals of the Decca studio opera recordings of the 1950s to 1990s was to capture the depth and perspective of the stage behind the orchestra and to position the singers according to the dramatic action at the time. In this way, it was seen as more akin to recording a play as it was performed and not simply recording actors reading from a playscript. Additionally, most of the theatrical effects included in the score and stage directions were included at the time of the recording. Things that would be added later were small effects such as clocks, bells, and the occasional sword fight, although offstage chorus might also be recorded separately (such as that for 'Nessun Dorma' from Puccini's *Turandot* (Decca, 1972); see Figure A1.9).

There are four elements to capturing staged opera as a recording session:

1 Stage set-up.
2 Chorus set-up.
3 Orchestral set-up.
4 Live offstage effects.

Additional practical considerations are communications between control room, conductor, and stage and the level of detail required to repeat stage positions for every take.

A1.1 The stage set-up

The stage might already be present in the hall, such as at Kingsway Hall, where around 80 chorus members could be accommodated behind the stage or on adjacent balconies if more singers were needed. Where there was no stage already in situ, a staging area was placed behind the orchestra, and it was designed to make replicating stage moves straightforward. During the later 1950s and early 1960s, a grid was marked out in squares on the stage floor, with each square measuring about 1.5–1.8 m (5′ to 6′) along each side. The stage measured at least five squares wide and three deep, and the singers had to move between the grid squares as part of the recording choreography. Vocal microphones were set up at three positions across the front of the stage in front of positions 1, 3, and 5. See Figure A1.1 for the grid layout and Figure A1.2 for a photo of an early recording with Kirsten Flagstad in which the stage grid is clearly visible.

From the mid-1960s onwards, Decca moved to a five-microphone stage (see Figure A1.3) to avoid singers losing focus if they were standing in areas 2 and 4 on the grid. Once the five-microphone set-up was in use, the floor grid was discontinued as the microphones were used as lateral reference positions instead.

The microphones were numbered 2, 4, 6, 8, and 10, and each microphone stand or its associated music stand was clearly labelled with a large reference number. Odd-numbered stage position references were located in between the microphones, and all choreographed stage positions were marked in the scores that were used by the conductor, producer, and engineer. Most of the stage directions were lateral movements, but in the event of needing a more distant sound, the

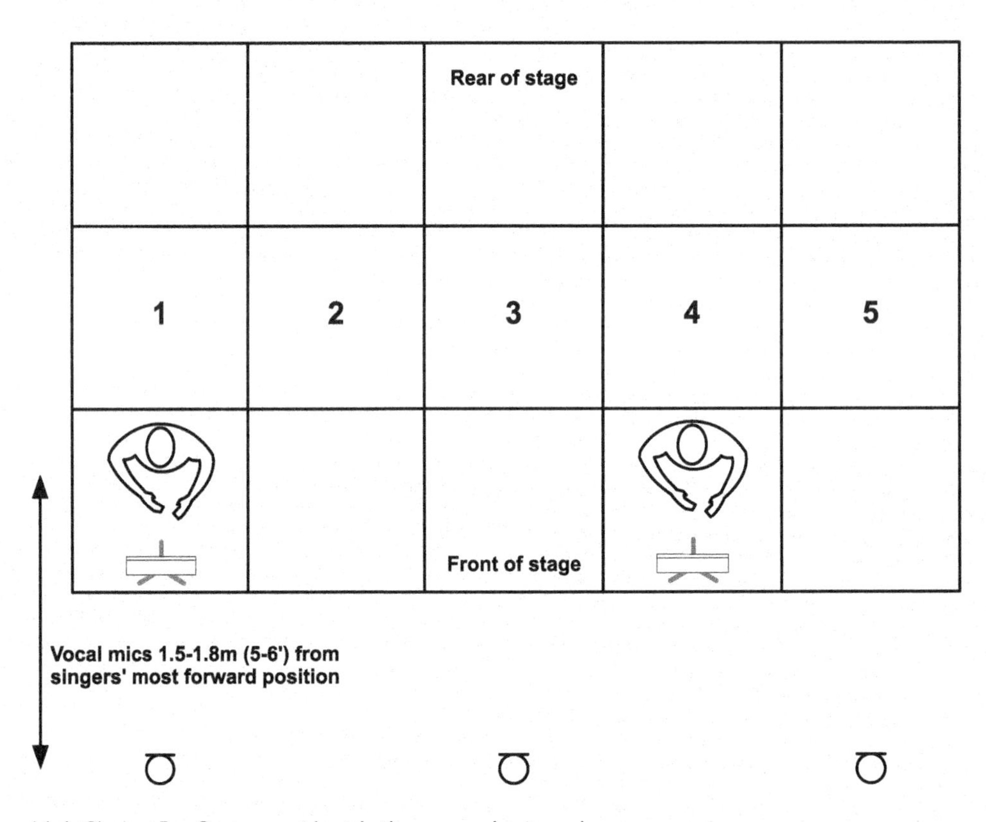

Figure A1.1 Shows 5 x 3 stage grid with three vocal microphones

Figure A1.2 The 1957 recording of *Die Walküre* in Sofiensaal, Vienna, with Kirsten Flagstad and the Vienna Philharmonic Orchestra conducted by Hans Knappertsbusch.[1] The stage grid layout is visible, along with the stage bar for mounting soloists' microphones and the experimental baffled tree.

Photo: Courtesy Decca Music Group Ltd.

singer would be asked to step back a bit or even to turn around and face away from the microphones. When standing closest to the front of the stage area, the singers were around 1.5–1.8 m (5′ to 6′) away from the microphones.

The assistant producer would spend the session on the stage area to make sure that the singers were always on the correct microphones, and became known as the 'collie' because of the parallel with herding sheep. The engineer followed the same cues in his own score in order to know which of the stage microphones needed to be faded up at any time; the five microphones were not left open at all times, as this would have confused the orchestral image. For large stage movements of singers, the engineer would have to crossfade between adjacent microphones, although some smaller movements could be simulated by panning the microphone while the singer stood still. Singers varied greatly in their ability to sing and move at the same time, with many only able to move first and sing afterwards.

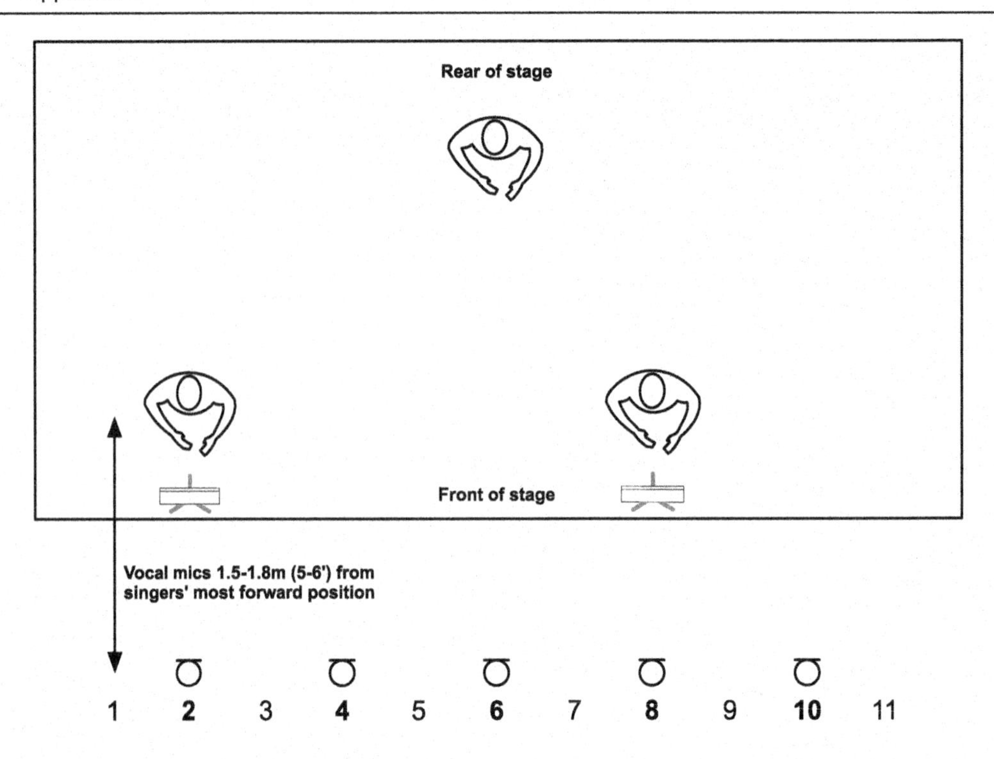

Figure A1.3 Five-microphone set-up with no stage grid. Microphone numbers are shown in bold (2, 4, 6, 8, 10) with the in-between stage positions taking the odd numbers.

The additional microphones were used to have an effect on the soloist's sound that was similar to the use of two spot microphones. When faded up in appropriate amounts, they produced some width and bloom around the singer so that he or she did not sound like a dimensionless point source. For a singer standing centrally, nearest to microphone 6, the relative gains and panning of the microphones would be roughly as shown in the table in Figure A1.4.

As can be imagined, the amount of fader moves needed to keep track of the singer (or singers) made this very intensive work for the engineer, particularly in the days of recording straight to stereo on session. Decca would use two balance engineers: one for the orchestra and one to concentrate on the soloists' microphones only. Recordings up to 1975 were recorded to four tracks: two tracks for the orchestra and two tracks for the chorus and soloists.

The vocal microphones used were M49s and KM56s in the early days, KM64 cardioids in the 1960s, and KM84s for most of the 1970s through the 1990s. In the Sofiensaal, Vienna, these were mounted on a single wide stage bar at a little above mouth height so the singers could still see the conductor, although this method meant they were unable to accommodate singers of very different heights. As noted in Chapter 16, positioning just below the mouth works well to reduce tonal variation where singers have a tendency to either look straight ahead or slightly down as

Mic No.	2	4	6	8	10
Level	−12 dB	−6 dB	0 dB	−6 dB	−12 dB
Panning	75% L	50% L	C	50% R	75% R

Figure A1.4 Gains and panning for a singer standing centrally on a five-microphone stage

they glance at the score. Figure A1.5 is from a session from the Sofiensaal that shows the bar-mounted microphones and also an acoustic screen behind the orchestra that was rarely used (see section A1.3).

An obvious development of the five-microphone opera stage would be to replace each of the three inner single microphones with a pair, and this forms the basis of the stage coverage technique for capturing live opera at the Royal Opera House outlined in Chapter 16.

Figure A1.6 shows the sessions for the 1993–1994 recording of Berlioz's *Les Troyens* with Dutoit and the OSM,[3] presenting the stage layout which uses seven soloists' microphones in this case.

A1.2 The chorus set-up

The chorus would be placed behind the soloists' stage area, and it was most effective when placed at least 10 m away from the orchestra. This meant that the spill between chorus and orchestra became very de-correlated, and they could be approached as separate sources that would later be laid on top of one another in the mix. Another row of microphones mounted on a bar was used to cover the chorus; in the early days, there would have been only three chorus microphones because of channel limitations on the mixing desk. This was later expanded to five microphones as for the soloists (see Figure A1.7). Figure A1.8 shows a session from *The Flying Dutchman* with the Chicago Symphony Orchestra that reflects this layout.

A1.3 The orchestra set-up

The orchestra set-up was very much as we might expect from the Decca orchestral recording tradition but with a few minor changes. Because there were also choir and soloist microphones open in the room, more indirect orchestral sound was picked up, and getting enough detail from the orchestra became more difficult. As a result, the orchestral microphones were placed slightly closer than for a purely orchestral session, with the main microphones mounted about 7.5 cm (3″) lower than the normal tree height.

During the period 1967–1972, Decca's opera recordings were made with the six-microphone tree array that is described in section 8.6; the first opera to go back to using the 'classic' Decca five-microphone tree was *Turandot* in 1972.[5] For this particular recording, screens were added behind the woodwinds to reduce orchestral spill into the soloists' stage microphones. These were about 2.25 m (7′4″) high from the orchestral floor, so the singers on stage could see over the top, but this arrangement was not very conducive to an enjoyable performance experience for the

Figure A1.5 The 1967 recording of *La clemenza di Tito* at the Sofiensaal, Vienna, with István Kertész conducting the Vienna State Opera Orchestra,[2] showing soloists' microphones on a bar at the front of the stage and microphone numbers beneath

Photo: Elfriede Hanak; courtesy Decca Music Group Ltd.

singers or to good communication between singers and conductor. The effect of the screens was also slightly audible on the voices, and their use did not catch on, although the reversion to the three-microphone tree endured. A copy of the set-up layout for Turandot is shown included in Appendix 3.

Figure A1.9 shows a view of the stage area from the Turandot recordings, with the chorus behind the soloists. The acoustic screens are just visible in the bottom right-hand corner.

Figure A1.6 *Les Troyens* sessions in Montreal, 1993–1994
Photo: Jim Steere; courtesy Decca Music Group Ltd.

A1.4 Offstage effects

Offstage effects were generally performed live, thus capturing a real sense of distance and not one created by artificial reverb or other processing. Part of the reason for this was because of constraints on track numbers and the lack of technology such as digital reverb; effects were often performed in a convenient stairwell to provide the right sense of distance or reverb. Effects performed in a different room or stairwell needed to be cued in at the right time in the score; the communications required to co-ordinate the operator are touched on in section A1.5. In old theatres such as Drottningholm, live theatrical effects could be captured as part of opera recording; the old-fashioned way of creating a thunder effect in a theatre was by rolling cannonballs across the stage. More unusual effects were created as required; a well-known example outlined in Culshaw's book *Ring Resounding*[6] was the use of a slowed-down recording of a bowling alley strike to represent the fall of Valhalla.

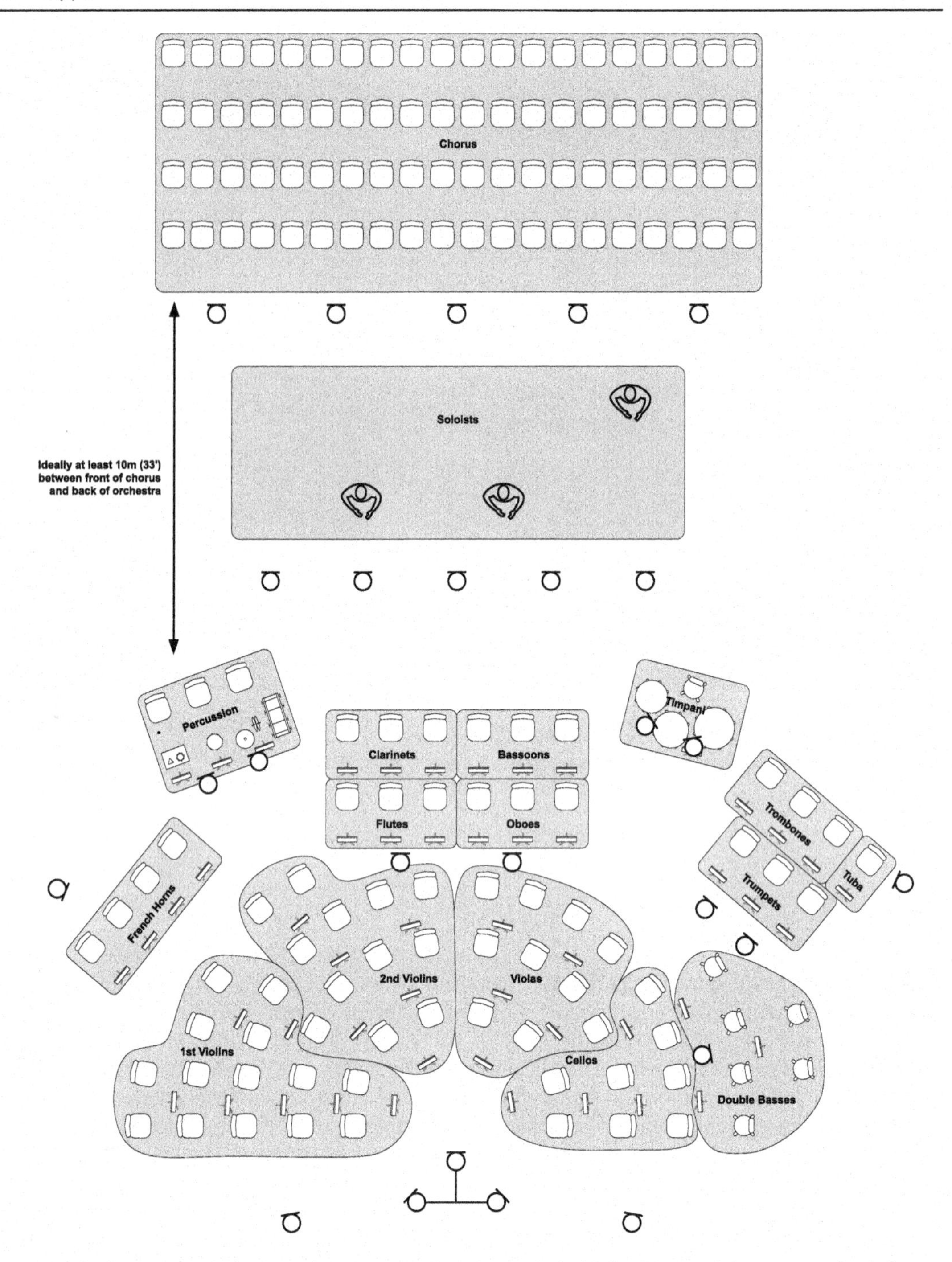

Figure A1.7 Orchestra plus soloists' stage area plus chorus behind. Five microphones on each of chorus and soloists.

Figure A1.8 A session from *The Flying Dutchman* with Sir Georg Solti and the Chicago Symphony Orchestra and Chorus in Medinah Temple in 1976.[4] The chorus is placed in the lower balcony, and a row of chorus microphones is clearly visible.

Photo: Courtesy Decca Music Group Ltd.

A1.5 Communications and control room management

Communications and logistics requirements for studio opera recording were complex and expensive, and in the days before mixing desks had assignable groups, an additional engineer was used to manage the large number of channels and active faders involved in mixing live on session. The nominally senior engineer usually dealt with the orchestra, any sound effects, and the overall balance, and the second engineer usually managed the soloists and chorus. This was not a rigid system, however, and was open to change according to personal preferences.

The main areas that needed to be linked by efficient communication systems were the control room, the conductor, and the stage area (where the assistant producer was stationed). Additional communications would be needed for any offstage performance, whether this was a chorus with an assistant conductor, a soloist, or sound effects operator. Phone links were set up between all areas, and cameras relayed live pictures of the stage area and the conductor back to the control room. The stage cameras would show a front view, but where possible a side view would be added as this made it easy to see how far back the singers were standing. The assistant producer

Figure A1.9 Ghiaurov, Pavarotti, Caballé, and Sutherland recording Turandot at Kingsway Hall, 1972. The white line is to keep the singers at the right distance from their microphones, and the acoustic screens at the back of the orchestra can just be seen at the bottom right.

Photo: Courtesy Decca Music Group Ltd.

on the stage area would also be equipped with headphones so that stage directions could be relayed quickly. The 'Pellowephone' was a mini-telephone exchange that enabled all the phones to talk to one another (see Figure A1.10).

Within the control room, two balance engineers and a producer had to sit so that everyone could hear, see the video monitors, and communicate with one another. On Decca recordings, the producer sat to one side of the centrally seated engineers rather than behind them in the best monitoring position, reflecting the non-hierarchical and collaborative way of working that had evolved at the company. Although as the most senior person he was nominally in charge, much of the decision-making and approach to the recording was down to discussion amongst the team. There was always a strong culture of sharing recording techniques and ideas between engineers at Decca, which went a long way to creating a familial atmosphere and the development of a certain consistency of house sound. If an engineer at any level discovered something that worked well, it would be shared with the others, and this provided a very fertile training ground for younger engineers and also kept the established engineers open to new suggestions. The need for two balance engineers to deal with live opera balancing was easily managed within this sort

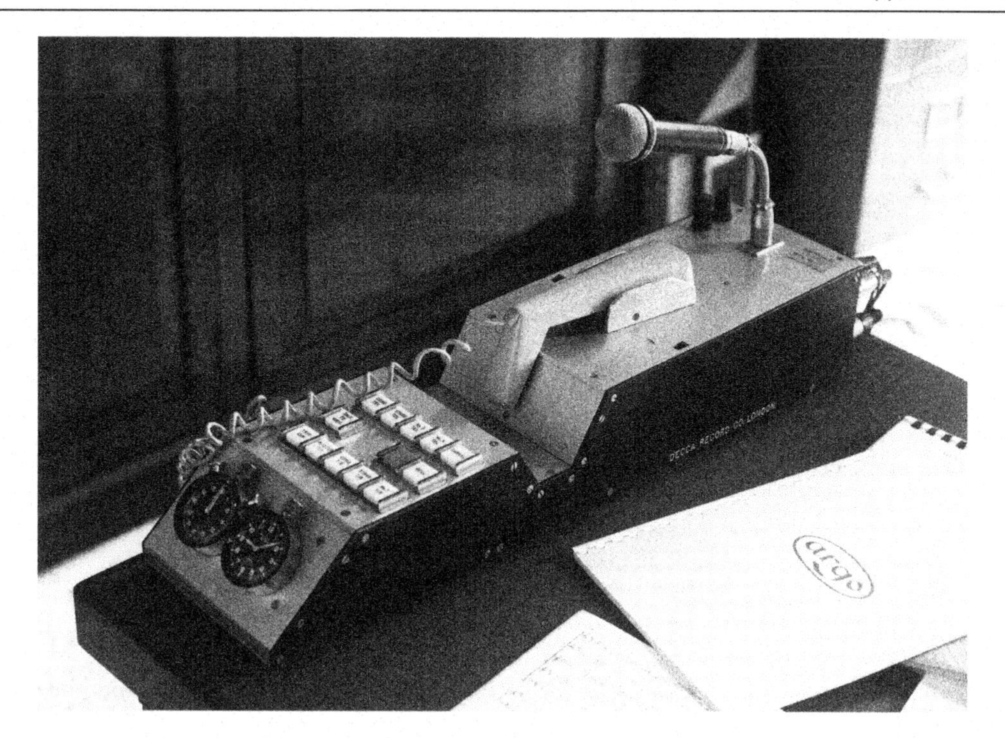

Figure A1.10 Communications equipment – the 'Pellowephone' telephone exchange in the control room
Photo: Mark Rogers.

of collaborative culture without the clash of egos that might have arisen between engineers elsewhere. Even when John Culshaw was producing, the producer's role was not seen as the great impresario set apart from the others, but as part of the team, so open to discussion and finding the best way forwards.

Notes

1 *WAGNER/Die Walküre/Vienna Philharmonic Orchestra/Knappertsbusch/Flagstad/DECCA (1959) CD (1990) 425 963–2*
2 *MOZART/'La Clemenza di Tito'/Vienna State Opera Orchestra/Kertész/Berganza/Krenn/Popp/DECCA (1967) CD (1990) 430–105–2*
3 *BERLIOZ Les Troyens/Montreal Symphony Orchestra & Chorus/Dutoit/Lakes/Voigt/Pollet/Quilico/Perraguin/Maurette/Ainsley/Dubosc/Schirrer/Carlson/DECCA (1994) CD 443 693–2*
4 *WAGNER/Der Fliegende Holländer/Chicago Symphony Orchestra and Chorus/Solti/Bailey/Martin/Kollo/Jones/Krenn/Talvela/DECCA (1977) CD (1985) 414 551–2*
5 *PUCCINI/Turandot/LPO/John Alldis Chor/Mehta/Sutherland/Pavarotti/Caballé/Ghiaurov/Krause/Pears/DECCA (1973) CD (1985) 414 274–2*
6 *Ring Resounding: The Recording of* Der Ring des Nibelungen
Author: John Culshaw
Publisher: Pimlico, 2012
ISBN: 978–1845951948

Appendix 2

Cheaper alternatives to classic microphones

Throughout this book we have made reference to the microphone types used by Decca, especially those made by Neumann, Schoeps, Royer, AKG, and Coles. Many of these models, such as the Neumann M50 used for the Decca Tree, are no longer available, and on the rare occasions that such microphones come onto the second-hand market, they command extraordinarily high prices. Those models that are still available new are usually very expensive as well.

One of the key messages of this book is that is perfectly possible to make excellent recordings with cheaper alternatives. There's no denying that a classic old Neumann in top condition costing £10,000 is a better-sounding microphone than a £100 mass-produced copy. But where the microphones are positioned and how they are balanced is at least as important as what the microphones are; a set of fabulous, expensive microphones put in the wrong places and badly balanced will make a worse recording than a set of reasonably good, cheap microphones positioned and balanced well.

A2.1 The importance of off-axis response when choosing alternative microphones

As we have shown, managing the quality of spill is a critical element in making a classical recording that works well. For example, for a successful violin and piano recording, the sound of the piano spill on the violin microphones needs to blend well with the sound of the piano on the piano microphones, and vice versa for the violin. Key to making this happen is the nature of the off-axis response of the microphones used. For spill to blend in well, the tone quality of sounds entering the side of a cardioid microphone should be as good as the tone quality of sounds entering at the front. This is not an easy thing for designers to achieve, and it is the reason why classic cardioid microphones such as the Neumann KM84 are so sought after – the off-axis sounds are of comparable tone quality to on-axis sounds, just lower in level. For some techniques, such as ORTF pairs, where most of the desired sound is entering the microphones off-axis, this is essential.

Almost all modern, cheaper condenser microphones sound good on-axis, but only some are good off-axis. Be sure to check the published polar plots of such microphones to see if they maintain a fairly consistent shape at different frequencies, and where they do vary, check that variation is smooth. If such information is unavailable, be more wary of making a purchase. Even better than relying solely on published data is to test them yourself. If you are able to borrow a pair and record a friendly musician with one microphone on-axis and the other off-axis, you will be able to compare the tone quality of the two signals in turn (after adjusting for level) and make a better assessment of the off-axis performance.

Note that budget omnidirectional microphones will usually be cheaper than budget cardioids as they are simpler to make. Producing a cardioid with a smooth off-axis response is more difficult.

A2.2 Acoustic pressure equalisers (spheres) for omnis

As shown in Chapter 8, the Decca Tree system works best with directional omnis, which can be made by mounting a pressure-operated microphone onto the surface of a sphere so that higher frequencies from off-axis are shielded by the sphere. Microphones such as the M50 were made this way, but it is possible to convert a cylindrical pencil microphone by adding a sphere – often called an acoustic pressure equaliser (APE). Branded APEs from the big manufacturers can be expensive, but for simple, cylindrical microphones without a tapered body, it is easy to make your own.

The microphone needs to be a single-diaphragm, pressure-operated omni (and not a variable directivity, dual-diaphragm design). For example, Figure A2.1 shows the excellent Studio Projects C4, which comes with both an omni and cardioid capsule. You can see that the cardioid

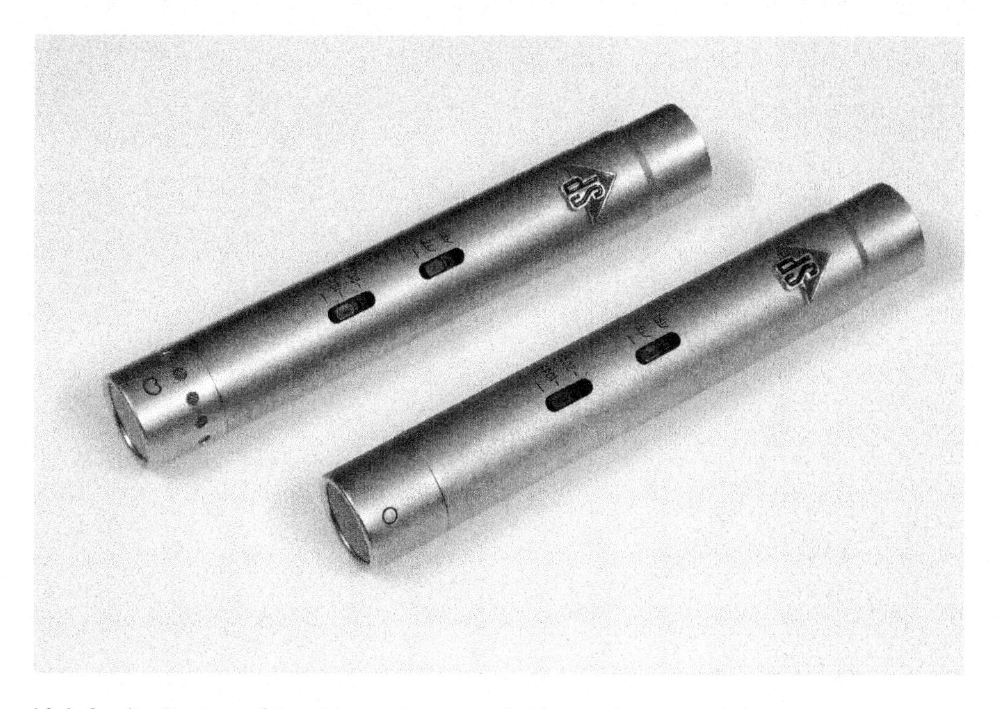

Figure A2.1 Studio Projects C4s with omni and cardioid capsules

Photo: Mark Rogers.

capsule has holes around the capsule near the top, which allow some sound to access the rear of the diaphragm and gives the microphone its cardioid directional characteristic. But the omni capsule is sealed and responds as a simple pressure transducer to sounds approaching from all directions. This kind of microphone, with a ball attached, will be ideal for use in a Decca Tree.

The C4, along with the Behringer B5, has a cylindrical diameter of 20 mm. Figure A2.2 shows a home-made 48 mm diameter sphere with a 20 mm diameter bore. The solid plastic ball was purchased for about £6 from the Precision Plastic Balls Company (www.theppb.co.uk) and is made of Polyamide Nylon 66 Zytel. It was mounted tightly in a vice clamped to a bench drill, making sure that it was centred as accurately as possible under the drill bit, and very slowly drilled through using a high-quality 20 mm flat-blade wood bit. With a little bit of filing to remove any burred edges, the final result works perfectly.

The ball will slide comfortably over the cylinder of the microphone, but how do you ensure that it doesn't fall off? Many of the commercially available spheres include one or more rubber O-rings mounted inside the bore, which provide a stable friction hold on the microphone body, but cutting the grooves where the O-rings will sit is a specialised task requiring a lathe, which is beyond the ability of many people. Fortunately, there's an easy way out: just use Blu-Tak!

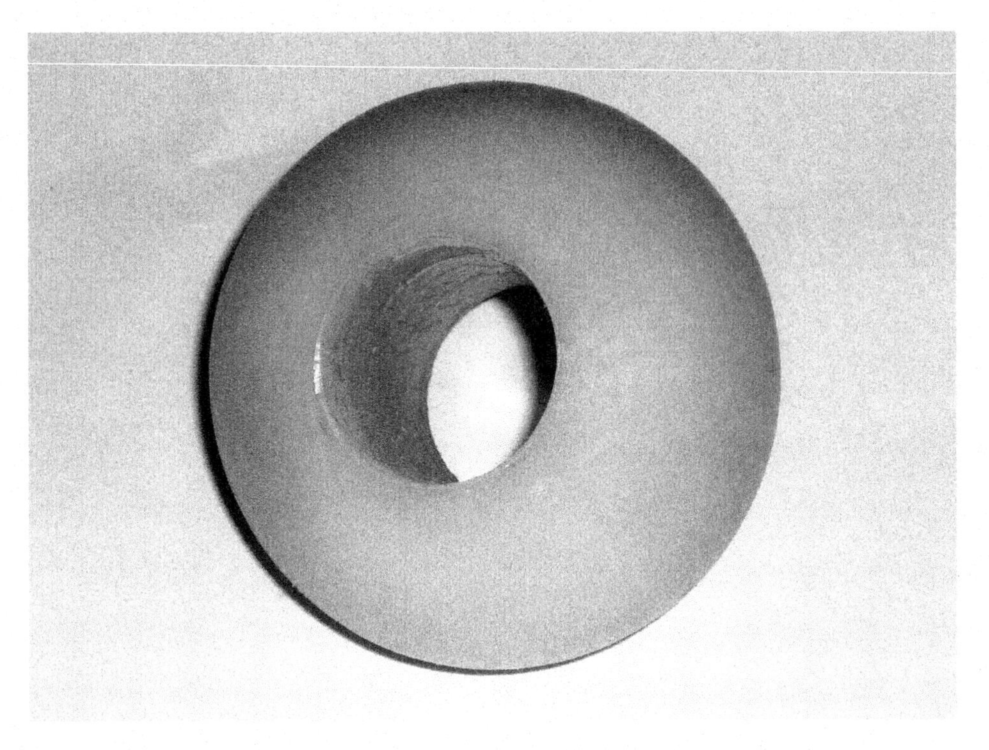

Figure A2.2 A home-made 48 mm microphone sphere

Photo: Mark Rogers.

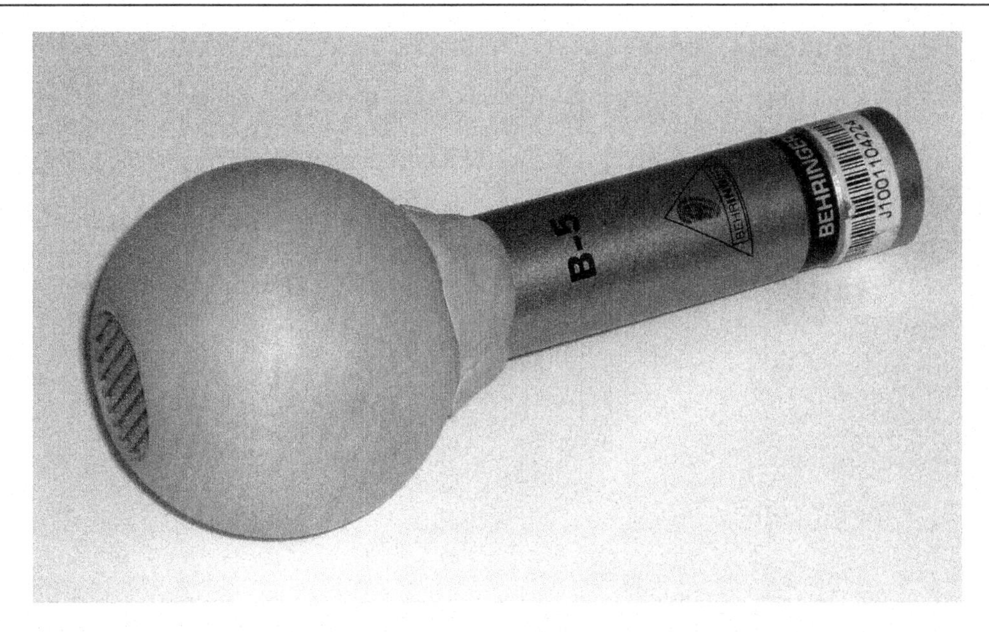

Figure A2.3 Home-made sphere attached with Blu-Tak to Behringer B5 microphone

Photo: Mark Rogers.

Figure A2.3 shows our home-made sphere attached to a Behringer B5 microphone with Blu-Tak. The Blu-Tak makes an excellent adhesive bond to the nylon ball and the metal body of the microphone, and the whole unit feels very solid and stable.

One other task is necessary if using 48/50 mm spheres with the Studio Projects and Behringer microphones. These microphones have their attenuation switches positioned quite high up the body, and they are in the way of the ball when it is mounted flush with the capsule. Pushing the ball into place will force the attenuation switch to be engaged. A small notch needs to be filed at one end of the ball's bore to accommodate the attenuation switch in either position.

The Behringer B5 is an extraordinary microphone for one in the budget range. It has a very even frequency response, low noise, and costs around £70 (2020 prices). So, for about £80, plus a few hours in a workshop, it is possible to make a Decca Tree–compatible microphone that compares quite favourably with classic microphones costing many thousands of pounds. There is no doubt that it is not as good as the real thing, but for those operating on a tight budget, realising maybe 90% of the performance for 3% of the cost is as good a deal as you are ever likely to get.

The Schoeps CMC series microphones (e.g. the MK2H or MK2S capsule attached to a CMC6 body) have a constant 20 mm diameter – which, by a fortunate coincidence, is exactly the same as the Studio Projects C4 and Behringer B5. So if you make a set of spheres for the cheaper microphones, they will still work if you are able to upgrade to the more expensive Schoeps microphones at a later date.

A2.3 High-frequency boost

A frequent criticism of modern cheaper microphones is that they are somewhat bright, with an exaggerated HF response. This can be a problem if they are used close to instruments, but for classical recording, where we are typically placing microphones much further away from the sound source, this can be a positive attribute. Many of the old classics, such as the Neumann M50, and more contemporary classics, such as the Schoeps MK2S, have intentional HF boosts to compensate for being used at a distance, where the air will have absorbed some of the HF (see Figure 8.10). This desired HF boost, combined with home-made APEs, make it possible to build a highly effective Decca Tree on a tight budget.

Appendix 3

Original session set-up sheets

Sketches and notes from John Dunkerley's personal archive

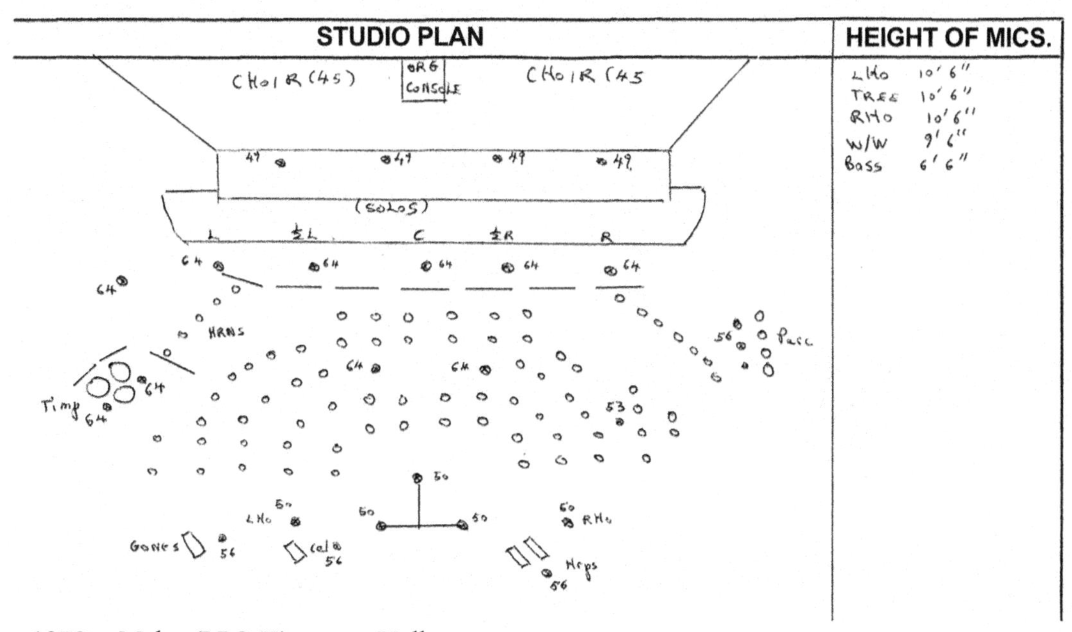

1972 Mehta/LPO Kingsway Hall
Puccini – *Turandot*
Orchestral set-up – *this one has been redrawn as the original document had become too faded to use*

ELECTRICAL RECORD OF SESSION

DATE: 13th Jan 77	STUDIO: Walthamstow	ENGINEERS: JD, JP	PRODUCER: JW
TITLE: Arnold – Cornish English and Scottish Dances	ARTIST: LPO – Arnold.		LABEL: Lyrita

CHANNEL	1	2	3	4	5	6	7	8	9	10	11	12	13	14	15	16	17	18	19	20	21	22	23	24	(E.R.I	E.I
LOCATION	LHo	TREE			RHo	Bass	WW		Hrns	Timp	Cel	Hp	Tuba	Brass	PERC											
TYPE MIC.	50	50	50	50	50	83	84	84	84	84×2	84	88	84	49	88	88	88									
TREB. FREQ.																										
TREB. DB.	+1	-	-	-	+1	+1	-	-	+1	+2	-	-	-	-	+1	+1	+1									
BASS FREQ.																										
BASS DB.	-	-	-	-	-	-	-	-	-	-2	-	-	+1	-	-	-	-									
PRESENCE																										
ECHO BF/AF.																										
PLATE/CHAMBER																										
CHAN. TO GROUP																										
GROUP TO TRACK																										
ECHO GAIN																										
ECHO PAN																										
PRE-SET GAIN	J	J	I	J	J	J	F	F	G	6	G	F	F	I	H	H	H									
SIG. PAN	L	8L	C	8R	R	R	5L	5R	7L	3L	3L	4L	8R	7R	6L	C	8R									
SIG. FADER	10½	10½	10½	10½	10½	12	12	12	15	12	15	9/12	12/18	18	15/18	15/18	15/18									

ECHO SEND	1	2	3	4	MASTER FADERS	1	2	3	4	5	6	7	8	ST.	GR.1	GR.2	GR.3	GR.4	L.S	REMARKS:
REVERB. TIME														15	9/12	12	15/18		half	Stretched

										TAPE MACHINES	
TREB. FREQ.					LIMITER	1	2	3	4		
TREB. DB.					THRESHOLD					TYPE	B62 ×2
BASS FREQ.					GAIN					REPLAY	NAB +4 DBM
BASS DB.					RECOV. TIME					SPEED	15 ips
FADER					POSITION					TAPE	206

1977 Arnold/LPO Walthamstow Town Hall
Arnold – *Cornish, English and Scottish Dances*
Electrical set-up

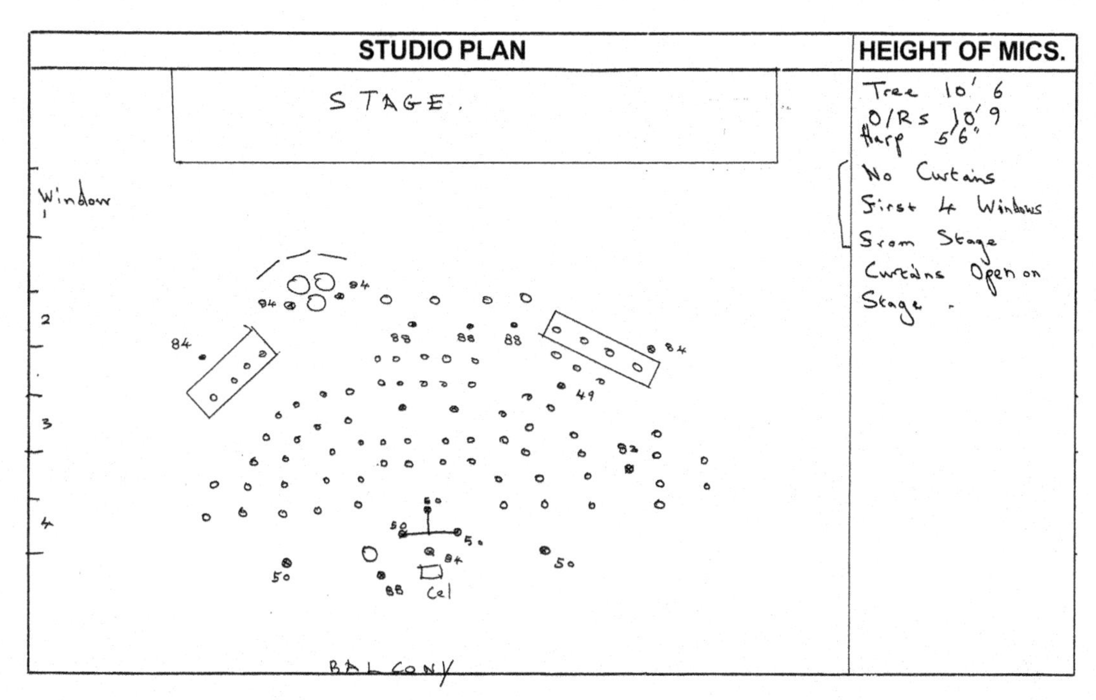

1977 Arnold/LPO Walthamstow Town Hall
Arnold – *Cornish, English and Scottish Dances*
Orchestral set-up

ELECTRICAL RECORD OF SESSION

DATE: 1st – 12th Sept 79. | STUDIO: St. Judes Hampstead. | ENGINEERS: JD, MA. DG. | PRODUCER: PW.

TITLE: Handel – Messiah. | ARTIST: Soloists, A.A.M., C.C.C, Hogwood Preston. | LABEL: L'OL.

CHANNEL	1	2	3	4	5	6	7	8	9	10	11	12	13	14	15	16	17	18	19	20	21	22	23	24	25	26	27	28	29
LOCATION	1st	Vla	Celli	Bss	Cb	Ob	Fag	Tpt	Timp	Org	–	Hns	Solos		←	CHOIR		→											
TYPE MIC.	50	50	50	50	83	84	84	49	84×2	84		84	84	84	49	49	49	49											
TREB. FREQ.																													
TREB. DB.					+1			+1	+2			+1			+1	+1	+1	+1											
BASS FREQ.																													
BASS DB.									-1								+1	+1											
CHAN. TO TRACK																													
GROUP TO TRACK																													
ECHO GAIN																													
ECHO PAN																													
PRE-SET GAIN	I	I	I	H	I	G	G	I	G	H		F	F	E	I	I	H	H											
SIG. PAN	L	7L	7R	R	5R	5L	2L	5R	9R	6L		7R	I	R	L	5L	5R	R											
SIG. FADER	10½	10½	10½	10½	12	12	12/15	15/18	18	21		15/18	9/12	9/12	9/12	9/12	9/12	9/12											

MASTER	1	2	3	4	5	6	7	8	ST.	GR.1	GR.2	GR.3	GR.4	GR.5	L.S
FADERS	10½	10½	10½	10½	10½	10½	10½	10½	10½	12	12	12	12		Various

TAPE MACHINES		REMARKS
TYPE	B62 ×2 (C1, C2). A80 ×1	8 TRK only used on Choral sessions
REPLAY	NAB +4 DBM. CCIR +4 DBM.	Balanced on Chartwells.
TAPE	BASF SPR 50.	

1979 Hogwood/AAM St Jude's, Hampstead
Handel – *Messiah*
Electrical set-up

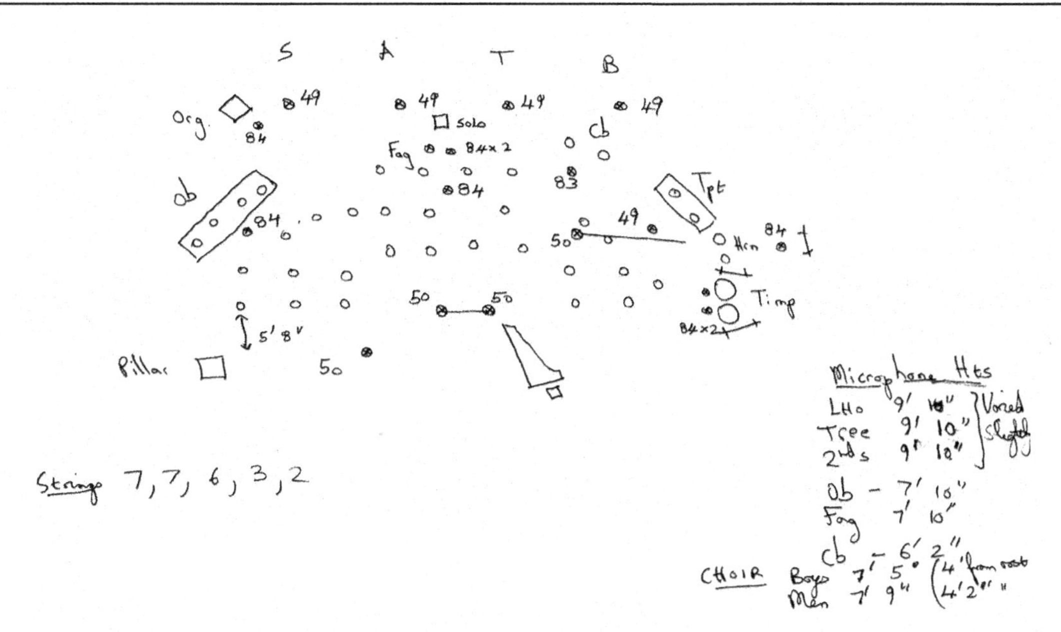

1979 Hogwood/AAM St Jude's, Hampstead
Handel – *Messiah*
Orchestral set-up

ELECTRICAL RECORD OF SESSION

DATE: July 1981. | STUDIO: St. Eustache Montreal. | ENGINEERS: JD, PC, mm. | PRODUCER: RM.

TITLE: De Falla / Ravel. | ARTIST: M.S.O. - Dutoit. | LABEL: Decca.

CHANNEL	1	2	3	4	5	6	7	8	9	10	11	12	13	14	15	16	17	18	19	20	21	22	23	24	25	26	27	28	29
LOCATION	LHo	TREE		RHo	Bass	Harp	WW	WW	Hns	Timp	Timp	Piano	Cel	BD	PERC 1	PERC 2	PERC 3												
TYPE MIC.	50	50	50	50	83	83	84	84	84	84	84	84	83	84	88	88	88												
TREB. FREQ.																													
TREB. DB.										+1	+1	+1																	
BASS FREQ.																													
BASS DB.		-1	-1		-1?					+1	-1																		
CHAN. TO TRACK																													
GROUP TO TRACK																													
ECHO GAIN																													
ECHO PAN																													
PRE-SET GAIN	J	J	J	J	J	I	G	G	H	H	H	G	H	I	H	H	H												
CH. PAN	L	7L	7R	R	R	6R	5L	5R	7L	2L	2R	6L	8L	5L	7L	C	7R												
CH. FADER	10½	10½	10½	10½	13½	13½	12/15	12/15	18	18	18	15	15	18	21	21	21												

MASTER	1	2	3	4	5	6	7	8	ST.	GR.1	GR.2	GR.3	GR.4	GR.5	L.S	REMARKS
FADERS									8½	7/12	12	12.				Mic Hts / Bass Cut on Bass mic
	TAPE MACHINES															LHo 10' 4'' / only on De Falla.
E																RHo 10' 6''
PLAY																TREE 10' 6''.
E																WW. 10' 2''.

1981 OSM/Dutoit St Eustache – Montreal
De Falla/Ravel
Electrical set-up

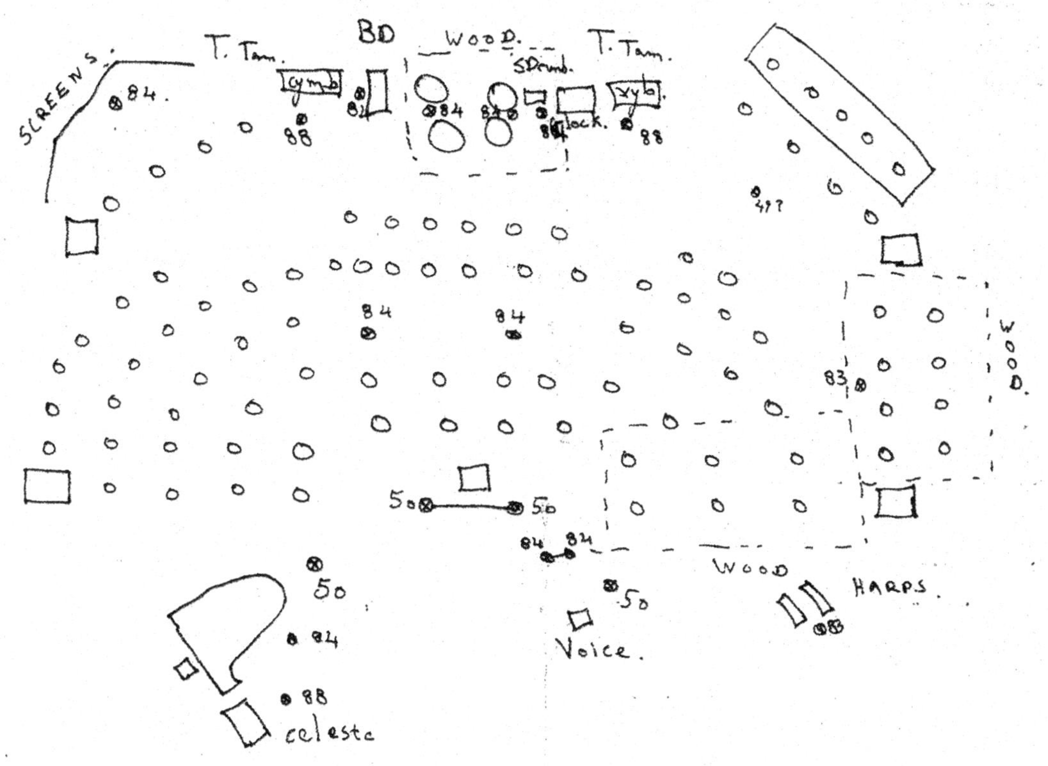

1981 OSM/Dutoit St Eustache – Montreal
De Falla/Ravel
Orchestral set-up

ELECTRICAL RECORD OF SESSION

DATE: 19th, 20th Sept 82 | STUDIO: Kingsway | ENGINEERS: JD, JP, PC. | PRODUCER: PW.

TITLE: Rachmaninov Concerto No.3. | ARTIST: Jorge Bolet / LSO / Fischer | LABEL: Dec

CHANNEL	1	2	3	4	5	6	7	8	9	10	11	12	13	14	15	16	17	18	19	20	21	22	23	24	25	26	27
LOCATION	LHO	TREE			RHO	Bass	WW		Hrn	Timp		-	Piano			SD	BD.										
TYPE MIC.	50	50	50	50	50	83	82	84	84	84×2			83	83		84	84.										
TREB. FREQ.																											
TREB. DB.										+1	+1						−1										
BASS FREQ.																											
BASS DB.		−1	−1	−1						−1	−1																
CHAN. TO TRACK																											
GROUP TO TRACK																											
ECHO GAIN																											
ECHO PAN																											
PRE-SET GAIN	K	K	K	K	K	K	H	H	H	I	I					H	H										
SIG. PAN	L	8L	C	8R	R	R	6L	6R	7L	3R	5R		L	R		8L	C										
SIG. FADER	13½	13½	13½	13½	13½	15	15	15	15/18	18	18		12/15	12/15		21	21										

MASTER	1	2	3	4	5	6	7	8	ST.	GR.1	GR.2	GR.3	GR.4	GR.5	L.S
FADERS									13½	12	12	10½			

TAPE MACHINES

TYPE	
REPLAY	
TAPE	

REMARKS

Mic Hts LHO 10′8″ TREE 10′ 8″
RHO 10′8″ WW 10′0
Piano 5′ 10″. Bass 6′
Bechstein No 174960

1982 LSO/Fischer/Jorge Bolet Kingsway Hall
Rachmaninov – *Piano concerto No. 3*
Electrical set-up

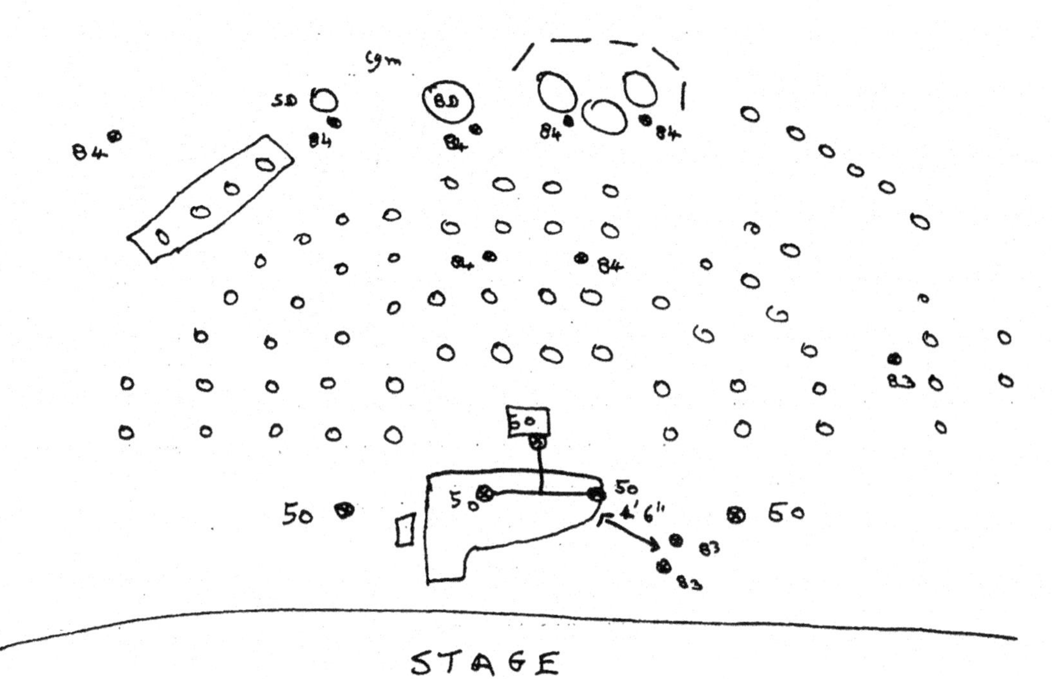

1982 LSO/Fischer/Jorge Bolet Kingsway Hall
Rachmaninov – *Piano concerto No. 3*
Orchestral set-up

ELECTRICAL RECORD OF SESSION

DATE: 16th – 19th August '82 | STUDIO: K.H. | ENGINEERS: JD, MA. | PRODUCER: P.M.

TITLE: Chants D'Auvergne. Vol 1. | ARTIST: Kiri Te Kanawa / E.C.O / Tate. | LABEL: Decca.

CHANNEL	1	2	3	4	5	6	7	8	9	10	11	12	13	14	15	16	17	18	19	20	21	22	23	24	25	26	27	28
LOCATION	LHo	← TREE →			RHo	bass	W	W	Hrn	Timp.		Piano	Solo	Voice	Perc/BD													
TYPE MIC.	50	50	50	50	50	83	84	84	84	84	84	84	84	84	84	84												
TREB. FREQ.																												
TREB. DB.												+1				−1												
BASS FREQ.			−2																									
BASS DB.																												
CHAN. TO TRACK																												
GROUP TO TRACK																												
ECHO GAIN																												
ECHO PAN																												
PRE-SET GAIN	J	J	J	J	J	J	H	H	H	I	I	H	G	F	H	H												
SIG. PAN	L	8L	C	8R	R	R	5L	5R	7L	8L	5L	6R	9L	9R	5R	7R												
SIG. FADER	12	12	12	12	12	12	15/18	15/18	15	18	18	15	12/15	12/15	21	21												

MASTER	1	2	3	4	5	6	7	8	ST.	GR.1	GR.2	GR.3	GR.4	GR.5	L.S	REMARKS
FADERS									12	12	12					

TAPE MACHINES	
TYPE	IVC ×2 B62 × 1
REPLAY	NAB +4DBM.
TAPE	461 SPR 50

1982 ECO/Tate/Kiri Te Kanawa Kingsway Hall
Chants d'auvergne
Electrical set-up

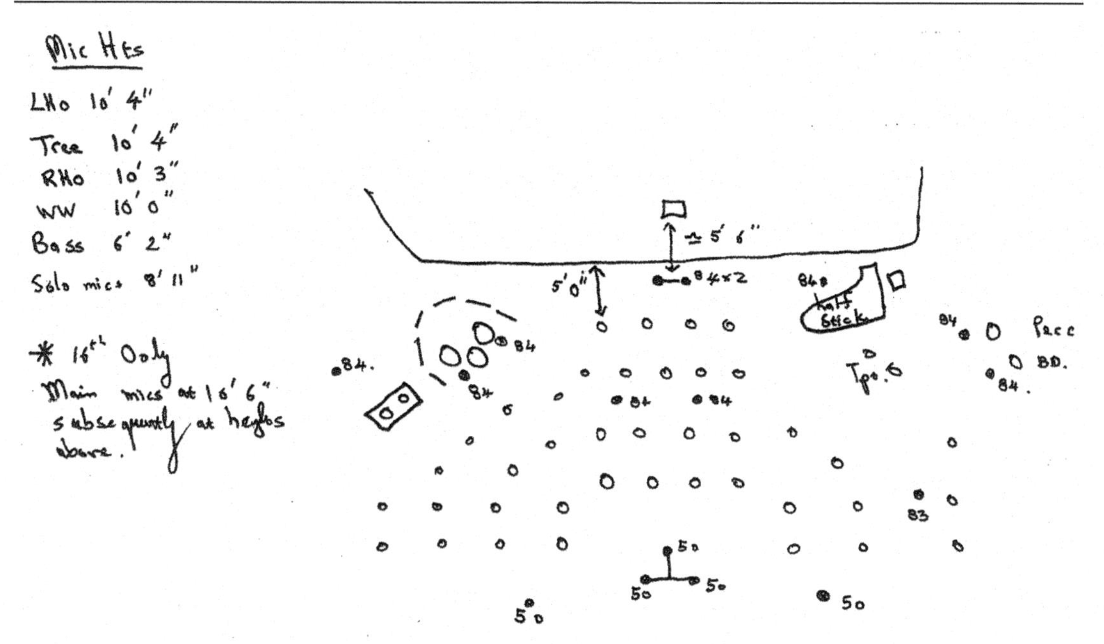

1982 ECO/Tate/Kiri Te Kanawa Kingsway Hall
Chants d'auvergne
Orchestral set-up

ELECTRICAL RECORD OF SESSION

DATE: 30th Sept. 82. | STUDIO: Kingsway | ENGINEERS: JD, P.C. | PRODUCER: A.C.

TITLE: Mussorgsky – Pictures at an Exhibition. | ARTIST: Philharmonia – Ashkenazy | LABEL: Decca

CHANNEL	1	2	3	4	5	6	7	8	9	10	11	12	13	14	15	16	17	18	19	20	21	22	23	24	25	26	27	28
LOCATION	LHo	← TREE		→	RHo	Bass	WW		Hn	Timp.		Harp	Cel	Brass	← 1	2 PERC	3	4	5	6 →								
TYPE MIC.	50	50	50	50	50	83	84	84	84	84	84	83	84	49	84	84	84	84	84	84								
TREB. FREQ.																												
TREB. DB.																												
BASS FREQ.																												
BASS DB.																												
CHAN. TO TRACK																												
GROUP TO TRACK																												
~~ECHO GAIN~~ LINE	1	2	3	4	10	11	19	20	21	29	30	12	7	31	22	23	24	25	28	26								
ECHO PAN																												
PRE-SET GAIN																												
SIG. PAN																												
SIG. FADER																												

MASTER	1	2	3	4	5	6	7	8	ST.	GR.1	GR.2	GR.3	GR.4	GR.5	LS	REMARKS
FADERS																
	TAPE MACHINES															
TYPE																
REPLAY																
TAPE																

1982 Philharmonia/Ashkenazy Kingsway Hall
Mussorgsky – *Pictures at an Exhibition*
Electrical set-up

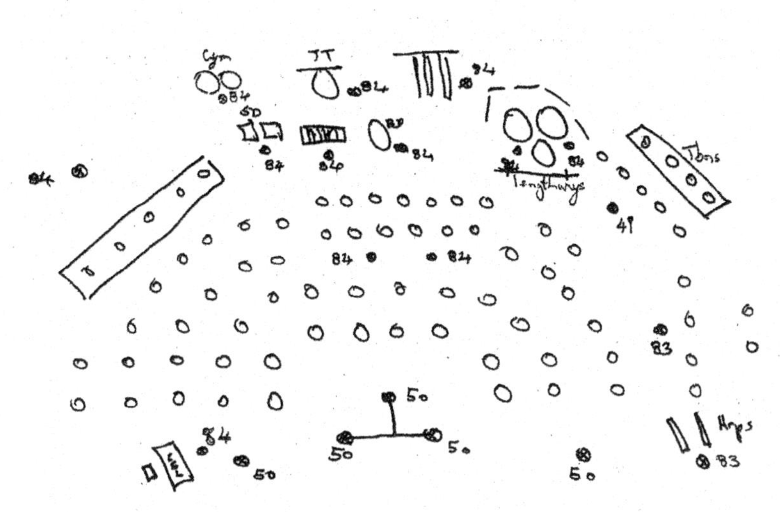

14, 12, 10, 8,
3, 3, 4, 3.
5, 4, 4, 1.
Timp
2 harps
Celeste.
xylo, glock,
2 As cymbs,
suspended cym, B

1982 Philharmonia/Ashkenazy Kingsway Hall
Mussorgsky – *Pictures at an Exhibition*
Orchestral set-up

ELECTRICAL RECORD OF SESSION

TE: 18th, 19th, 24, 25th June 82 | STUDIO: St. Eustache. | ENGINEERS: JD, JL, MM, | PRODUCER: RM.

LE: Respighi – Pines/Fountains Rome, Festa Romane / Ravel Piano Conc | ARTIST: Rogé – OSM – Dutoit. | LABEL:

ANNEL	1	2	3	4	5	6	7	8	9	10	11	12	13	14	15	16	17	18	19	20	21	22	23	24	25	26	27	28	29
CATION	LHO	← TREE →			RHO	Bass	WW	WW	Hcn	Timp	Timp	Harp	Piano		Cel	COND MIC	Brass	Nightingale	P 1	E 2 R	C 3	4		T/A					
'E MIC.	50	50	50	50	50	Sch MK2	84	84	84	84	84	83	MK4	MK4	84	83	49		84	84	84	82							
:B. FREQ.																													
:B. DB.	-	-	-	-	-	-	-	-	-	+1	+1	-	-	-		+1	-		-	-1	-	-							
:S FREQ.																													
:S DB.	-	* -1	-1	* -1	-	-1	-	-	-	-1	-1	-			-	-1	-		-	-	-	-							
AN. TO TRACK																													
OUP TO TRACK																													
IO GAIN																													
IO PAN																													
-SET GAIN	K	K	K	K	K	L	I	I	I	I	I	I	J	J	H	H	H	-	H	I	H	H							
PAN	L	8L	C	8R	R	R	5L	5R	7L	3L	3R	6L	8L	8R	8L	C	7R		8L	4L	4R	8R							
FADER	12	12	12	12	12	15	15	15	15/18	18	18	15/18	18/21	18/21	27		21/24		24	24	24	24							

STER	1	2	3	4	5	6	7	8	ST.	GR.1	GR.2	GR.3	GR.4	GR.5	L.S
DERS									15	12	12	12	12		

TAPE MACHINES

E

LAY

E

REMARKS

Mic Hts LHO TREE RHO = 10' 4"
WW = 10' – 10' 2"
* Bass cut on L + R Tree for Piano Concs only.
Strings 17, 14, 12, 10, 8

1982 OSM/Dutoit/Pascal Rogé St Eustache – Montreal
Respighi – *Pines of Rome*/Ravel – *Piano concerto*
Electrical set-up

ELECTRICAL RECORD OF SESSION

DATE: 29th May – 2nd June 86. | STUDIO: St. Eustache – Montreal | ENGINEERS: JD, SG, MM. | PRODUCER: PM, CH

TITLE: Mendelssohn Overture/M. N. Dream/ Holst – Planets. | ARTIST: MSO – Dutoit. | LABEL: Decca

CHANNEL	1	2	3	4	5	6	7	8	9	10	11	12	13	14	15	16	17	18	19	20	21	22	23	24	25	26	27	28
LOCATION	LHO	TREE		RHO	Bass	Hcp	WW		Hrn	Timp 1		Timp 2	Brass	1	PERC 2	CBO 3	4	Tuba	CHOR					T/A				
TYPE MIC.	50	50	50	50	MK2	MK2	84	84	84	84	84	84 x2	49	84	84	84	84	84	MK3	MK3								
TREB. FREQ.										+1	+1	+1				−1		+1										
TREB. DB.																												
BASS FREQ.					[illegible]	−1				−1	−1	−1							−1	−1								
BASS DB.																												
CHAN. TO TRACK																												
GROUP TO TRACK																												
ECHO GAIN																												
ECHO PAN																												
PRE-SET GAIN	K	K	K	K	M	K	I	I	I	J	J	H	I	J	J	J	J	I	J	J								
SIG. PAN	L	6L	6R	R	R	6L	5L	5R	7L	3L	3R	4R	7R	4L	5L	2L	5R	8R	L	R								
SIG. FADER	12	12	12	12	15	18/21	12/15	12/15	15	18	18	18	18/21	21	22	18	21	21	15	15								

MASTER	1	2	3	4	5	6	7	8	ST.	GR.1	GR.2	GR.3	GR.4	GR.5	L.S	REMARKS
FADERS	13½	13½	13½	13½					12	12	12	12	12			2 × V800 Amps in bridged [illegible]

	TAPE MACHINES
TYPE	2 × 2TRK, 2 × 4TRK (Neptune only)
REPLAY	
TAPE	

1986 OSM/Dutoit St Eustache – Montreal
Holst – The Planets
Electrical set-up

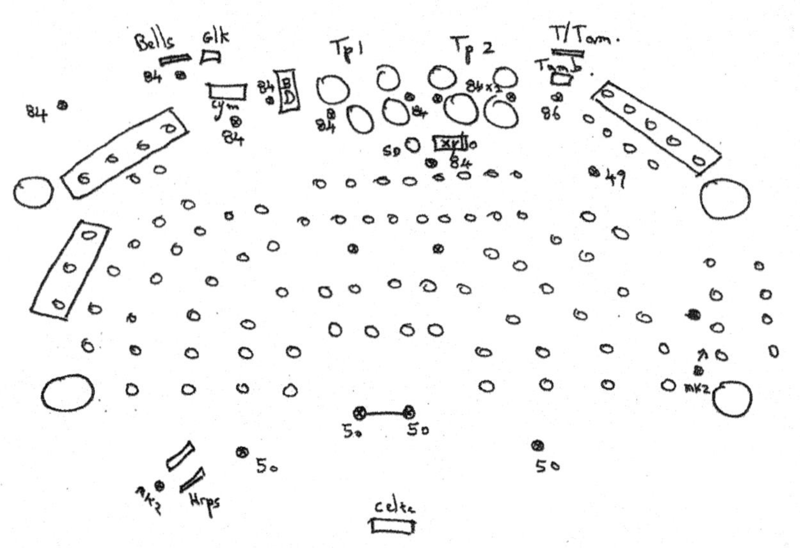

Mic Hts
LHo 10' 5"
TREE 10' 4"
RHo 10' 5"
WW 10' 2"
Bass 6' 5"

Perc 1 SD, xylo
Bells, Glk
Perc 2 cymb.
Perc 3 BD (10d
pa
Perc 4 Tamb, T/T

Chor in Top
Balcony with
Curtain at front
of Balcony.
Chor facing
backwards.
Mics – 2x mk3
not used on STere
only on 4TRK.

1986 OSM/Dutoit St Eustache – Montreal
Holst – *The Planets*
Orchestral set-up

ELECTRICAL RECORD OF SESSION

DATE: 28th, 29th Oct 87. | STUDIO: Methuen M. M. Hall, Boston | ENGINEERS: JD, RH, MM. | PRODUCER: PW.

TITLE: Mozart - Gran Partita. | ARTIST: Amadeus Winds - Hogwood | LABEL: L'O Lyre

CHANNEL	1	2	3	4	5	6	7	8	9	10	11	12	13	14	15	16	17	18	19	20	21	22	23	24	25	26	27	28	29
LOCATION	T	RE	E	Bsn	B.Hn	Horns		Bass																					
TYPE MIC.	50	50	50M	84	84	TL	MS	MK2.																					
TREB. FREQ.																													
TREB. DB.																													
BASS FREQ.	47 Hz		Cut on All Chs.																										
BASS DB.		-2																											
CHAN. TO TRACK																													
GROUP TO TRACK																													
ECHO GAIN																													
ECHO PAN																													
PRE-SET GAIN	38	38	38	41	41	41	41	35																					
SIG. PAN	L	C	R	3L	3R	5L	5R	8L																					
SIG. FADER	+1	-[illegible]	0	-5	-5	-9	-9	-10.																					

MASTER	1	2	3	4	5	6	7	8	ST.	GR.1	GR.2	GR.3	GR.4	GR.5	L.S	REMARKS
FADERS									0	+5					12	Nave + St T Accessories
	TAPE MACHINES															H.H Gain 0
TYPE																
REPLAY																
TAPE																

Area Required for String Orch. 32' W 16' Deep = 16 Sheets of 8x4 Hi[illegible]

1987 Hogwood/Amadeus Winds Methuen Masonic Hall, Boston Mozart – 'Gran Partita' Electrical set-up

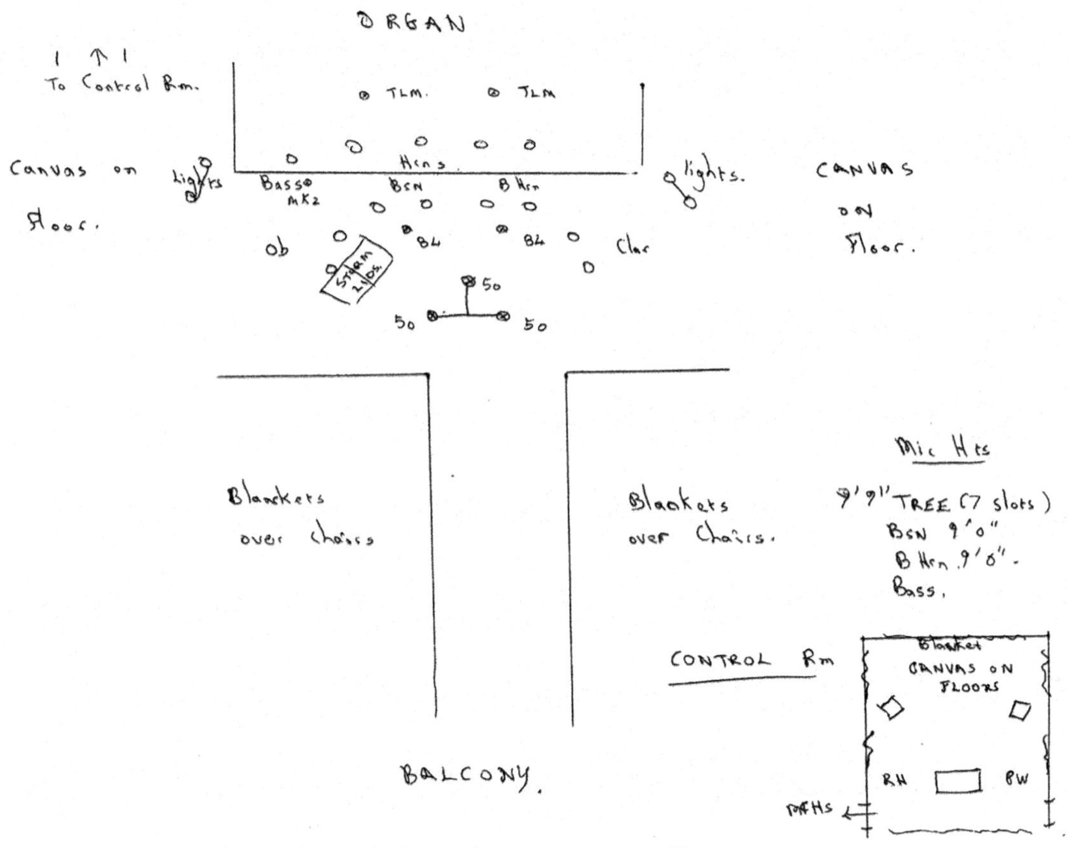

1987 Hogwood/Amadeus Winds Methuen Masonic Hall, Boston
Mozart – 'Gran Partita'
Ensemble set-up

Appendix 4

Orchestral layout notation

This is a commonly used method of indicating the number of players in session notes and layout instructions. The numbers of players in each section of the orchestral are listed in turn, usually starting with the strings, in this order: strings/woodwinds/brass/percussion.

First Violins. Second Violins. Violas. Cellos. Double basses/Flutes. Oboes. Clarinets. Bassoons/ Horns. Trumpets. Trombones. Tuba/Timpanist plus other percussion players/other instruments.

Only the numbers of players are given, without any instrument names. Instruments are listed in abbreviation if necessary for clarification in the 'other instruments' category. A common variation to this layout is to put the strings at the end; it is the only section with five component parts so can be readily identified.

An example layout listing looks like this:

12.10.8.6.4 / 3.2.2.2 / 4.2.2.0 / T+1 / Hp, Cel, Pno

Which means:

12 first violins
10 second violins
8 violas
6 cellos
4 double basses

3 flutes (this will usually mean 2 flutes and a piccolo)*
2 oboes
2 clarinets
2 bassoons

4 horns
2 trumpets
2 trombones
0 tubas

Timpani
1 other percussion player

Harp
Celeste
Piano (in the orchestra)

* Wind section variations can be notated in many ways, as some players might double (i.e. play two instruments), and the notation is not always consistent:

2+picc OR 2+1 means 3 players, 2 flautists plus piccolo player
3d1 = 3 players, one of whom plays both flute and piccolo ('d' stands for 'doubling')

2+cor OR 2+1 means 3 players, 2 oboists plus cor anglais player
3d1 = 3 players, one of whom plays both oboe and cor anglais

2+bass OR 2+1 means 3 players, 2 clarinettists plus bass clarinet player
3d1 = 3 players, one of whom plays both clarinet and bass clarinet

2+contra OR 2+1 means 3 players, 2 bassoonists plus contra-bassoon player
3d1 = 3 players, one of whom plays both bassoon and contra-bassoon

The four principal wind players sit in the centre of the wind section with the piccolo, bass clarinet, cor anglais, and contra-bassoon at the outer edges (see Figure 7.3).

Bibliography and further reading

Ring Resounding: The Recording of *Der Ring des Nibelungen*

Author: John Culshaw
Publisher: Pimlico, 2012
ISBN: 978–1845951948

The Birth of Decca Stereo

Author: Michael H. Gray
Association for Recorded Sound Collections (ARSC) Journal – Vol. 18, No. 1–2–3 (1986)
www.arsc-audio.org/journals/v18/v18n1-3p4-19.pdf

Two-Channel Pathfinder

Author: Malcolm Walker
International Classical Record Collector (ICRC) – Vol. 3, No 10 (Autumn 1997) pp52–62

Lyrita, 40 Years On

Author: Andrew Achenbach interviewing Richard Itter
International Classical Record Collector (ICRC) No 18 (Autumn 1999) pp34–40

Recording the Classics: Maestros, Music and Technology

Author: James Badal
Publisher: Kent State University Press, 1996
ISBN: 978–0873385428

Great British Recording Studios

Author: Howard Massey
Publisher: Hal Leonard Corporation, 2015
ISBN: 978–1458421975

Eargle's The Microphone Book (Audio Engineering Society Presents)

Author: Ray Rayburn, John Eargle
Publisher: Focal Press, 2011
ISBN: 978–0240820750

The Musician's Guide to Acoustics

Author: Murray Campbell, Clive Greated
Publisher: OUP Oxford, 2011
ISBN: 978–0198165057

The Physics of Musical Instruments

Author: Neville H. Fletcher and Thomas D. Rossing
Publishers: Springer
ISBN 978-0-387-98374-5

Fundamentals of Musical Acoustics

Author: Arthur H. Benade
Publisher: Dover
ISBN 0–486–26484-X

Mastering Audio: The Art and the Science

Author: Bob Katz
Publisher: Focal Press, 2014
ISBN: 978–0240818962

Index

Note: Page numbers in *italics* indicate a figure.